W9-CFO-350

TELECOMMUNICATIONS
AND THE CITY

Telecommunications and the City provides the first critical and state-of-the-art review of the relations between telecommunications and all aspects of city development and management.

Drawing on a range of theoretical approaches and a wide body of recent research, the book addresses key academic and policy debates about technological change and the future of cities with a fresh perspective. Through this approach the complex and crucial transformations underway in cities in which telecommunications have central importance are mapped out and illustrated. Key areas where telecommunications impinge on the economic, social, physical, environmental and institutional development of cities are illustrated by using boxed extracts and a wide range of case study examples from Europe, Japan and North America.

Rejecting the extremes of optimism and pessimism in current hype about cities and telecommunications, *Telecommunications and the City* offers a sophisticated new perspective through which city–telecommunications relations can be understood. It will be of interest to students and researchers in urban studies, planning, urban geography, sociology, public administration, communications and technology studies.

Stephen Graham and **Simon Marvin** are both Lecturers at the Centre for Urban Technology, Department of Town and Country Planning, University of Newcastle. They can be contacted on e-mail (s.d.n.graham@ncl.ac.uk and s.j.marvin@ncl.ac.uk) or World Wide Web:http://www.ncl.ac.uk:80/~ncut/.

TELECOMMUNICATIONS AND THE CITY

electronic spaces, urban places

STEPHEN GRAHAM and SIMON MARVIN

LONDON AND NEW YORK

First published 1996
by Routledge
11 New Fetter Lane, London EC4P 4EE

Simultaneously published in the USA and Canada
by Routledge
29 West 35th Street, New York, NY 10001

© 1996 Stephen Graham and Simon Marvin

Typeset in Garamond by Solidus (Bristol) Limited
Printed and bound in Great Britain by Biddles Ltd, Guildford
and King's Lynn

British Library Cataloguing in Publication Data
A catalogue record for this book is available from the British Library

Library of Congress Cataloguing in Publication Data
A catalogue record for this book has been requested

ISBN 0–415–11902–2
0–415–11903–0 (pbk)

CONTENTS

PLATES

FIGURES

BOXES

PREFACE

This book explores the complex and poorly understood set of relationships between telecommunications and the development, planning and management of contemporary cities. It provides a new interdisciplinary and international perspective on how remarkable advances in telecommunications affect all aspects of urban development: social, economic, physical, environmental, geographical and governmental. This book represents the first attempt to provide such a broad and synoptic approach to fill the gap left by the long neglect of tele-communications in urban studies and policy-making.

Because of this neglect, the book's 'journey' through the most important types of city–telecommunications relations is analogous to an early expedition into largely uncharted territory. This journey is assisted only by a highly imperfect map; there are many gaps and areas of poorly understood territory. This is because the study of telecommunications in cities remains so immature, but it is also due to the extremely rapid rate of change in the subject. This book develops a new framework to analyse the diverse range of policy and research that is emerging on telecommunications and cities.

We aim to stimulate more sophisticated debate and research on city–telecommunications relationships. We also aim to assist teaching by providing a book that draws together a diverse and eclectic range of material which is presented in accessible form. However, we remain unable to provide answers to all questions about this embryonic subject; inevitably, this book raises as many questions as it answers.

This book was developed because of the problems we have experienced in developing a course on telecommunications and urban development for town planning students. We and our students have all been confused by the range, complexity and diversity of material on the subject. We have also been frustrated by the difficulties often involved in tracking down literature and obscure 'grey' material on the subject. We found that in the literature on the subject profound pessimism coexists with utopian optimism but there is very little actual empirical study of how telecommunications relate to cities. At the same time, however, we have been disappointed and surprised that no coherent book exists on the subject which brings the diffuse, specific and specialised material together to introduce how cities and telecommunications are related.

Hence we have written this book. In it we emphasise and illustrate the complex relationships which exist between telecommunications and cities by covering neglected subjects such as the urban environment, urban government and urban utilities as well as the more familiar ground of socioeconomic development, transport and urban form. We set out the debates between dystopian and utopian theorists and establish a framework for considering the range of relationships between cities and telecommunications. We link these theories to debates about the social, economic, geographical, political and environmental development of contemporary cities, and bring out the technological dimensions of each. Finally, we consider questions of urban management, planning and policy integrally with our wider considerations of urban development and telecommunications.

The book will appeal to students of urban studies, local government studies, geography, planning and technology and communications studies who are interested in new technologies and the city. It will also interest urban policy-makers who are keen to inform themselves about state-of-the-art research and policy in this burgeoning and increasingly important area. The book has been designed to act as a set text for advanced specialised courses in telecommunications and cities. It is also suitable as a basis for exploring specific issues and topics, as each section includes a context-setting introduction and an up-to-date guide to further reading on each subject.

ACKNOWLEDGEMENTS

Many people have contributed to the development of this book. Within the University of Newcastle, we have received excellent support and encouragement from John Goddard, James Cornford, Allan Gillard, Patsy Healey and Simon Guy. The Department of Town and Country Planning, particularly John Benson, gave vital financial support to the writing of the book. Within the Department, colleagues in the Centre for Research in European Urban Environments and the Centre for Urban Technology have provided an exciting and stimulating work environment. Next door, the Centre for Urban and Regional Development Studies has provided a lively research environment, which has helped us enormously in the writing of this book. Helen Price in the Town Planning office provided superb secretarial support. Elsewhere in the University, we must thank the librarians in the interlibrary loans section at the Robinson Library and Mick Sharp in the Audio Visual unit for help with the diagrams.

A wide variety of reviewers have given invaluable advice at all stages of the preparation of the book – Peter Hall, Mike Batty, Dave Wield, Ian Miles, Mitchell Moss and Ralph Negrine. Many thanks for your time and useful suggestions. Obviously, we must take the responsibility for the book's contents. Third year students on our Telecommunications and the City option course also provided useful feedback on the material used in the book.

Thanks also to our Editor, Tristan Palmer, and his colleagues, Matthew Smith and Caroline Cautley, for their advice and support. On the copy-editing front, Penelope Allport had the patience of a saint in dealing with our references; Connie Tyler provided an excellent index.

Grateful thanks to Annette Kearney and Nicola Turner – writing this book consumed far too many evenings and weekends. Thanks for your tolerance and encouragement!

Thanks to the following for their generous permission to reprint material in the book from:

Box 3.1 'The Vanishing City', by Anthony Pascal, *Urban Studies* (1987), vol. 24, pp. 597–603.

Box 3.2 'The electronic cottage', from *The Third Wave* by Alvin Toffler. Copyright © 1980 by Alvin Toffler. Used by permission of Bantam Books, a division of Bantam Doubleday Dell Publishing Group, Inc.

Box 3.3 'Electronic spaces: new technologies and the future of cities', by K. Robins and M. Hepworth. This article was first published in *Futures* (April 1988), vol. 20, no. 2, pp. 155–176 and is reproduced here with the permission of Butterworth-Heinemann, Oxford, U.K.

Box 3.4 'The politics of citizen access technology: the development of public information utilities in four cities', by K. Guthrie and W. Dutton, *Policy Studies Journal* (1992), vol. 20, no. 4, pp. 574–597.

Box 4.1 'Telecommunications, world cities and urban policy', by M. Moss, *Urban Studies* (1987), vol. 24, pp. 534–546.

Box 5.1 'The overexposed city', by Paul Virilio, *Zone* (1987), vol. 1, no. 2. Urzone Inc.

Box 6.2 *The New Urban Infrastructure – Cities and Telecommunications*, J. Schmandt *et al.* (eds), pp. 107–110 *passim*. Praeger Publishers, an imprint of Greenwood Publishing Group, Inc., Westport, CT, 1990. Copyright © 1990 by the University of Texas at Austin. Abridged and reprinted with permission of Greenwood Publishing Group, Inc. All rights reserved.

Box 6.3 'Using computers for the environment', by J.E. Young, in L.R. Brown *et al. State of the World 1994* (1994). Worldwatch Institute, Norton New York.

Box 6.4 *An Enhanced Urban Air Quality Monitoring Network: A Feasibility Study*, by Environmental Resources Limited (February 1991). Department of the Environment, Air Quality Division.

Box 7.1 'New information technology and utility management', *Cities and New Technologies* (February 1992), pp. 51–76. OECD, Paris.

Box 7.2 *Information Horizons: The Long-Term Social Implications of New Information Technologies*, by I. Miles *et al.* (1988), pp. 119–121. Edward Elgar Publishing Ltd.

Box 7.3 'Confusing signals on the road to nowhere', by J. Whitelegg, *The Times Higher Education Supplement* (19 November 1993), pp. x–xi.

Box 7.4 'The intelligent city: utopia or tomorrow's reality?', by J. Laterrasse, in F. Rowe and P. Veltz (eds), *Telecom, Companies, Territories* (1992). Presses de L'ENCP.

Box 8.1 'Foresight and hindsight: the case of the telephone', in I. de Sola Pool

(ed.), *The Social Impact of the Telephone* (1977), pp. 140–145. MIT Press. ©
I. de Sola Pool 1977.

Box 8.2 'Transportation and telecommunications networks: planning urban
infrastructure for the 21st century', by R.E. Schuler, *Urban Studies* (1992),
vol. 29, no. 2.

Box 8.3 'Communications technologies and the future of the city', by A.
Gillespie, in M.J. Breheny (ed.), *Sustainable Development and Urban Form*
(1992), pp. 67–78. European Research in Regional Science v 2, Pion
Limited, London.

Stephen Graham and Simon Marvin
Newcastle upon Tyne
May 1995

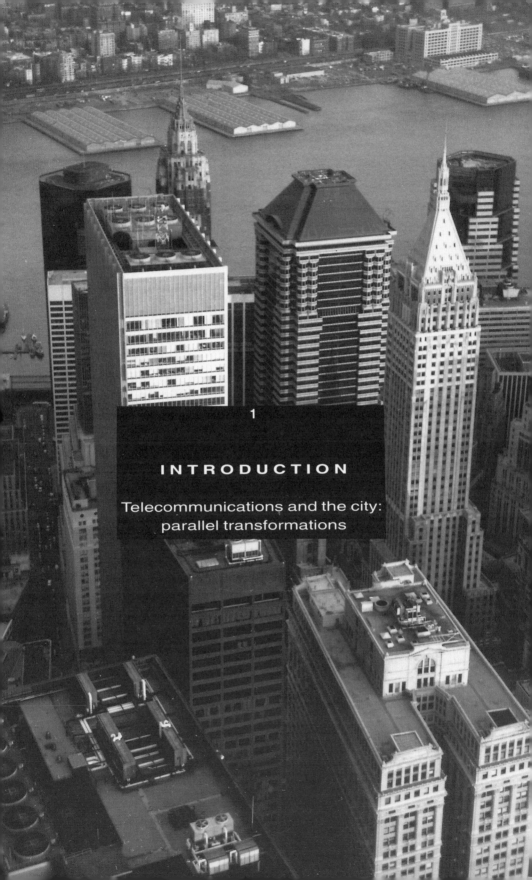

1

INTRODUCTION

Telecommunications and the city:
parallel transformations

TELECOMMUNICATIONS AND URBAN TRANSFORMATIONS

A rapid transformation is currently overtaking advanced industrial cities. As we approach the verge of a new millennium, old ideas and assumptions about the development, planning and management of the modern, industrial city seem less and less useful. Accepted notions about the nature of space, time, distance and the processes of urban life are similarly under question. The boundaries separating what is private and what is public within cities are shifting fast. Urban life seems more volatile and speeded up, more uncertain, more fragmented and more bewildering than at any time since the end of the last century.

Apparently central to this transformation, according to nearly all commentators, are remarkable leaps in the capability and significance of telecommunications. Much of contemporary urban change seems to involve, at least in part, the application of new telecommunications infrastructures and services to transcend spatial barriers instantaneously. Telecommunications – literally communications from afar – fundamentally adjust space and time barriers – the basic dimensions of human life (Abler, 1977). They connect widely separated points and places together with very little delay – that is, in ways that approach 'real time'.

As telecommunications themselves become digital and based on micro-electronics, they are merging with digital computer and media technologies. These are diffusing into a growing proportion of homes, institutions, workplaces, machines and infrastructures. The result of this merging is a process of technological convergence and a wide and fast-growing range of so-called 'telematics' networks and services. Following the French word *télématique*, coined in 1978 by Nora and Minc (1978), 'telematics' refers to services and infrastructures which link computer and digital media equipment over tele-communications links. Telematics are providing the technological foundations for rapid innovation in computer networking and voice, data, image and video communications. It is increasingly obvious that telematics are being applied across all the social and economic sectors and functions that combine to make up contemporary cities. It is also clear that telematics operate at all geographical scales – from within single buildings to transglobal networks. As William Melody argues, 'information gathering, processing, storage and transmission over efficient telecommunications networks is the foundation on which developed economies will close the twentieth century' (Melody, 1986).

As part of this transformation, cities are being filled with what Judy Hillman calls 'gigantic invisible cobwebs' of optic fibre, copper cable, wireless, microwave and satellite communications networks (Hillman, 1991; 1). The corridors between cities, whether they be made up of land, ocean or space, are in turn developing to house giant lattices of advanced telecommunications links. These connect the urban hubs together into global electronic grids. Such grids now encircle the planet and provide the technological basis for the burgeoning flows of global telecommunica-tions traffic: voice flows, faxes, data flows, image flows, TV and video signals. Instantaneous electronic flows now explode into the physical spaces of cities and buildings and seem to underpin and cross-cut all elements of urban life.

Clearly, then, contemporary cities are not just dense physical agglomerations of buildings, the crossroads of transportation networks, or the main centres of economic, social and cultural life. The roles of cities as electronic hubs for telecommunications and telematics networks also needs to be considered. Urban areas are the dominant centres of demand for telecommunications and the nerve centres of the electronic grids that radiate from them. In fact, there tends to be a strong and synergistic connection between cities and these new infrastructure networks. Cities – the great physical artefacts built up by industrial civilisation – are now the powerhouses of communications whose traffic floods across global telecommunications networks – the largest technological systems ever devised by humans.

Many have argued that these shifts are part of a wider technological and economic revolution which seems to be underway within advanced industrial societies and within which both the development of telecommunications and urban change hold central significance (see Miles and Robins, 1992). A wide and sometimes confusing range of analytical perspectives have developed that try to chart this transformation from an industrial, manufacturing dominated society to one dominated by information, communications, symbols and services.[1] Because western societies are fundamentally urban societies – with between 60 and 90 per cent of their populations living in towns and cities – cities are at the front line of this revolution. Cities are the dominant population, communication, trans-action and business concentrations of our society. This makes them the central arenas within which we would expect the effects of current telecommunications innovations to be felt. As we move towards an urban society based more and more on the rapid circulation of messages, signs and information via global electronic networks, it would therefore be hard to pinpoint a more important set of technology–society relations than those which link cities to telecommunications.

THE URBAN 'IMPACTS' OF TELECOMMUNICATIONS

But what are the implications of these shifts? What becomes of cities in an era dominated by electronic flows and networks? What fate lies in store for our urban areas in the world where 'virtual corporations', 'virtual communities' and the abstract 'electronic territory' of 'cyberspace' are developing, based fundamentally on the use of telematics as space and time transcending technologies?

The growing use and significance of telecommunications throws up many profound and fundamental questions which go to the heart of current debates about cities and urban life both today and in the future. For example, how do cities and urban life interrelate with the proliferation of electronic networks in all walks of life and at all geographical scales? What happens to cities in the shift away from an economy based on the production and the circulation of material goods to one based more and more on the circulation and consumption of symbolic and 'informational' goods? (Lash and Urry, 1994). How are cities to sustain themselves economically given that more and more of their traditional economic advantages seem to be accessible, 'on-line', from virtually any location? Are cities being affected physically by advances in telecommunications as many claim they were in previous eras by the railway and the automobile? How does the movement from physical, local neighbourhoods to specialised social communities sustained over electronic networks – such as those on the Internet – affect the social life of cities? How are social power relations and the traditional social struggles within cities reflected in the new era of telecommunications? What is the relevance of telecommunications for burgeoning current debates about the 'environmental sustainability' of industrial cities? And what do all these changes imply for the ways in which cities are planned, managed and governed?

Such questions have recently stimulated much speculation and debate about the future of cities and the role of advances in telecommunications in urban change. Many commentators excitedly predict very radical changes in the nature of the city and urban life as advanced telecommunications, telematics and computers weave into every corner of urban life and so 'impact' on cities. Arguments that this will mean the dissolution of the cities and the emergence of decentralised networks of small-scale communities or 'electronic cottages' are widespread. In fact they are so common that visions of the end of cities seem almost to have reached the status of accepted orthodoxy within some elements of the popular media. Here, speculations abound surrounding the apparently revolutionary importance of the 'communications revolution', the 'information age', the 'information superhighway', 'cyberspace' or the 'virtual community' for the future of cities.

Unfortunately, however, these debates tend to be heavily clouded by hype and half-truth. They have generated much more heat than light. Such debates often tend also to be extremely simplistic, relying on assumed and unjustified assumptions about how telecommunications impact on cities. Many accounts of city–telecommunications relations amount to little more than poorly informed technological forecasts. Often, these are aimed at attracting media attention or generating sales and glamour for technological equipment. As a result, remarkably little real progress has been made in debates about telecommunications and cities. Amidst all the general hype about telecommunications and cities, remarkably little real empirical analysis of city–telecommunications relations exists.

This leaves the terrain open to extremes of optimism and pessimism. On the one hand, utopianists and futurologists herald telecommunications as the quick-fix solution to the social, environmental or political ills of the industrial city and industrial society more widely. On the other, dystopians or anti-utopians paint portraits of an increasingly polarised and depressing urban era dominated by global corporations who shape telematics and the new urban forces in their own image. Meanwhile, the increasing importance of telecommunications in cities has stimulated urban policy-makers, managers and planners to begin to get involved in the development of telecommunications within their cities. But they, too, often remain confused about how their cities are really affected by developments in telecommunications. This, and the need to be seen to be successful means that they themselves can become prone to hyping up their urban telecommunications policies in the language of the quick technical fix.

The immaturity and neglect of urban telecommunications studies means that there has been a tendency to approach the whole subject without trying to justify the theory or methodologies adopted. In the excitement to address these neglected and important areas, Warren (1989) notes what he calls a 'candy store effect':

The topic [of telematics and urban development] creates a 'candy store' effect by providing license to deal with a range of phenomena. The result is an effort to cover far too much with no logic or theory offered to explain why some consequences are discussed and others are not and why some evidence is presented and other findings are not. . . . We are left with an analysis which lacks any theoretical base and an explicit methodology, gives more attention to marginal than primary effects of telematics, and, in many instances, is in conflict with a significant body of research.

(Warren, 1989; 339)

THE NEGLECT OF TELECOMMUNICATIONS IN URBAN STUDIES

This 'candy store' effect is one symptom of the wider immaturity and neglect of telecommunications issues in both urban studies and urban planning and policy-making. In many ways, cities can be thought of as giant engines of communication – physical, social and electronic (Meier, 1962; Pool, 1977). We might therefore expect technologies that allow communication over distances – that is *tele* communications – to be a central focus of disciplines which aim to understand the city and professions involved in urban planning and management. This is especially so given that telecommunications are absolutely central to current innovation and restructuring all of the activities that combine to make cities: in manufacturing, transportation, consumer and producer services; in leisure, media and enter-tainment industries; in education, urban government, public services and urban utilities; and in social and cultural life.

But, despite these two points, telecommunications remain far from being a central focus in urban studies or urban policy-making. The subject of tele-communications and cities is a curiously neglected and extremely immature field of policy and research. Urban telecommunications studies remains perhaps the most underdeveloped field of urban studies. Telecommunications is also one of the least developed areas of urban policy (Mandlebaum, 1986). Recently Michael Batty argued that 'interest and insights into the impact of communications patterns on the city with respect to information flow have . . . been virtually non-existent' (Batty, 1990a; 248) and that current 'understanding of the impacts of information technology on cities is still woefully inadequate' (ibid.; 250).

Urban studies and policy remain remarkably blind to telecommunications issues. Compared to the enormous effort expended by urban analysts and policy-makers on, say, urban transportation, urban telecommunications have received only a tiny amount of attention. Vast libraries and many professional bodies and dedicated journals now exist in the field of urban transport issues; only a handful of books have directly looked at telecommunications and the city.[2] Things have not greatly changed since Bertram Gross argued in 1973 that 'urban planners seem most comfortable when dealing with urban problems in terms of *transportation*. Indeed, the most advanced techniques and the most "scientific" body of knowledge readily available to such decision makers are those of transport. . . . Urban planners . . . must become aware of the problems and possibilities of tele-communications' (Gross, 1973; 29). 'Urban analyst' or 'commentator' could easily replace 'planner' here. At most only about a dozen urban commentators in the Anglo-Saxon world have directly researched the relationships between telecommunications and urban development since Gross made that statement. Only rarely have these had much impact of the urban disciplines.

This relative neglect means that the field has been left open to other non-urban specialists who have developed very influential speculations on how cities might relate to telecommunications. Importantly, though, these speculations have not been based on any particular understanding or analysis of cities *per se*. Instead, they have tended to start with rather simplistic and utopian approaches. Often, new technologies have been seen unproblematically as technical-fix-style solutions for the perceived social and environmental inadequacies of the industrial city. Often, these ideas have been directly fuelled by interests in computing and tele-communications industries, keen to foster positive public images to new technologies as a stimulus to the growth of markets (Slack and Fejes, 1987). Mark Hinshaw was one of the first to diagnose the link between the so-called 'utopianist' approaches that were then increasingly influential and the neglect of telecommunications by urban planning and urban studies. He remarked that:

Many planners may well feel that communications technology will have little or no effect on urban development. Virtually any recognition at all of the relationships between urbanism and communications has come from academics and professionals outside the fields most directly involved in urban analysis and policy development. Most of the literature coming from such sources, however, treats communication and information-generating hardware as the *means of solving most of the urban problems with which we are presently confronted*.

(Hinshaw, 1973; 305. Emphasis added)

THE NEED FOR MORE SOPHISTICATED APPROACHES TO CITY—TELECOMMUNICATIONS RELATIONS

While recent efforts to understand city–telecommunications relations have grown markedly (Brunn and Leinbach, 1991), it is still clear that urban tele-communications researchers and policy-makers are still fighting an uphill battle. Facing them are the overwhelming invisibility of the subject, the long legacy of neglect, and the powerful influence of the utopianists and futurists who have tended to fill the vacuum left by the neglect of telecommunications in urban studies. We believe that these problems are significant enough to challenge the paradigms underpinning urban studies and policy. They mean that – while they are increasingly numerous – references to telecommunications in both the policy and urban studies literature still tend to be general and speculative rather than specific or grounded in real analysis. Conceptual sophistication still tends to be rudimentary.

As with much social research on technology, literature on telecommunications and cities still tends to invoke what Gökalp (1988) calls 'grand metaphors' of the nature of telecommunications-based change in cities. Invariably, modern tele-communications are seen as a 'shock', 'wave' or 'revolution' impacting or about to impact upon cities. Technological determinism is common: current or future urban changes are often assumed to be determined by technological changes in some simple, linear cause and effect manner. The use of simple two-stage models to describe changes in cities and society is common. Cities are seen to be placed in a new age in which telecommunications increasingly have a prime role in reshaping their development. Most usual here are notions that capitalism is in the midst of a transformation towards some 'information society' (Lyon, 1988) or 'post-industrial society' (Bell, 1973), or that a more general 'communications revolution' (Williams, 1983) or 'third wave' (Toffler, 1981) is sweeping across urban society.

Most often, because of the general inability to analyse real change and the influence of futurology, analysis centres on speculating the impacts of tele-communications on future cities in a general and vague way. Actual telecommunications-based developments in real contemporary cities are rarely analysed in detail. Even when they are, because they are so intangible and difficult to untangle that they are often described using physical analogies with the more comprehensible elements of the industrial city. Thus, the satellite ground station becomes the 'tele*port*', the highly capable trunk network becomes the 'information super *highway*'; the computer conferencing system becomes the 'virtual *community*' or the 'electronic *neighbourhood*'; the local community electronic bulletin board is labelled the '*Public Square*'. The wide range of such metaphors and grand scenarios which have now been offered up to describe the increasingly telecommunications-based city is shown in Figure 1.1. This lists the various telecommunications-related labels and metaphors that have been used to describe the contemporary city.

A related tendency is to assume that the 'impacts' of telecommunications on cities are all the same, and are seen to be relatively simple, homogeneous, linear and one-directional (Gökalp, 1988). The difficulty of undertaking empirical studies of such impacts however, means that they tend to remain assumed rather than being tested empirically. Many commentators, for example, have predicted that, because they allow instantaneous communications, telecommunications across distance will automatically undermine the spatial 'glue' that concentrates all large cities (see, for example, Martin, 1978; Toffler, 1981). But usually these expectations remain just that: forecasts of some future urban state rather than empirical analyses of real change. In fact, evidence points to a wide range of

---

The 'invisible city' (Batty, 1990)
The 'informational city' (Castells, 1989)
The 'weak metropolis' (Dematteis, 1988)
The 'wired city' (Dutton *et al.*, 1987)
The 'telecity' (Fathy, 1991)
The 'city in the electronic age' (Harris, 1987)
The 'information city' (Hepworth, 1987)
The 'knowledge-based city' (Knight, 1989)
The 'intelligent city' (Latterasse, 1992)
The 'virtual city' (Martin, 1978)
'Electronic communities' (Poster, 1990)
'Communities without boundaries' (Pool, 1980)
'Electronic cottage' (Toffler, 1981)
The city as 'Electronic spaces' (Robins and Hepworth, 1988)
The 'overexposed city' (Virilio, 1987)
The 'Flexicity' (Hillman, 1993)
The 'Virtual Community' (Rheingold, 1994)
The 'non-place urban realm' (Webber, 1964)
'Teletopia' (Piorunski, 1991)
'Cyberville' (Von Schuber, 1994, quoted in Channel 4, 1994; 1

Figure 1.1 Metaphorical characterisations of the contemporary city

experiences in city–telecommunications relations; a complex set of new processes is leading to a new type of 'telegeography' (Staple, 1992). This is based on the degree to which nation states, regions, cities, rural areas, neighbourhoods and households are the foci of investment in telecommunications or are switched into the new globally driven dynamics of telematics-based change.

Such technological determinism and forecasting does little to foster more sophisticated views of city–telecommunications relations in contemporary cities. In fact, what little evidence there is suggests that these approaches are far too simplistic. In this book we show how the effects of telecommunications on cities seem to be far more ambiguous and complex than many would have us believe. Rather than revolutionising cities by suddenly disinventing them – spreading their contents equally across regions and nations – telecommunications and telematics are intimately involved in complex and diverse incremental urban changes across all areas of urban life. Their impacts on cities are not all the same; they are not even all in the same direction. In fact, when one starts to scrutinise the relationships between cities and telecommunications in more detail, a wide range of complex relationships emerge. These defy easy description and make the use of crude 'shock', 'wave' or 'revolution' labels extremely unhelpful. Tele-communications are intimately involved in many of the social, economic, environmental and geographical changes that make up the urban restructuring process. Invariably, however, the precise nature of this involvement is subtle and difficult to disentangle.

Given the immaturity and neglect of urban telecommunications studies, this book is an attempt to develop a more sophisticated and considered approach to analysing the complex relations between cities and telecommunications. Adopting an interdisciplinary and international perspective, the book aims to help to overcome the divorce which exists between the urban and telecommunications studies communities, so allowing an integrated and socio-technical understanding of tele-mediated urban change to be developed. In other words, we want to explore the complex interactions between technologies and the social, economic, cultural and political change underway in contemporary cities. We aim to avoid the pitfalls of the extremes of optimism and pessimism, of crude technological or social determinism, and of the simple recourse to some all-explaining grand metaphor. Rather, we aim to ground our analysis in a comparative evaluation of the theoretical approaches available, to build on empirical evidence where it is available and to synthesise work from a wide variety of disciplines and sources on the full range of key issues which arise at the complex interface between cities and telecommunications. Critical social science is the perspective towards which we aspire.

In the remaining chapters of the book, we trace how telecommunications are emerging to challenge the prevailing paradigms underpinning urban understanding and policy-making. We explore the theoretical perspectives that can be adopted to explore telecommunications developments in cities. We then go on to review some of the key aspects of city–telecommunications relations – economic, social, environmental, infrastructural, physical and governmental. These themes provide 'windows' through which we can start to explore the complex relations between cities and telecommunications. Although far from perfect, we believe that this broad cross-cutting perspective allows us usefully to construct a more complete picture of these relations than has been built up before.

Before we can do these things, however, we need to introduce the context by summarising the current forces transforming both cities and telecommunications; we do this in the rest of this chapter. First, we shall look at the remarkable technological and regulatory transformations currently underway in telecommunications. Second, we shall briefly review the profound economic, political and social changes that have radically restructured advanced industrial cities over the past fifteen years or so.

THE TRANSFORMATION OF TELECOMMUNICATIONS: FROM THE 'PLAIN OLD TELEPHONE SERVICE' (POTS) TO TELEMATICS

In less than two decades, the telecommunications industry has moved from a slow-moving and largely ignored sector to an important force which is increasingly involved in the current transformation of capitalist society. Already, the telecommunications industry is on the verge of becoming the world's largest; it is certainly the world's fastest growing (*Business Week*, 1994). For example, within Western Europe telecommunications accounted for 2 per cent of Gross Domestic Product in 1984; by the year 2000 this is expected to be 7 per cent and 60 per cent of all jobs will be supported either directly or indirectly by telecommunications (Mulgan, 1991).

STANDARDISATION AND EQUALISATION: THE 'PLAIN OLD TELEPHONE SERVICE' ERA

Incredibly, only fifteen years ago, telecommunications were virtually synonymous with one service – the basic telephone or Plain Old Telephone Service (POTS). At this time, telecommunications meant, effectively, telephones with some minor flows of telexes, telegraphs and data communications (and, of course, the broadcasting services necessary for TV and radio). Between the nationalisation of the first telephone systems at the start of the twentieth century and the mid-1970s, all western nations except the United States maintained a state monopoly over their telecommunications networks through their Postal Telegraph and Telephone (PTT) authorities, who also ran national postal systems (see Figure 1.2). In the United States, AT&T operated a private monopoly in a similar fashion. These monopolies were maintained in order to roll out basic telephone systems, known as Public Switched Telecommunications Networks (or PSTNs) that were universally accessible within and between the cities of the national urban system. PSTNs were based on the use of analogue signals (where the voice was transmitted as an electrical wave that was its direct analogy), electromechanical telephone exchanges (which had physical moving parts to connect lines) and copper wires for transmitting signals. When high capacity was needed coaxial copper cable was laid; when only one telephone was to be linked up, a narrow, twisted-pair copper cable was strung to the house or office. During this period, the telephone was extended from an élite service for perhaps 15 per cent of the population to a service for the majority (60–75 per cent of the population). Telephone services were often seen to be a quasi-public good where a single, universal network was necessary because of the vast costs of developing a network through all parts of the nation state and the need for less affluent users and areas to gain access to the telephone.

Fundamentally, then, the emphasis in this so-called POTS era was on standardisation and social and geographical equalisation. Tariffs to services were kept the same despite very wide disparities in the costs of serving people. Underpinning this were systems of cross-subsidy from the lucrative trunk routes and heavy users (cities and businesses) to rural and lower income users. This was not surprising as PTTs and the universal service concept were developed as part of the elaboration of wider Keynesian welfare states during this period (Lüthe, 1993). 'The arrangement served the important goal of interconnecting society and operated as a means of redistribution' (Noam, 1992; 3). Often, PSTNs were

	'Plain Old Telephone Service' POTS era	Telematics era
Regulation	Single, national monopoly (either public or private), under Postal, Telegraph and Telephone Administration (PTT).	Liberalised competition.
Organisations delivering telecommunications services	PTT	Wide variety: privatised or market-driven PTT, new private entrants, other utilities.
Technologies involved	Analogue Public Switched Telephone (PSTN) network. Mainly copper cables and elecromechanical switches.	Telephone, telex and vast range of data, image and video communications services
Services involved	Telephone, telex, some data, plus separate ratio and television services.	Wide variety: updated digital PSTN interlinked with competing systems and overlays such as radio, microwave plus local cable systems.
Geographical characteristics	Desired universal access to, and diffusion of, telephone: spatial equalisation.	'Cherry picking' of lucrative customers on increasingly international basis focused on big business centres: spatial polarisation.
Urban policy relevance	Negligible. National regulation ensured relatively equal access. Policy-making highly divorced from city level.	Substantial and growing: new importance of access to networks plus growing unevenness of 'telegeography'. New policy opportunities and growth of local social and economic strategies.

Figure 1.2 General characteristics of the transformation from the 'Plain Old Telephone Service' era (POTS) to the 'telematics era'

also laid out in advance of demand as part of the broader effort at state level to stimulate regional equalisation and national economic development. In sum, 'for a century, telephony throughout Europe had been a ubiquitous, centralized, hierarchical network operated by a monopolist' (ibid.; 3).

In general, the distant, centralised nature of PTTs meant that POTS and the PSTN tended to be ignored at the level of city authorities. But the low urban salience of POTS did not mean that the elaboration of national telephone grids had no urban effects during this time. As Pool (1977) has demonstrated, the development of these extensive and standardised PSTN networks had important effects on the development of cities. They integrated national urban systems. They supported the development of central business districts and skyscrapers (as offices could separate from factories and still control them at a distance from central business districts). They allowed social networks to be continued in widely dispersed suburbs. They encouraged the planned zoning of cities, because phone companies came to rely on the predictability that zoning gave their own network expansion plans. Finally, PSTNs supported a whole new range of phone-based business practices (Pool, 1977). More generally, as telephones became diffused through the economic and social fabric of society, they provided an important boost to the elaboration of the mass production and mass consumption system of capitalist society, based on individual households who were able to act at a distance in 'real time' by communicating with each other and a wide range of service providers.

TECHNOLOGICAL CONVERGENCE AND POLITICAL LIBERALISATION: TOWARDS THE TELEMATICS ERA

Since the late 1970s, the world of telecommunications has been in continuous turmoil. Radical technological and regulatory change has been a constant feature. The previously separate areas of telecommunications, computing and media technologies are now converging around a core group of digitalised technologies (see Figure 1.3). Essentially, these new technologies allow all types of information – voice, data, sounds, images and video signals – to be processed and transmitted in the form of the 'bit' streams of binary code used in computers – as a series of zeros and ones. This is the basis behind the much-vaunted multi-media technologies and services, which allow sounds, images, voice and data to be

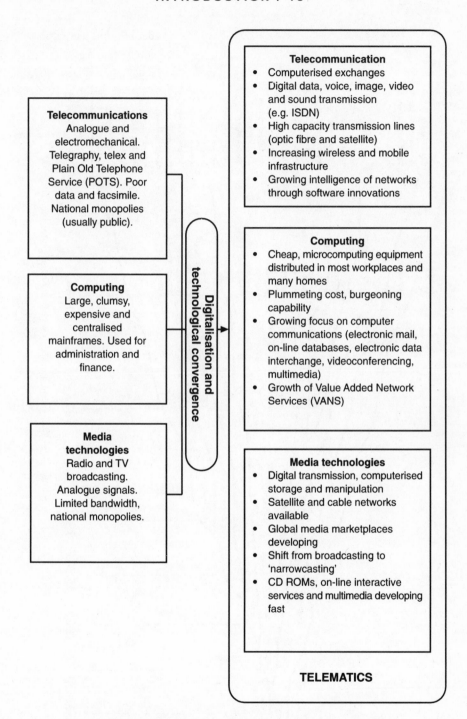

Figure 1.3 Technological convergence and the development of telematics

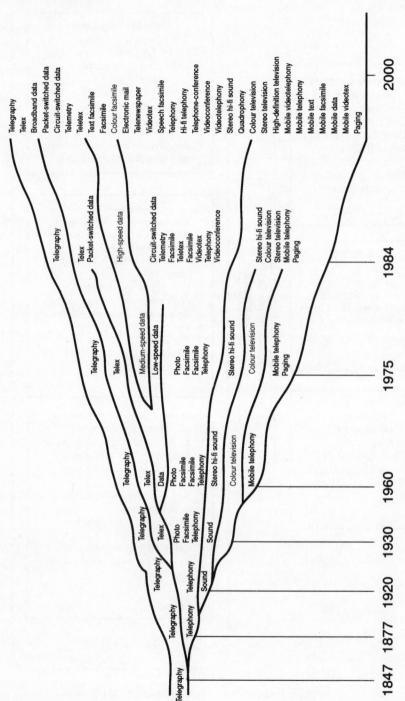

Figure 1.4 The expanding range of telecommunications services: prospects for the year 2000

Source: Adapted from Ungerer, 1988

managed and transmitted in an integrated way. Digital telecommunications are more capable, more accurate, more flexible and often less costly than analogue telecommunications. The result has been an explosion of technologies and services and a remarkable rate of change (see Figure 1.4). A multitude of telecommunications networks and services have been proliferating within and between cities ever since and as this innovation tends to be most advanced and rapid in cities, it is centrally important to urban development.

Unfortunately, this dazzling variety and speed often defies understanding of the technological and regulatory dynamics at work. The simplest way to try and understand the radical shift in telecommunications is to split it into its four interrelated elements. These are:

* changes in the types of switching used in telecommunications networks;
* changes in the way that the transmission of communications signals occurs between the switches;
* changes in the terminal equipment that are the sources and destinations of communications flows;
* shifts in the ways in which telecommunications are regulated.

Digital switching and intelligent networks

On the switching side, now that telecommunications networks are largely digital, their switches increasingly consist of highly sophisticated computers rather than sets of electromechanical equipment. The national telecommunications infra-structures of the POTS era are increasingly being upgraded to digital standards. This makes them capable of dealing with higher volumes of traffic more cheaply and with greater accuracy. Video, image and data signals can be handled as easily as voice flows – provided the networks have the bandwidth (transmission space) and switching capabilities to accommodate these. New, more capable tele-communications systems for carrying this broader range of services are also being developed (for example, the Integrated Services Digital Network (ISDN) which offers seamless communication of voice, data, image and sound over a basic phone line). New advances here in using computers to compress signals mean that much more can be squeezed out of even the basic copper-wire telecommunications link to the telephone.

At a more advanced level, completely new computer-controlled switching systems are being deployed; an example of this is the so-called 'Intelligent Network'. Here the control of service flows and the management of vast telecommunications networks rest with a few centrally located computers and the sophisticated software which runs on them. This means that the flexibility and capability of networks is much enhanced and can be upgraded simply by reprogramming computers rather than relaying networks or replacing switches.

Increasingly, such networks operate at the global scale, geared to the sophisticated communications needs and the lucrative markets of transnational corporations (TNCs). The $10 billion per year market for the leading 2,500 TNCs drives technological innovation, as the TNCs attempt to develop single networks on a global basis through which their flexibility, speed of response and competitiveness can be improved (Keen, 1986; Valovic, 1993). The scale of such private, corporate networks, which weave many localities and cities together in 'real time', often now surpasses that of many national telecommunications infrastructures. IBM's own network, for example, has 800,000 users in 90 countries (Roche, 1993).

New telecommunications networks and infrastructures

On the transmission side, the traditional copper and coaxial cable links are increasingly being supplemented or replaced by optic fibre, wireless, microwave and highly efficient satellite systems. The capabilities of optic fibres are now legendary and have entered popular consciousness: each hair-like strand can now accommodate up to 60,0000 simultaneous telephone calls (as opposed to 6–7,000 for a much wider coaxial cable). This technology has a particular lustre which attracts technological forecasters. Many utopian forecasts about telecommunications and cities centre on the idea of a universally accessible optic fibre grid to all homes. Cost and technical barriers, however, still mean that this is far from being realised: optic fibre remains concentrated as the trunks of national and global electronic grids or the local links to very large users of telecommunications such as office blocks. Often, the largest cities, as the main centres of demand for advanced telecommunications, now have substantial and fast-growing optic fibre infrastructures at the metropolitan level. This is demonstrated in Figure 1.5, which shows the current growth in Melbourne's optic fibre network. Increasingly, these metropolitan networks are being connected through international and

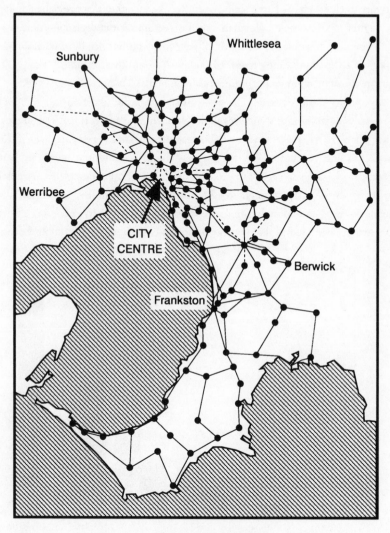

------ Existing optic fibre
network, 1989
———— Planned optic fibre
network, 1994

Figure 1.5 The expanding optic fibre network in Melbourne, Australia

Source: Adapted from Newton, 1991

intercontinental optic fibre networks. The growth of transoceanic optic fibre networks across the Atlantic and Pacific has been especially crucial in underpinning the shift towards a global information economy. Between 1986 and 1996, for example, the number of optic fibre 'voice paths' across the Atlantic is expected to rise 66 times from 22,000 to 1.45 million (Staple, 1992). This dramatic growth is shown diagrammatically in Figure 1.6.

Advances in both switching and transmission have allowed four new ranges of telecommunications infrastructures to develop which are largely centred on cities. First, there are wireless and mobile communications systems which link telephones and computers by radio signals to fixed telephone networks. Such systems are laid out in 'cells' to build up coverage of a building, city or region. Others work directly through satellite connections. Wireless networks allow telephone and data communications to operate on the most flexible basis and are being widely heralded as the next revolution as communications systems become

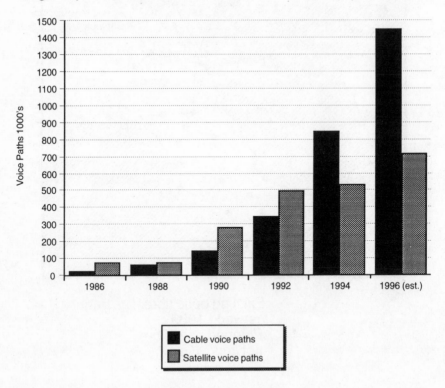

Figure 1.6 The number of voice paths across the Atlantic by satellite and transoceanic cable

Source: Adapted from Staple, 1992

mobile and personal rather than tied to fixed buildings, as with the traditional PSTN network. Wireless systems are currently the fastest growing area of telecommunications and this rapid growth has led some to proclaim that 'when the history of communications in the 20th century is written, the 1990s will be remembered as the decade of wireless communications' (Spector, 1993; 403). Annual rates of increase in the numbers of mobile phones globally approach a remarkable 50–60 per cent, as technology has improved, networks have been rolled out to wider and wider areas, costs have been slashed, and competition between new private mobile operators has grown (*Financial Times*, 1994). New technologies and lower costs are now allowing wireless phones to begin entering the mass market. Physically vulnerable groups and blue-collar workers who face customers are now using mobile phones as well as the stereotypical 'yuppies' who dominated their use in the 1980s (Wood, 1994).

However, because of the costs to private companies of rolling out these systems they tend to be centred on cities and urban corridors where demand is most concentrated. Figure 1.7 shows the way in which the new digital mobile networks within Europe known as GSM are developing along the main business corridors while ignoring the more rural and remote areas.

Second, there are *broadband cable networks*. These are overwhelmingly urban networks, built primarily to deliver Cable TV (CATV) services to residential consumers. Cable is the basis for distributing much larger numbers of TV channels that have traditionally been accessed via terrestrial radio-based TV transmission. Increasingly, however, these networks are becoming more sophisticated and capable, with optic fibre trunks and new digital switching technologies complementing or replacing the coaxial copper cable of previous systems. The result is a slow growth of other interactive services, from telecommunications to 'pay per view', teleshopping and other value added services. In some nations, cable networks are allowed to compete directly with the PSTN in mainstream telecommunications services, as in the UK, where many cable companies make larger revenues from telecommunications than television services.

Third, there are a new generation of *satellite* infrastructures. In 1990 there were over 200 communications satellites in orbit, from eighteen nations or communications agencies (Kellerman, 1993; 46). Demands for satellite services are growing fast, particularly in Europe and Asia. As the skies become more crowded with a widening range of powerful satellites, a plethora of new services – from broadcasting through high-speed data communications to paging, navigation and mobile phones – are developing based on satellite communications. The proliferation of satellite TV, known as Direct Broadcasting by Satellite (DBS)

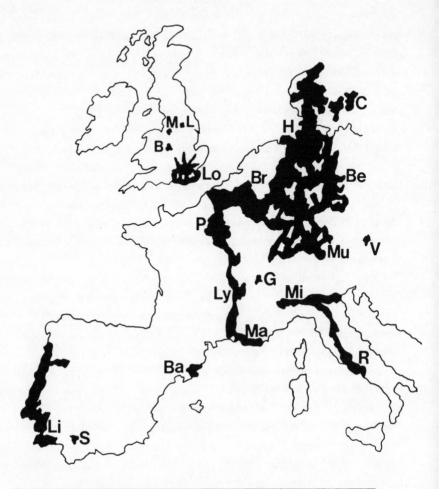

Key to cities

B	Birmingham	**Ly**	Lyon
Ba	Barcelona	**M**	Manchester
Be	Berlin	**Ma**	Marseilles
Br	Brussels	**Mi**	Milan
C	Copenhagen	**Mu**	Munich
G	Geneva	**P**	Paris
H	Hamburg	**R**	Rome
L	Leeds	**S**	Seville
Li	Lisbon	**V**	Vienna
Lo	London		

Figure 1.7 An example of the urban bias of market-led telecommunications development: the availability of GSM digital mobile telecommunications in Europe, 1992

Source: Adapted from Arnbak, 1993

is the best-known example here, but there are many other satellite services available geared towards voice, data and video communications.

The ground stations needed to link up with satellites used to require major developments and special sites. Often this rare example of telecommunication being very visible and requiring major physical developments led to large-scale planning and urban development projects like teleports through which cities ambitiously tried to develop their economies. Now, however, miniaturisation and technological changes are reducing these requirements. Satellite phones and transmission equipment can now be fixed to individual buildings or even be portable by humans. This is particularly so with a new range of Very Small Aperture Terminal (VSAT) networks, which can provide a flexible and cheap substitute to terrestrial corporate networks because they rely on small micro ground stations that can be attached easily to individual buildings. Thus the great advantages of satellite communications – its flexibility, the pervasive coverage of satellite 'footprints' and the relative ease of deployment – are becoming even more important for specialised applications.

Finally, there are *microwave systems*. These support short-distance, point-to-point and line-of-sight transmission for voice, data or TV/video services. They are especially useful in congested city centres where they link heavy tele-communications users with the nodes of terrestrial networks, so bypassing the often inadequate cables in the local loop between nodes and customers' premises. When many microwave towers are strung together, whole national and international telecommunications networks can de developed.

The layering of urban telecommunications infrastructure

With this proliferation of infrastructures beyond the basic PSTN, many telecommunications networks are now being superimposed within and between cities where only one – the PSTN – tended to exist previously. This layering of the telecommunications networks within a typical western city is illustrated in Figure 1.8, which shows schematically how the older, modified PSTN now tends to be superimposed with other competing telecommunications networks (where liberalisation has occurred), which centre on the most profitable areas. It also shows the layouts of cable and mobile systems and satellite ground stations. In fact, this itself is an oversimplification. Such are the fast-moving dynamics in many

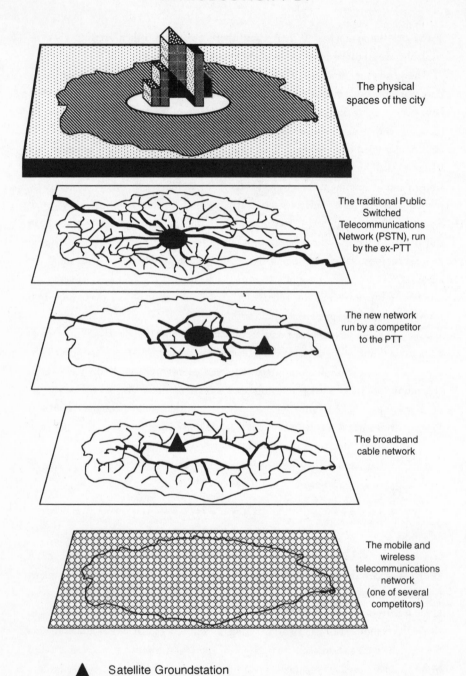

The physical spaces of the city

The traditional Public Switched Telecommunications Network (PSTN), run by the ex-PTT

The new network run by a competitor to the PTT

The broadband cable network

The mobile and wireless telecommunications network (one of several competitors)

▲ Satellite Groundstation

Figure 1.8 The superimposition of different telecommunications infrastructures within a typical medium-sized western city

larger cities that up to a dozen different telecommunications networks can be superimposed within the centres of cities like London and New York, based on complex combinations of hard infrastructure, private leased lines, and 'resold' capacity on general telecommunication lines.

The transformation of customer premises equipment (CPE)

Paralleling these changes in transmission and infrastructure, the terminal equipment which subscribers use to communicate with has been completely transformed in the past fifteen years. This change has been most dramatic in computers. The bulky, expensive and slow computers of the 1960s that used to sit squat in the basements of office blocks are now museum pieces. The radical improvement in the performance and cost of microelectronics has meant that smaller, cheaper and far more powerful items of computerised equipment now inhabit virtually every workplace and a growing proportion of homes. Personal computers (PCs), in particular, have diffused through all economic and social sectors and are now an accepted fixture of everyday life. The processing power of PCs now doubles every two years while their costs are actually falling. Increasingly, these lowering costs and increased capabilities mean that personal computers are shifting from stand-alone workstations to communications and work terminals, linked via the improved telecommunications systems discussed above. In many professional circles, an electronic mail address is an increasingly expected element of a person's contact details.

A crucial development here has been the improvement of computer network technologies. The most important twin advances are the development of broadband computer networks within small sites or a small group of buildings known as Local Area Networks (LANs) and the linking together of computers, computerised equipment and LANs across wider geographical distances via telecommunications lines such as the networks used by transnational corporations known as Wide Area Networks (WANs). In addition, many computer networks and broadband infrastructures at the scale of the city are developing called Metropolitan Area Networks (MANs).

Some WANs, having reached a critical mass of users, are growing with startling rates. For example the Internet – the network of Wide Area computer networks that links universities, research institutions and thousands of organisations around the world – is now estimated to have between 20 and 35 million electronic mail users and to be growing at around 10–20 per cent per month. Figure 1.9 shows the exponential recent growth of the Internet, which is now expanding into many new commercial areas, driven by the increasing involvement of media conglomerates who are trying to take advantage of the network's commercial potential.

Many other types of terminal have also developed to complement the phone and PC as entry points on to telecommunications networks. Facsimile machines are now extremely common. Pagers are increasingly being used to keep in touch. Mobile computers and Personal Digital Assistants (PDAs) are providing sophisticated platforms for managing complex business schedules on the move. Videotex terminals (like the French Minitel) have diffused widely (although unevenly). Digital entertainment and media systems within the household are proliferating, such as CD-ROMs and interactive CDs. Some of the latest of these use 'virtual reality' technology in which people are immersed in 'virtual environments' constructed by computers. For these a wide range of what Mitchell (1995) calls 'exoskeletal devices' – 'datagloves', helmets, visual and audio sensors and

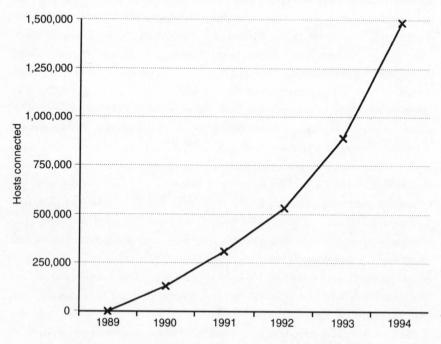

Figure 1.9 The exponential growth of the Internet 1989–1994

Source: Ogden, 1994

surrounding screens – are being developed which link into computers and allow this real-time sensory immersion to take place. In addition, microchips now control many phones and an increasing range of domestic and industrial appliances – TVs, stereos, utility meters, even cookers and heating systems – as well as office, transport and factory equipment. All of these are already geared, or are being redesigned, to orient towards inter-communication across distance via telecommunications networks. In addition, many pieces of terminal equipment are now being developed which aim to take advantage of the new switching and transmission capability of telecommunications: videophones, ISDN terminals, highly capable interactive and high definition televisions and multimedia personal computers, to name but a few.

On the financial front, networked electronic fund transfer (EFT) and point-of-sale terminals (EFTPOS) are replacing traditional shop tills. Together with electronic credit cards and so-called 'smart cards' (which have a microchip embedded in them for storing information and money), these are allowing an increasingly diverse range of transactions to occur electronically and cash free, through transfers of money between bank accounts. In the manufacturing and logistics areas, Electronic Data Interchange (EDI) terminals are stringing out across supply chains, providing computerised systems for linking firms into complex transactional chains. Automatic Teller Machines (ATMs) are complementing bank cashiers. Computerised production machinery is replacing mechanical equipment.

Such is the crucial significance of these widening arrays of high technology equipment that the US Commerce Department recently estimated that fully 38 per cent of all economic growth since 1990 has derived from such sales (*Business Week*, 1994; 36). If we consider the global set of telecommunications networks to be a single set of global networks, it is possible to trace the growth in the number of terminals attached – wired telephones, mobile telephones, fax machines, data modems and videoconferencing sets. This is shown in Figure 1.10.

Not surprisingly, paralleling these trends in switching, transmission and terminal equipment has been an explosion of innovation and a proliferation of telematics services supporting the communication of data, sounds, images and video between these diverse sets of computerised equipment – as well as advanced forms of human voice communications. Figure 1.4 (p. 16) shows how this convergence is allowing a rapid growth in the range of telecommunications available. For the first time, such telematics services, which are embraced under the umbrella term Value Added Network Services (VANS), allow many different

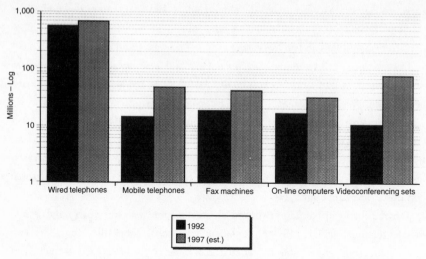

Figure 1.10 Equipment plugged into the world's telecommunications networks
1992–1997 (estimated)

Source: Staple, 1992

types of computer-mediated information flows (e.g. on-line databases), computer-mediated communications (e.g. electronic mail), and computer-mediated transactions (e.g. Electronic Data Interchange or Automatic Teller Machines) to be supported both within and between cities. Such services increasingly merge voice, data, image and video signals in complex mixtures of multimedia telematics flows.

Just as important, though, are the plummeting costs of telematics hardware, software and services. As each year passes, more information can be processed and sent longer distances for lower costs. Technological change, the growth of markets and increasing competition is leading to rapid cost reductions while performance and capability are continually enhanced. The example of France, which is summarised in Figures 1.11 and 1.12, is typical here. While personal computers in France are less than one third of the real costs they were in 1988 (Figure 1.11), the real costs of telecommunications are also falling dramatically (Figure 1.12).

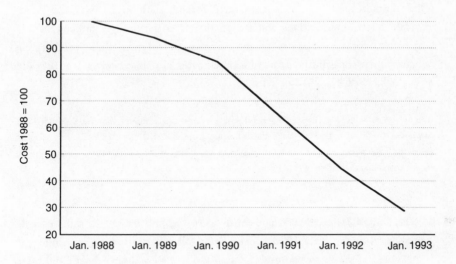

Figure 1.11 The plummeting cost of personal computers: the French case 1988–1993
Source: Adapted from Volle, 1994

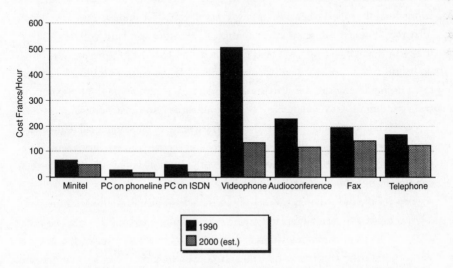

Figure 1.12 The plummeting costs of telecommunications services: the French example
Source: Volle, 1994

Changing regulatory regimes

These inter-related changes are, in turn, forcing shifts in the way that states regulate telecommunications and the media. It is increasingly difficult, if not impossible, for nation states to monopolise control over these proliferating telematics networks in the name of a single national public interest – as they did with the basic POTS/PSTN service:

The idea of collective solutions, of a general public interest that is more important than individual interests, has found it hard to make the transition from the relatively standardised services of telegraphy, telephony and television to the more complex worlds of the fax, electronic mail, videotex and high-definition television.

(Mulgan, 1991; 243)

The United States, Britain and Japan were the first nations to shift towards competition in the early 1980s; increasingly, all western nations are finding it impossible not to do so because of the risk of business disinvestment and poor competitiveness in this key sector. Pressures from international business interests to deliver and benefit from a global, commercialised marketplace in telematics are becoming increasingly powerful. John Sale, a network strategist for Rank Xerox, summarises the approach of TNCs and their lobbying for more liberalisation in continental Europe, where shifts away from the PTT monopolies have been relatively slow. He admits that his company's 'position is that if you don't allow it in [continental] Europe, we will build it in the U.S. or get the permission to do it in the U.K.' (quoted in Schenker, 1994; 12). National political movements towards neoliberalism are also pushing privatised and marketised regimes of telecommunications regulation as the route towards faster innovation and national economic competitiveness in these vital infrastructures. Supranational bodies such as the European Union, the G7 and the General Agreement on Tariffs and Trade (GATT) are also becoming powerful advocates of global, liberalised markets in communications and telematics services.

As a result, most nation states are now liberalising and often privatising their telecommunications regimes and turning them into marketplaces for a multiplicity of different communications and information services. Usually, this process starts by relaxing controls over who can deliver telematics services or develop new telecommunications infrastructures like mobile. Although national approaches vary considerably, partial or full privatisation of the PTT often follows along with new regulations allowing other entrants to compete in telecommunications markets. When this occurs, privatised PTTs such as BT in the UK often emerge as some of the largest and most powerful private corporations within national and local economies.

These complex changes mean that telecommunications and telematics markets are extremely complex and fast-moving. Within them, incumbent PTTs (for example, France Télécom) fight it out to serve lucrative users with many new market entrants. Included here are new private sector operators (like MCI, BT and AT & T), specialised global data network operators (for example, IBM), basic infrastructure providers who are increasingly involved in telecommunications (utility and transport companies such as British Rail), private IT suppliers (like Motorola) and smaller entrepreneurs (like Hutchinson Telecom) (Kok, 1992). Increasingly, these organisations are forging global alliances so that they can extend to cover the lucrative and booming markets for global telecommunications on a one-stop basis, focusing on TNCs (Cooke and Wells, 1991).

Thus, companies that were only a few years ago privileged as public monopolies within nation states now face many foreign competitors on their home soil, as well as the imperative to extend their own efforts to other nations. The result is a frantic process of globalisation, with telecommunications services and infrastructures driven more and more by an international logic and international flows of capital. To make matters even more complex, the technological convergence between computing, telecommunications, information and broadcasting industries is also stimulating a flurry of mergers, acquisitions and alliances as tentative efforts are made to make the most of the blurring of boundaries. Cable companies are joining telecom companies; newspaper and media conglomerates are buying into cable companies; equipment companies are forging alliances with entertainment and telecom companies. Telematics and media giants are therefore emerging which are truly global and well placed to try and engineer and benefit from the growing multi-media synergies between telecommunications, computing, media and services within global, private markets. AT&T, for example, has recently announced a World Partners Program, which links it to several PTTs; BT and MCI have an alliance to build global networks; France Télécom and Deutsch Telekom have developed a partnership; and Microsoft, the software giant, and TCI, the cable company, recently formed an alliance to develop interactive television.

These extremely rapid processes of globalisation and restructuring are leading to a more international and open set of liberalised markets which are linked closely into international flows of technology, services and capital. But, ironically, this convergence in national regulation is actually associated with an increasing heterogeneity in terms of the spatial and social development of telecommunications infrastructures and services. The old POTS-style certainties of standardisation and equalisation are replaced by fragmentation and polarisation. Not surprisingly, market-based regimes for telecommunications tend to encourage much greater extremes and contrasts between the most favoured groups and areas and the least favoured. This tends to mean that cross-subsidies supporting universal services tend to decline and much more variegated patchworks of telecommunications infrastructures roll out within and between cities. Recent experience in Britain shows that while companies compete ruthlessly to 'cherry pick' from lucrative areas and markets such as those dominated by large financial services companies, poor areas with low demand are left with inferior and older infrastructures (Graham and Marvin, 1994). Sometimes, the most marginal users of the telephone actually drop off the basic networks, driven by higher line rents and tariff rebalancing (a process through which rates are cut for the higher and longer distance users and increased for the marginal users who benefited most from the universal service idea). This polarised pattern in access to, and use of, telecommunications, is even more extreme when the more advanced services are considered. The result is a complex and uneven electronic landscape within and between cities.

But the stakes here go way beyond telecommunications policy and the development of cities. Industrial, innovation and trade policy are also crucially affected by these global shifts, stimulating Japan, Europe and the USA to construct 'information superhighway' policies with which to boost their economic positions and their strength in export markets for hardware, software, consultancy support and services (Lanvin, 1993). In the United States, the frenzy of alliance formation between the 'Baby Bells' – the regional telecom monopolies – and cable and entertainment companies effectively consists of a jostling for position to make the profits that seem likely with the development of the much-vaunted information superhighway or national information infrastructure (NII). In post-Maastricht Europe, the aim is to replace the fragmented set of national monopoly telecommunications and telematics systems with a single market made up of one 'European Information Space' (European Union, 1994). All these shifts also have important implications for cities and urban development because cities are the centrepieces of these national and increasingly international telematics marketplace (Graham, 1993).

THE TRANSFORMATION OF CITIES: TOWARDS PLANETARY URBAN NETWORKS

The rest of this book will be concerned with exploring in detail how these technological and regulatory developments affect cities and urban development. Nevertheless, it is also necessary here to examine briefly the remarkable shifts in cities that have paralleled these radical changes in the telecommunications sector. A key argument of this book is that the complex development of telematics and their infusion into cities cannot be divorced from considering the parallel crisis and restructuring that is underway in western cities themselves. The last twenty years have witnessed remarkable changes in the economic, social and geographical makeup of western capitalist cities – and in their political and cultural dynamics (Moss, 1987; Healey *et al.*, 1995).

ECONOMIC RESTRUCTURING, TELECOMMUNICATIONS AND CITIES

The end of the long post-war boom in western capitalist society has triggered a massive restructuring which has radically altered cities. Globalisation and the intensification of global competition have torn away the traditional industrial fabric of many western cities through 'deindustrialisation' (Dicken, 1992). Huge transfers of manufacturing activity have focused on less developed and newly industrialised countries (LDCs and NICs) creating a global division of labour. The vertically integrated manufacturing giants of the so-called 'Fordist' era are everywhere being replaced by more responsive and flexible networked corporations operating across these global distances and tending to buy in goods and services from small firms. In western cities, information, high-tech manufacturing, service and leisure industries have grown (albeit patchily), forcing great changes in urban labour markets and urban socioeconomic dynamics. Political shifts towards liberalisation and the growth of investment markets have led to a remarkable boom in financial services. This has fuelled the growth of the larger cities which are placed at the hubs of the global electronic and financial services networks.

The result of these shifts is that economic activity involving processing and adding value to knowledge and information now dominate the economies of western cities as never before (Knight and Gappert, 1989). Even commodity-based industries such as retailing and manufacturing are increasingly information rich. With the unprecedented turbulence, competitiveness and volatility of markets, higher inputs of knowledge and information are being used to reduce uncertainty and improve responsiveness. Because information has such a central place in production in all sectors, it, too, is emerging as a key commodity to be bought, sold, traded and exchanged in markets. This is made possible by the capabilities that telematics bring for processing, storing and controlling vast flows of electronic information on a continuous and real-time basis. In short, as a result of all these shifts, industrial cities have been transformed into 'post industrial' (Savitch, 1988) or 'information' cities (Hepworth, 1987), dominated by consumption industries and the processing and circulation of knowledge and symbolic goods rather than physical goods. A corresponding shift has gone on in the labour markets of cities, with 60–70 per cent of new jobs typically now concerned with some form of information processing, distribution or production. Investment in telematics – the basic information infrastructures of cities – now

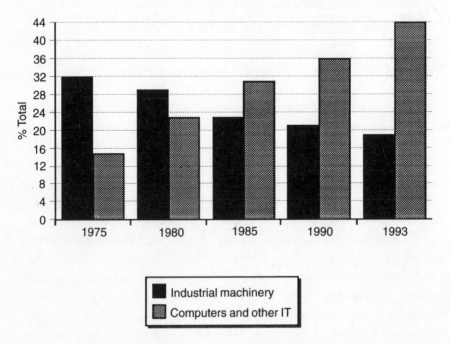

Figure 1.13 Industrial investment in the USA in telematics and other industrial machinery

Source: *Business Week*, 1994

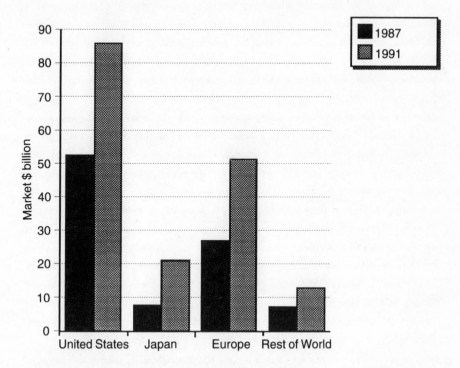

Figure 1.14 The global market for computing and communications services 1987–1991
Source: Gille and Mathonnet, 1994

surpasses investment in other industrial machinery (see Figure 1.13 for the example of the United States). Because telematics make information so easy to move around, these shifts increasingly take a global complexion, tied in with the wider shifts towards globalisation and the growing power of TNCs. The result of the pervasive application of telematics across all economic sectors is that markets for computers and communications equipment are growing rapidly. As shown in Figure 1.14, this growth is dominated overwhelmingly by the three urban-industrial blocks which are the powerhouses of global capitalism: North America, Western Europe and Japan.

This economic transformation in cities, however, has been associated with the growth of structural unemployment. The growth of well-paid, knowledge intensive jobs has been dwarfed by the loss of manufacturing jobs and the growth of poor quality jobs in retailing, leisure and tourism. Globalisation and the application of telematics also seem to be associated with the fracturing and disintegration of city economies, as they become 'exposed' to global telematics networks that invisibly and silently cross-cut them (Virilio, 1987). The result is that cities are being restructured from internally integrated wholes to collections of units which operate as nodes on international, and, increasingly, global economic networks (Dematteis, 1994). It is increasingly impossible to understand the forces that are shaping the restructuring of cities from a purely local perspective; contemporary city economies can only be understood through their relations to global economic, political and technological changes. The instantaneity of telematics networks is a key facilitator of this linking of the 'local' into the 'global', largely through the construction of corporate telematics networks and global transport networks (Mazza, 1988). As a result, cities are now being tied together with a new level simultaneity. The interactions between and within cities now approaches real time – or at least operates with an unprecedented velocity. The global urban world now operates as a vast set of international systems based on electronic telecommunications-based flows of information, money, services, labour power, commodities and images as well as by advanced transport networks. Dematteis speaks of the emergence of a 'planetary metropolitan system' (Dematteis, 1988).

But the fortunes of cities are very uneven within these shifts. The world financial capitals have emerged as key command and control centres where the best jobs are located. Certain smaller, usually non-industrial cities have managed to specialise in advanced manufacturing, research and development or high technology services. Others have emerged as key tourist centres. At the same time, however, many older industrial cities have had to compete even for low order service jobs such as those in back offices, branch plant manufacturing and shopping centres. In this context, because of the speed of these systems and the erosion of the attachment to place, city economies are more turbulent and face very uncertain futures.

URBAN SOCIAL AND CULTURAL CHANGE

Such economic restructuring and the telematics-based globalisation of cities has been associated with profound urban social and cultural change. On the social front, the urban economic crisis of the 1980s precipitated the end of the post-war welfare–Keynesian consensus and the ascendancy of neo-liberal approaches to urban management (Harvey, 1989; Healey et al., 1995). In concert with the effects of economic restructuring, these changes have forced social and geographical polarisation within cities.

The ways in which new telecommunications and telematics innovations are involved in the social life of cities tends to both reflect and support this polarisation. While affluent and elite groups are beginning to orient themselves to the Internet and home informatics and telematics systems, other groups are excluded by price, lack of skills or threaten to be exploited at home by such new technologies. Advanced telecommunications and transport networks open up the world to be experienced as a single global system for some. But others remain physically trapped in 'information ghettos' where even the basic telephone connection is far from a universal luxury.

The growing divisions between affluent and poor areas can lead to rising fear of crime, the 'fortressing' of neighbourhoods through electronics surveillance systems, and an increasingly home-based urban culture where people's working, shopping, access to services and social interaction may become mediated more via telematics rather than by social interaction in the public spaces of cities. The parallel shift toward market-based telecommunications regimes has added further momentum to this polarisation and growing unevenness in the social landscapes of cities. It is clear that while many speak of 'time–space compression' through new technologies, it is important to differentiate between the diverse social experiences of these processes of change for different groups of people.

A major area of debate currently centres on the degree to which telematics can be used to support socially liberating and progressive changes within cities – by overcoming the isolation of disabled groups for example. Whilst there are certainly some examples of such liberating applications, for example with 'virtual communities' for marginalised and housebound groups, critical commentators stress that, on the whole, such technologies may in fact be a basis for exacerbating further the social and geographical polarisation within urban places. Current evidence certainly suggests that the future of those in cities who are at the margins of the information society seems grim. The hyped-up promises of technological fixes in the new telematics era threaten to have a distinctly hollow ring when considering the position of the most disadvantaged groups within the contemporary city.

These changes in the social dynamics of western cities are in turn interwoven with the *cultural* dimensions of globalisation, a process which is closely tied into wider shifts towards a global, 'post-modern' urban culture (Lash and Urry, 1994). For example, advances in telecommunications and telematics are, along with the liberalisation of media regulations, helping to support the emergence of truly global cultural and media industries. These feed into support social and cultural change in international systems of cities (Morley and Robins, 1995). There is little doubt that the human experience of place and the social construction of cultural identities by groups and individuals are being radically altered in these 'global times'. This is because new advanced telecommunications act as conduits for flows of images, knowledge, information and symbols which integrate places and people into the global cultural system in 'real time'. Thus, traditional national mass broadcasting systems are giving way to a broadening range of global systems of mass communications (cable and satellite TV) and interactive personal communications (such as electronic mail and the World Wide Web on the Internet). Through these systems a growing proportion of social interaction and cultural flow take place.

But interpreting the implications and effects of these shifts remains highly contested. To some, shifts towards a more participative and interactive media culture through the explosion of the Internet and 'cyberspace' represent the empowerment and liberation of individuals and groups who were simply passive consumers of media in the past. To critics, though, the globalisation of urban culture is little more than the ruthless commodification of all information by fast-growing media conglomerates such as Rupert Murdoch's News International corporation. It is argued that this erases local differences; it superimposes western cultures over non-western ones; it polarises social access to information; and it leads to a cacophony of signs which often alienates urban inhabitants within an 'uprooted' and bewildering global culture.

Whether one stresses the positive or critical interpretations, or some blend of the two, there is little doubt that, in this movement-dominated post-modern world, the whole cultural idea of cities is being redifined. This transformation is creating a 'global sense of place' which challenges all previous ideas about what it means to live in a region, nation or city. But this does not simply erase the attachment of people to urban places. Instead, a complex interaction between telemediated cultural exchanges in electronic spaces and place-based ones in urban places is emerging. Networked communities of interest now span the globe, based on specialised foci of interest or various combinations of ethnicity, gender, sexuality, profession, etc. These mesh and interact with place-based communities, but in ways which we have only begun to understand.

URBAN ENVIRONMENTS

These economic and social shifts have led to a growing concern amongst urban planners and policy-makers to address the *environmental* dimensions of their cities. Planners are trying to address the legacies of pollution and dereliction from the industrial era as well as the side-effects of burgeoning traffic congestion. The need to compete as an attractive business environment is joining with wider social awareness to force environmental issues to the top of the urban agenda. Concern now centres on the need for environmentally sustainable urban futures. Once again much attention here has turned to the potential role of telecommunications and telematics for contributing towards the development of more sustainable cities (Gillespie, 1992). Most often, however, this analysis has been hampered by the assumption that telematics-based flows of electronic information can be used directly to substitute for the environmentally damaging flows of physical transportation. More critical scrutiny of such ideas casts doubt on such claims, however, because they rely on oversimplified conceptions of the relationships between urban environments and telecommunications.

Rather than simply substituting for transportation, telecommunications have a wide range of contradictory linkages. Telecommunications can help to stimulate more travel as cheaper and more accessible forms of communication generate new demands for the physical movement of people and goods. At the same time new services such as road information systems and auto route guidance can help drivers to overcome the uncertainties of traffic congestion, thereby improving the attractiveness of the road network. Although telecommuting and teleworking initiatives may be able to help reduce levels of peak-time congestion they have a multitude of second-order effects. The teleworker has to heat and power the home during the day, the road space created by the teleworker can quickly induce new traffic on to the road network while weakening the need to live in close physical proximity to work can encourage urban sprawl. Similar contradictions also occur in the energy, water and waste sectors. Telecommunications do not necessarily lead to reductions in material flows through cities. While the new control capabilities of telematics do have the potential to shape the level and/or time of resource consumption the current supply-oriented logic of network management does not necessarily provide much incentive to infrastructure to reduce total flows.

URBAN TRANSPORT AND INFRASTRUCTURE

Telecommunications are radically transforming the management and provision of urban transportation and infrastructure networks. Old ideas about the role of monopolistic, standardised and universally available infrastructure networks available to meet all demands for movement, mobility, heat, power and water supply are being challenged. Increasingly telecommunications are supporting the splintering of infrastructure services to facilitate competition on what were previously considered to be monopolistic networks. Telecommunications enable infrastructure providers more effectively to control their networks by identifying the costs of servicing different types of customers. At the same time these capabilities provide the opportunity of providing premium, enhanced and value added services to particular groups of customers.

The application of telecommunications technologies in the management of networks is developing in tandem with shifts towards the liberalisation and privatisation of infrastructure networks. Privatisation has transformed service provision from monopolistic, universally available, standardised systems into complex new patterns of service provision. Increasing levels of choice for large users mean that incumbent operators are forced to remove cross subsidies from large to small users to compete with new entrants who cherry pick their most valuable customers. Telematics play a central role in these new logics of infrastructure management. For instance, Geographical Information Systems (GISs) provide the tools for allocating costs to different types of customer, allowing new entrants to target the most lucrative and profitable customers. The new control capabilities of telematics networks being fitted over what were previously 'dumb' infrastructure networks provide much more sophisticated control and operational systems. These can significantly improve the efficiency and profitability of networks by more effectively balancing demand and supply of services. New smart metering technologies enable premium customers to have increasing levels of choice while prepayment metering technology based on smart cards allows utilities to socially dump expensive, marginal and poor customers.

Telecommunications are also having profound implications within the transportation sector. Historically, there have always been close linkages between communication technologies and transportation. The telegraph and later the telephone enabled railway companies to standardise time while monitoring and controlling the movement of trains across their rail networks. In the contemporary city these linkages have significantly strengthened as the bundle of applications within the road transportation informatics technologies lead to important changes in the use and management of transportation networks. Electronic road pricing systems have been demonstrated in cities across Europe allowing transport authorities potentially to charge for road space according to levels of congestion in real time. The development of freight logistic systems linking together production and distribution sites electronically is enabling companies drastically to cut warehouse stocks and centralise warehouse and distribution functions, thereby more closely utilising the transport network as part of the production process. These companies are also relying more heavily on transportation informatics to maintain real time contact with drivers to route deliveries according to congestion levels and the demands of retailers. Public transport operators are also utilising telematics to provide real time information to users and to make efficient use of expensive transport capacity by more efficiently meeting demands.

Consequently, telematics are facilitating radical transformation in all types of infrastructure networks. New information and telematics technologies are helping to create new markets in infrastructure services, introduce competition on to networks, differentiate between particular types of customers and provide a wide range of value added services. But these new logics of network management hardly figure in the urban policy literature. With the exception of tele-communications networks themselves, there has been relatively little academic or policy interest in how the new capabilities of telecommunications are transforming the management and provision of urban networked services.

URBAN PHYSICAL FORM

Reflecting the economic, social and political restructuring of western cities, the urban *landscape* and *physical urban forms* of advanced industrial cities are in turn being radically reshaped. Global economic forces are taking over local property markets. Derelict or decaying old industrial spaces are being transformed into post-modern urban developments as foci of global consumption and culture (Relph, 1987; Knox, 1993). The sprouting of new telecommunications equipment, and the infusion of many new telecommunications infrastructures into the old fabric of the city, have been an essential part of this transformation. In addition the deconcentration of many cities, and the emergence of the multicentred urban area – what Jean Gottman (1983) called 'megalopolis' – has been, at least in part, facilitated by the new capabilities of telecommunications and telematics for supporting dispersed economic activities away from urban cores (Hall, 1993). Suburban office complexes, business and technology parks, out-of-town shopping malls and, increasingly, whole 'Edge Cities' are reshaping the physical layout of urban areas, using the combined decentralising power of automobiles and telematics (Garreau, 1988). Core cities are being turned into extended urban regions; these themselves now blend into wider megalopoli; at the final level megalopoli merge into the planetary metropolitan system. Instead of single centres linked into some single central place hierarchy, then, very complex networks between cities are emerging based on complex complementary relationships. This trend is particularly advanced in North America. But this decentralisation does not represent the end of cities as we know it. Complex combinations of both decentralisation and centralisation are occurring simultaneously, with the world cities in particular the focus of new pressures for more centralisation because of telematics.

URBAN PLANNING, POLICY AND GOVERNANCE

The overwhelming importance of the *economic* imperative in cities means that the increasing emphasis of urban governance is on public–private partnerships oriented towards an explicit economic development agenda rather than the social, redistributional one that characterised the post-war period (Healey *et al.*, 1995). City authorities have been plunged into a new competitive era in which they act as 'urban entrepreneurs' in increasingly global 'marketplaces' for investment from multinational corporations, public agencies, media, sport and leisure corporations and tourists (Harvey, 1989).

Because city economies today operate as fragmented collections of nodes on global networks, this fight for an improved nodal status is intense and very competitive. Increasingly, elected local governments work in corporatist ways with non-elected public agencies and local firms, utilities and property interests to fight for the regeneration of their cities. Telecommunications companies of all types are, once again, involved here as growing players in such local coalitions – because they are dependent upon long-term revenues from the infrastructure they have, quite literally, sunk into the physical fabric of their home cities. Therefore they have much to gain from supporting the local growth of information-intensive economic sectors (Logan and Molotch, 1987). The importance of telecommunications and telematics to the image and competitiveness of urban areas mean that they are a key focus of such entrepreneurial policy. These new approaches to urban government reflect the emergence of truly international systems of interlinked cities, where urban policy-makers need to consider the role of their city as a node on urban networks, mediated by advanced tele-communications and global transport systems. To parallel the globalisation of their markets, telecommunications companies are increasingly being set up on a global basis so as to meet the needs of multinational corporations for private, global networks (Graham and Marvin, 1995).

Urban governance and public services are also being transformed through the application of telematics. The wider shift from Keynesian to individualistic and conservative welfare regimes is driving cost cutting, privatisation and restructur-ing in urban welfare services at the same time as increased polarisation is increasing demands for these services. In these circumstances, the talk is now to reinvent government more along business lines and to use telematics innovations as the new mechanism for delivering services with minimum costs and maximum flexibility. As a result, telematics-based restructuring in urban governments is burgeoning, often with the result of replacing physical, staffed service delivery offices with virtual and electronic ones. Many routine functions are also being outsourced to distant, even less developed country, locations.

THE STRUCTURE OF THE BOOK

The rest of this book is structured around nine chapters, summarised in Figure 1.15. In Chapter 2 we try to address the growing sense that many current ways of looking at cities are becoming obsolescent because of the neglect of many telecommunications-based changes underway in cities and urban life. We explore in detail how the invisibility of telecommunications-based change in cities combines with their space and time-adjusting qualities to challenge fundamentally many of the assumptions that still underpin much urban analysis and most urban planning and policy-making. Attempts to grapple with the need for new conceptual frameworks which map and explain telecommunications-based chan-ges in cities are then explored.

In Chapter 3 we go on to critically evaluate the competing theoretical perspectives that can be used to address city–telecommunications relations. Four of these are identified: technological determinism, utopianism and futurism,

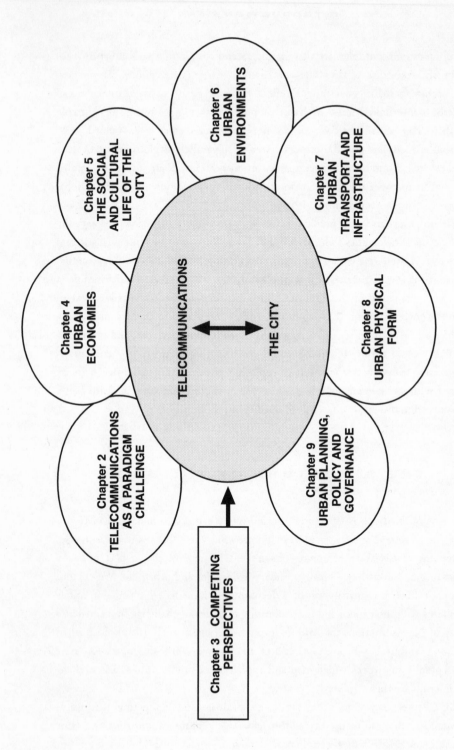

Figure 1.15 The structure of the book

TELECOMMUNICATIONS

THE CITY

Chapter 5
THE SOCIAL
AND CULTURAL
LIFE OF THE
CITY

Chapter 6
URBAN
ENVIRONMENTS

Chapter 7
URBAN
TRANSPORT AND
INFRASTRUCTURE

Chapter 4
URBAN
ECONOMIES

Chapter 2
TELECOMMUNICATIONS
AS A PARADIGM
CHALLENGE

Chapter 8
URBAN PHYSICAL
FORM

Chapter 9
URBAN PLANNING,
POLICY AND
GOVERNANCE

Chapter 3 COMPETING
PERSPECTIVES

urban political economy and the social construction of technology (SCOT) approach. Examples of these are given through extracts from the work of the leading analysts that have adopted these perspectives. This analysis is used to develop a new approach to city–telecommunications relations which provides the framework for the rest of the book.

Each of the following six chapters then goes on to use this approach as the basis for analysing a key aspect of city–telecommunications relations. These are the development of urban economies (Chapter 4), the social and cultural life of cities (Chapter 5), the urban environment (Chapter 6), urban transport and infrastructure (Chapter 7), the physical form of cities (Chapter 8), and urban planning, policy and governance (Chapter 9). Finally, in the concluding chapter, we pull out some key issues which emerge from the approach and attempt to assess the implications of telecommunications for the reality of socioeconomic and cultural life, development and policy within the turbulent and uncertain world of modern cities.

NOTES

1. At least six of these can be identified: the information economy approach (Hepworth, 1989), the post-industrial and information society theses (Bell, 1973), the technoeconomic paradigm approach (Hall and Preston, 1988), the post-Fordist approach (Amin, 1994) and the post-modern approach (Harvey, 1989).

2. Included here are the books by Meier (1962); Dutton et al. (1987); Castells (1989); Brotchie et al. (1991); Schmandt et al. (1990); OECD (1992); and Mitchell (1995). More peripheral accounts are by Hepworth (1989) and Gottman (1990).

2

**TELECOMMUNICATIONS AS A
PARADIGM CHALLENGE FOR
URBAN STUDIES AND POLICY**

INTRODUCTION

A key argument of this book that the current burgeoning importance of telecommunications fundamentally challenges the paradigms underpinning urban studies and policy-making. Even the recent growth of research on urban telecommunications has largely failed to reach beyond the specialist audience to help shift telecommunications from the margins to the centre of contemporary understanding of cities. As a result, the conceptual and policy-making frameworks built up since the nineteenth century to deal with the physical, geographical, social and environmental aspects of the industrial city still tend to underpin – at least implicitly – a large proportion of urban analysis and policy-making. These approaches to understanding cities seem less and less able to cope with current urban change, as it becomes more and more mediated at the most fundamental level by telecommunications. The remarkable neglect of telecommunications in urban studies means that current very rapid telecommunications-based changes in cities demand a corresponding shift in the analysis and understanding of cities and the processes of urban development. But because these have yet to emerge fully, we argue, many urban analysts and policy-makers still see cities through analytical lenses which actually have less and less to do with the real dynamics of telecommunications-based urban development. There is, in short, a threatening 'paradigm crisis'.

In this chapter, we explore this paradigm challenge in detail. We look at the reasons behind it and the ways in which it is manifesting itself. We also try to point towards new ways of thinking about cities and urban space that are more appropriate to tackling the increasingly tele-mediated nature of urban life. This discussion then feeds into the next chapter which analyses the various theoretical approaches that can be taken to studying city–telecommunications relations. As firms, individuals and organisations become increasingly reliant on tele-mediated interactions, no longer is city growth fuelled solely by the need to concentrate physical markets, the application of labour, the exchange of money and commodities, and face-to-face communications within the highly concentrated area of a traditional metropolitan core. Instead, telecommunications and fast transport systems are being used to recombine the urban world in new ways which bring the physical and social aspects of cities into continuous and constant interaction with the electronic worlds of telecommunications.

These transformations are not just occurring with remarkable speed; their nature can be profoundly disorienting to those used to the old rhythms and regularities of modernist urban life. These processes operate in ways that can stretch our ability to perceive and understand cities, distance, space and time, the nature of what is public and what is private to breaking-point. Through telecommunications and innovations in rapid transportation, relations between the physical and locational aspects of cities and the operation of social and economic systems are being fundamentally loosened and reworked, often on a global basis. Many electronic fluxes now explode continuously into the physical spaces of cities, providing profoundly new systems through which urban life can be reorganised and remade. In a very real sense, current telematics systems provide technical networks inside which new spaces and times are being created in all areas of urban life. These networks 'bind together' places in many different spatial and temporal positions in the form of 'real time' networks (Gillespie and Williams, 1988; 1317).

The danger of this paradigm challenge is that these new spaces and new times will develop to radically affect cities while being largely ignored by the very disciplines which purport to understand and manage the city. Clearly, our abilities to comprehend and understand the changing nature of cities becomes threatened by this growing mismatch between the paradigms we use and the nature of urban reality.

TELECOMMUNICATIONS AS A PARADIGM CHALLENGE

Three inter-related components can be identified within the broader threat of a telecommunications-based paradigm challenge for urban studies and policy-making. These we call: the challenge of invisibility; the conceptual challenge; and a challenge to urban planning.

THE CHALLENGE OF INVISIBILITY

Urban studies and policy tend to be dominated by a concern with the visible, tangible and perceivable aspects of urban life. This applies particularly to geography and with its tradition of seeing the 'world as exhibition' (Gregory, 1994) and to urban planning, with its ancestry in the technical and design sciences. Gillian Rose, for example, noted recently that urban geography tends to give a 'privileged role . . . to the visual as a means of accessing knowledge'. This reliance on the eyes in turn shapes the 'particular interpretation of space that results' (Rose, 1993; 70). She continues, 'when geographers gaze at social space, the space economy, urban space and so on, their claim to know and so understand rests on the notion of space as completely transparent, unmediated and therefore utterly knowable' (Rose, 1993; 90. See Gregory, 1994; 16).

Given this visual preoccupation, it is easy to diagnose the virtual invisibility of telecommunications in cities as a key reason for the curious neglect of telecommunications issues in cities. When a new road is built in a city, the unpriced positive and negative externality effects are all too salient and unequivocal. Increased access and travel time savings accrue to road users and noise, pollution and danger accrue to those who lose out. It is hard for people to escape these effects. The new road network and the uses to which it is put are patently obvious to all because of the physical disruption involved in construction and the very obvious tangibility of road networks and road traffic. Because of their greedy use of urban space and direct physical support to urban development, roads also directly affect the distribution of land and property values and rights. Roads therefore impinge very directly into local politics and the urban planning and policy-making process; these have evolved to manage the sorts of physical and locational disputes which roads and other developments trigger as they become incorporated into the land use and environmental patterns of cities.

In stark contrast, new telecommunications networks tend to be largely invisible and silent or, at most, relatively hard to discern. Most weave unseen through the fabric of cities, using very little space. Urban telecommunications networks consist of underground networks of ducts and cables or aerial lattices of wires over which near speed-of-light flows of electrons or photons carry information. Radio and satellite-based telecommunications networks rely on the truly invisible flows of electromagnetic radiation across space between antennae, transmitters and aerials. The positive and negative externalities of telecommunications networks are often extremely enigmatic (Moss, 1987). Only rarely do they erupt to stimulate the major conflicts in urban politics that continually afflict other developments and infrastructures such as roads, railways and property development. In fact efforts often have to be made to increase the visual and physical impact of telecommunications in cities, as when prominent satellite dishes are developed to boost the image of high-tech office development and teleports. In one case, for example, such a dish has been proposed purely for cosmetic reasons, even though no satellite facilities were actually technically required (IBEX, 1991).

Because local politics and planning is so dominated by the tangible externalities of hard infrastructure, land and development, urban telecommunications systems tend largely to be concealed to local politicians and planners. Local political salience for telecommunications has historically tended to arise only in reaction when hard and physical impacts actually arise – such as digging up the roads for a new cable network or erecting a new microwave antenna. This has been exacerbated by the fact that, traditionally, telecommunications networks have been developed and operated by distant, technical, public bodies, such as PTTs, who remained divorced from local policy-makers.

Although, as we shall see later, they are increasingly common, it is still rare for proactive urban telecommunications policies to be developed aimed at positively shaping the use of these key technological infrastructures underpinning cities. By contrast, many western cities maintain elaborate and costly transport policies aimed at securing some form of balance between the competing demands for environmental sustainability, economic development and/or socially equity. Although things are now slowly changing, Seymour Mandlebaum's argument that 'communication is not usually treated . . . as part of the technical infrastructure of urban life' often still tends to hold sway (Mandlebaum, 1986; 132). So, despite the many excited general technological debates which currently rage in the media and popular press, telecommunications still have a comparatively low salience in local politics and planning as well as in urban studies. Many city planners and managers do not even know what the telecommunications infrastructure is in their cities; very few have the power, influence, or conceptual tools to reshape it to have desired impacts. Mitchell Moss captures this extremely well when he argues that:

The telecommunications infrastructure – which includes the wires, ducts and channels that carry voice, data and video signals – remains a *mystery* to most cities. In part this is due to the fact that key components of the telecommunications infrastructure, such as underground cables and rooftop microwave transmitters, are not visible to the public. Unlike airports and garbage disposal plants telecommunications facilities are not known for their negative side effects, and until recently, have not been the source of public disputes or controversy.

(Moss, 1987; 535. Emphasis added)

This fundamentally hidden nature of telecommunications distinguishes them from virtually all other aspects of urban development. Michael Batty suggests that 'much of that [telecommunications-based] change has a degree of invisibility which does not characterise traditional economic and social activity' (Batty, 1990a). Instead of directly influencing virtually every urban resident in very tangible, direct ways, the development and application of telecommunications therefore tends to escape the attention of all but the specialist. As a result, telecommunications developments in cities tend to be intangible, abstract and dominated by arcane terminology and concepts.

Because of this, the development of telecommunications can be secret and private to a degree that other infrastructures cannot approach. Telecommunications are silent, stealthy and incremental in themselves and in their effects upon cities; this makes them an extremely difficult subject both for urban research and urban policy development. To make matters even more difficult, the current shift away from public telecommunications monopolies towards the competitive telecommunications operators and the rapid growth of private, corporate telematics networks make accessing information on the nature and use of urban telecommunications infrastructures even harder. Urban telecommunications networks and the information that flows on them tend now to be intentionally kept firmly in the private domain, away from the prying eyes of urban policy-makers and analysts alike. The drive to accrue the maximum commercial and competitive benefits to the corporations and institutions who use and have control over them is increasingly powerful. The invisibility and intangibility of the networks makes this especially easy: if urban residents, urban commentators and urban politicians rarely understand what their city's telecommunications infrastructure consists of, they are unlikely to understand what they are used for and by whom. Michael Batty captures this by arguing that 'most of these networks will never form part of the public domain. Restricted access is a feature of their development; indeed their very purpose is to "lock out" part of the population from their information' (Batty, 1990a).

This private orientation and secrecy compounds the invisibility problem and further prevents empirical work and policy development on telecommunications and cities. It makes telecommunications even more mysterious to local policy-makers. Finally, it also tends to support highly polarised social patterns of access to these networks and the information on them within cities. Increasingly, telecommunications and information services in cities are allocated purely as market goods with access dependent on the ability of the customer to pay. Attempts to deliver telecommunications services as public or quasi-public goods accessible to all are becoming more and more difficult (see Pinch, 1985). As this book demonstrates, the tension between commercially driven secrecy and public, democratic access and accountability runs right through current research and policy debates surrounding telecommunications and cities.

These factors help to explain the apparent paradox where massive amounts of general speculation in the media about telecommunications and society coexists with a continuing low profile for telecommunications within urban studies and urban policy-making. Despite much rhetoric and hype about the current 'communications revolution' (Williams, 1983), the 'wired society' (Martin, 1978) or the 'information age' (Dizard, 1982) into which cities are supposed to be embedded, telecommunications are only just beginning to emerge as a focus of either serious urban empirical analysis or urban policy development.

THE CONCEPTUAL CHALLENGE: TIME, SPACE AND CITIES

The challenge of invisibility and the lack of research about real telecommunications developments in specific cities is bound up with a broader analytical problem. This is a general inability to develop appropriate conceptual frameworks which capture and help explain the ways in which telecommunications relate to cities. It has been suggested that this conceptual problem threatens the future momentum of urban studies and urban policy and planning, as it becomes less and less relevant to its subject cities (Batty, 1990a). To Michael Batty, for example, 'our observation of the way cities work is becoming increasingly more difficult. Cities are becoming invisible to us in certain important ways and it seems that this invisibility is increasing at a faster rate than our ability to adapt our research methods to these new circumstances' (Batty, 1990a; 130). Mark Hepworth adds that urban studies has 'made little progress in coming to terms with the major research and policy issues raised by the diffusion of information technology in cities. Much of the literature on this topic is woefully out-of-date and anticipatory' (Hepworth, 1989; 183).

Without new conceptual frameworks which help understand city–telecommunications relations, of course, the nature of these relations will continue to remain largely invisible and mysterious: in the absence of very obvious and observable effects, the conceptual deficit becomes all the more crucial because subtle and hidden effects are likely to go unnoticed.

To understand the origins of this conceptual challenge, we must explore the origins of urban studies and planning and the way they tend to conceptualise space and time. Both are rooted in the development and management of the industrial city. Their original concerns related to the dynamics of manufacturing employment and the transportation, raw material or locational advantages that first stimulated urban growth. At the core of their analyses were the social, economic, spatial and environmental aspects of cities dominated by manufacturing and the physical distribution of physical goods, services and people. These issues surrounded the emergence of, first, the initial nineteenth-century urban industrial societies, and later, after World War I, the so-called Fordist mass production and mass consumption societies.

Time and space as the external containers for urban life

Crucially, from our point of view, was the general treatment of space and time in these approaches. Urban studies traditionally approached cities and social structures as though they developed within the external environments of an 'objective' time and an external, physical and 'Cartesian' space (Lefebvre, 1984; 1). After Descartes, space was treated as an absolute object within which human life was played out. These ideas in turn stemmed from the dominance in social theory of the Newtonian approaches to space and time which originated in the natural sciences. As Anthony Giddens argued, 'neither time nor space have been incorporated into the centre of social theory; rather, they are ordinarily treated more as "environments" in which social conduct is enacted' (Giddens, 1979; 202). Such views make it difficult, if not impossible, to fully appreciate the importance of telecommunications, not only as space and time transcending technologies, but as technological networks within which new forms of human interaction, control and organisation can actually be constructed. As Bruno Latour argues:

Most of the difficulties we have in understanding science and technology proceeds from our belief that space and time exist independently as an unshakable frame of reference *inside which* events and place would occur. This belief makes it impossible to understand how different spaces and different times may be produced *inside the networks* built to mobilise, cumulate and recombine the world.

(Lafour, 1987; 228. Original emphasis)

Urban studies, particularly urban geography, have, historically, tended to take these views of space and time. As Ed Soja argues, until very recently, 'geography . . . treated space as the domain of the dead, the fixed, the undialectic, the immobile – a world of passivity and measurement rather than action and meaning' (Soja, 1989; 37). There was an objective 'Euclidean' geometry to space which could be observed and drawn, and which acted as the empty area within which the spatial relations and spatial processes that shaped cities were enacted (Lefebvre, 1984; 1; see Massey, 1992). Cities were special portions of this space, bounded, enclosed and separated from rural areas by the frictional effects of distance and the time it took to travel (Emberley, 1989). Meanwhile, the importance of time in social life was seen to echo classical Newtonian Physics, to be a simple, invariant, container for urban and social life (Lash and Urry, 1994; 237).

Acting in parallel with the invisibility of telecommunications, these ideas led to a predisposition in both urban studies and urban analysis against taking much direct account of telecommunications (Nicol, 1985). This is true despite the major paradigm shifts in the analytical approaches to cities and urban policy that have occurred in the past two decades.

From the late 1950s to the mid-1970s, for example the 'scientific' approach to urban studies gained ground. Analysts of cities and urban systems became preoccupied with the use of neoclassical economics to build up models explaining the physical structure and functioning of cities within this external and objective time–space framework (Herbert and Thomas, 1982; 15). To do this, simplified theoretical frameworks were built up such as Christaller's central place theory and Webber's industrial location theory (see Lloyd and Dicken, 1982). These were used to help develop an understanding of how cities and urban systems grew. Such approaches placed an overwhelming priority on the unavoidable need to transport people and heavy, physical goods as the limiting factor defining how cities and urban systems developed. The central notion here was the idea of the 'frictional effects of distance': travelling across the Euclidean plain was expensive in both money and time so larger distances discouraged interaction. The theories 'grounded explanation primarily in social physics, statistical ecologies and narrow appeals to the ubiquitous friction of distance' (Soja, 1989; 37). People, industries and firms were seen to behave and locate in economically rational ways. These minimised the costs of money and time needed physically to access the places where services, raw materials and markets were located. In particular, cities and activities became patterned across an idealised Euclidean spatial plain, shaped by the friction of distance.

Thus, services of various orders – from the local shop to the international stock market – spread out over space concentrated in different sized settlements so that consumers could most efficiently access them. Factories located where they could best access markets as well as raw materials. It was assumed that complex patterns of 'distance decay' – decreasing interaction with larger distances – therefore shape how cities and urban systems are structured. These fundamentally underpinned the physical and locational structures of neighbourhoods, cities, regions and systems of cities. The result was an urban landscape of central places evenly distributed across nations and regions so that the aggregate frictional effects of distance on physical movement were minimised. Most famously, attention centred on the degree to which actual national systems of cities met with the perfect central place hierarchy predicted by Christaller, within which different sized cities developed functional specialisations (see Figure 2.1).

Crucially, from our point of view, communications by any other means than travelling tended to be ignored or neglected in most of these approaches (although see Abler, 1974, 1975, 1977; Goddard, 1975; Tornqvist, 1968, 1974, for notable exceptions). In these models, the only way goods and services could be consumed, or communications could take place, was through physical transportation. Telecommunications as a means of moving information, services, capital and labour power around was virtually ignored as a factor influencing cities. In fact, one of the key simplifying assumptions of such approaches was often that producers and consumers had perfect information about the choices open to them: the services they could consume and how, the changing nature of markets and organisations and legal, technological and regulatory developments (Nicol, 1985). Thus what economist call 'transaction costs' – the costs of running the economic system, of searching and gathering information and making transactions – were assumed – wrongly – to be zero. As Lionel Nicol argues, 'transaction costs – the costs of running the economic system – are often assumed away when, in reality, they represent a major impediment, being in many instances an insuperable obstacle to the formation of markets' (Nicol, 1985; 196). This completely negated the importance of telecommunications, and information flows across distance. And because of the objective approach to space and the invariant notion of time, the effects of telecommunications networks in producing

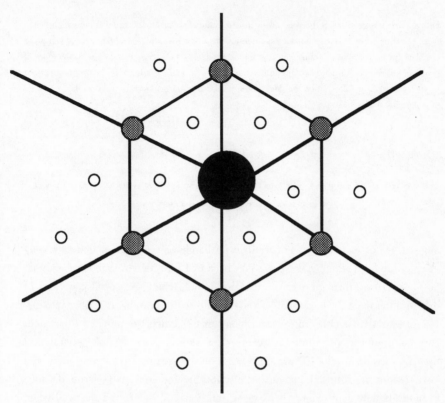

Figure 2.1 A Christallerian urban hierarchy
Source: Adapted from De Roo, 1994; 14

'different spaces and different times' inside which wide ranges of fundamentally new social and economic relations could be constructed was utterly ignored.

But this conceptual challenge creates problems even when telecommunications are considered. This is because they are often described as 'distance shrinking' technologies or 'electronic highways' which are laid out across space and operate in a way analogous to transport networks. Andy Gillespie and Howard Williams argue that these approaches act as 'a "strait-jacket" in preventing an adequate conceptualisation of the role telecommunications is coming to play in restructuring space and spatial relations' (Gillespie and Williams, 1988; 1317). This is because:

The idea of telecommunications as 'distance-shrinking' makes it analogous to other transport and communications improvements. However, in so doing the idea fails to capture the essential essence of advanced telecommunications, which is not to *reduce* the 'friction of distance' but to render it entirely meaningless. When the time taken to communicate over 10,000 miles is indistinguishable from the time taken to communicate over 1 mile, then 'time–space' convergence has taken place at a fairly profound scale. Because all geographical models and our contemporary understanding of geographical relationships are based, implicitly or explicitly, on the existence of the friction imposed by distance, then it follows that the denial of any such friction brings into question the very basis of geography that we take for granted.

(Gillespie and Williams, 1988; 1317)

From national urban hierarchies to 'hub' and 'spoke' urban networks

Suggestions of such a looming paradigm challenge or crisis in our understanding of cities are far from new, however. Way back in 1968, Melvin Webber urged that 'we are passing through a revolution that is unhitching the social processes of urbanization from the locationally fixed city and region' (Webber, 1968; 1092–3). He foresaw that as cities move from being manufacturing-dominated to services and communications-dominated centres, a radical new set of geographical dynamics would develop. These are centred on the use of telecommunications and fast transportation to link producers, distributors and consumers across distance in radically new ways. Pierre Beckouche and Pierre Veltz capture this well when they argue that:

On the whole, the old geography, which linked businesses to the sources of raw materials and to consumer markets, is being thrown into confusion at the expense of a more complex geographical arrangement where the production–distribution system can fight it out in space using the length of the infrastructure and communication networks on a national, even planetary level.

(Beckouche and Veltz, 1988)

These trends mean that the single, functional hierarchies of cities within nation states that were described so painstakingly within central place theory are breaking down. They are being combined and remade as more interconnected networks linking specialised urban economies across international boundaries via highly capable transport and telecommunications networks. Giuseppe Dematteis notes 'the passage from a functional organisation in which the centres are graded with a multi-level hierarchy (as in the models of Christaller and Lösch) to interconnected networks organised on the basis of the corresponding com- plementarities of the nodes and the synergies produced' (Dematteis, 1994). To some, these trends mean that the old territorial identity of the city

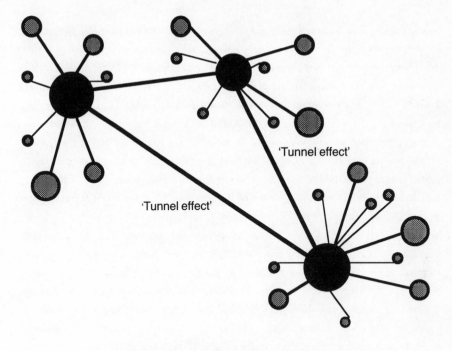

Figure 2.2 A 'hub' and 'spoke' urban network
Source: Adapted from De Roo, 1994; 14

economy as the heart of its hinterland has been totally lost; instead the city is 'divided into as many fragments as the networks which traverse it' (Dematteis, 1988). The use of telematics allows multinational firms to operate a wide variety of production units in many different functional sites while integrating them together in real time – with little respect to the traditional barriers of space and time through telecommunications and telematics. The result is that 'cities have become places where the large corporation chooses to locate segments of its operation' (Mazza, 1988). The implication of these trends is that telematics seem to be involved currently in an important reorientation in the relationships between economic activity and the locations within which it takes place. This has profound implications for all city economies. Luigi Mazza suggests that 'the city seems to have become trapped by a new economic environment in which the traditional location factors, which always linked activities to a community and a place, seem to have lost their power of attraction' (Mazza, 1988). According to Castells, this new 'global geometry' of production, consumption and information flows 'denies the specific productive meaning of any place outside its position in a network whose shape changes relentlessly in response to the messages of unseen signals and unknown codes' (Castells, 1989; 348).

This new emerging type of urban network, where the position of cities is not merely determined by population size, is shown schematically in Figure 2.2. This shows how telecommunications and fast transport networks now interconnect cities into systems of 'hubs' and 'spokes' across wide distances. Distance decay and the friction of distance are remodelled or are no longer of overwhelming importance. Rather, there is a set of so-called 'tunnel effects'. These are caused by the warping of time and space barriers by the advanced telecommunications and transport infrastructures. Often, these link nodes and city centres together into networks while excluding much of the intervening spaces from accessing the networks, because the networks pass through these spaces without allowing local access. Good examples of such tunnel effects can be found in the advanced telematics systems that link New York, London and Tokyo into a single global 'virtual' financial marketplace and the fast train or TGV networks that link up the major French cities while excluding smaller, intervening centres from having access to the networks. It is networks of cities rather than single, one-centred cities in functional hierarchies, that are the key to understanding today's knowledge-intensive urban economies (Buijs, 1994). Within these networks or 'multiple hierarchies', as Jaeger and Dürrenberger (1991) call them, one city may dominate another in, say, financial services but be subordinate to it in legal services. The results are close interdependencies within networks of cities. As David Batten argues, 'in today's global network economy, knowledge corridors link major knowledge-intensive nodes or hubs to form network cities. European examples include the Cambridge–London and Stockholm–Uppsala corridors as well as the Ranstaad Holland. The Kansai region in Japan in another example of a network city' (Batten, 1994; 2).

The broader intellectual and policy challenge

But this paradigm challenge applies not just to cities and urban policy-making. Old ideas in many intellectual and policy areas are equally being challenged by the current transformation. It is clear that this new era is characterised by the subversion of the concepts of linear, Euclidean and physical space that were built up during the industrial age by concepts of abstract and logical spaces based within electronic telecommunications networks. Geoff Mulgan, for example, argues that:

Computers have done much to spread familiarity with the idea of logical rather than physical space, with their topological representations on flow diagrams, branching trees and other patterns. In a culture based more on images and less on the linear logic of texts there seems to be a need for topographical and visual metaphors to replace older metaphorical systems based on such concepts as equilibrium and order. . . . Just as the physical machinery of networks and computers creates new logical spaces of configuration, this visual, metaphorical language transforms how the world is seen.

(Mulgan, 1991; 20)

In this context, nation states, as well as cities, are being fundamentally challenged. They 'become powerless confronted with the ability of capital to circulate globally, of information to be transferred secretly, of markets to be penetrated or neglected, . . . and of cultural messages to be marketed, packaged, recoded and beamed in and out of people's minds' (Castells, 1989; 349). Global telematics networks under the private control of corporations represent important tools which undermine the ability of nation states to impose tax, control capital flows and regulate their economies (Irwin and Merenda, 1989). Once again, Geoff Mulgan brilliantly captures this sense that our analytical capabilities and urban research traditions threaten to be left high and dry by the bewildering telecommunications-based changes underway in modern cities. He writes:

A world built on networks challenges traditional categories and intellectual structures on many fronts. It calls into question older conceptions of space and power. Where the early market economies grew out of the temporal and spatial regularities of city life, today's are built on the logical or 'virtual' regularities of electronic communication, a new geography of nodes and hubs, processing and control centres. The nineteenth century's physical infrastructures of railways, canals and roads are now overshadowed by the networks of computers, cables and radio links that govern where things go, how they are paid for, and who has access to what. The physical manifestation of power, walls, boundaries, highways and cities, are overlaid with a 'virtual' world of information hubs, databases and networks. Buildings are redefined in terms of their position in networks as the electronic office and the automated factory are joined by the 'smart house', and as national boundaries are by passed by flows of data and television images.

(Mulgan, 1991; 3)

A CHALLENGE TO URBAN PLANNING

The evolution of urban planning and policy was heavily influenced by the theories of urban development that emerged during the 1950s and 1960s. Filled with the zeal to apply the new scientific theories to practice, urban planners tried between the 1960s and 1970s to apply them to real cities. At this time, the public health crises and philanthropic drives that first stimulated planning movements had eased. Planners now came to concern themselves with economic and social modernisation and the need to improve the locational efficiency of cities and urban regions. They tried to plan their cities to approximate to the perfect physical systems predicted by theory. To many, this was seen to be a technical, rational process that would be socially neutral in its effects, helping to secure a single 'public interest' (Rydin, 1993). To improve locational efficiency, the frictional effects of distance needed to be minimised to improve quality of life and land uses needed to be laid out to improve economic efficiency and environmental quality. In some extreme cases, for example County Durham in the UK, planners actually suggested the destruction of villages and towns so that settlement systems accorded better with the theoretical perfection of central place theory.

At this time, planners assumed that cities were physically integrated places amenable to local land use and development policies that would then go on to solve their economic, social and environmental problems. As with urban studies, space and time were seen to be little more than the external containers for urban life. Planners had developed an individual remit and professional identity centred on dealing with the problems of rapid urbanisation and the dilemmas associated with the allocation of urban space (for public health, sanitation, infrastructure development and industry). The population, social structure, and economic and environmental dynamics were all seen to be very closely related within the city, closely underpinned by the city's physical built form. Cities were therefore seen to behave effectively as unitary objects amenable to physical intervention at the local level. 'In both the urban sciences and in urban planning', wrote Melvin Webber, 'the dominant conception of the metropolitan area and of the city sees each as a *unitary place*' (Webber, 1964; 81). These ideas were supported by notions of physical and environmental determinism: human life was shaped by the environment and location within which it occurred. The frictional effects of distance on transportation severely limited external influences. The economic and social lives of city residents was assumed to equate unproblematically with this almost completely physical and locational conception of cities and urban life. Webber continued, arguing that 'most planners share a conviction that the physical and locational variables are key determinants of social and economic behavior and of social welfare' (Webber, 1964; 85).

These ideas focused the attention of planners on the needs to improve the physical mobility of people and goods within cities, and to locate physical facilities and land uses with maximum efficiency within a unitary, integrated city (Webber,

1968; 1093). New environments would lead to new, improved patterns and processes of urban life. This was reflected in the major preoccupation of policies: new urban motorway programmes; the massive redevelopment of city centres and slum areas; developing integrated 'neighbourhood units' within New Towns; locating new facilities at their 'optimum' physical sites (see Hall, 1988). Phil Cooke describes this approach, where 'policies are *ipso facto* constructed in response to areal problems and take the form of providing physical solutions' (Cooke, 1983; 39). The basic assumption was therefore that 'by changing the *physical and the locational environments at the places* in which families live out their lives and in which groups conduct their business, the lives and the businesses can be improved' (Webber, 1964; 84. Emphasis added).

Once again, though, telecommunications were almost completely ignored in these considerations. Their intangibility meant that they had never stimulated the urban political movements that led to urban planning in other areas. The all-pervasive nature of communications meant that they never led to the development of a coherent professional or institutional fora within cities. As Mandlebaum argues, 'the concept of an urban communication infrastructure was not expressed in local institutions comparable to those which realized public concerns with education, environmental quality, housing or transportation' (Mandlebaum, 1986; 134). Responsibility for regulating and developing slow-moving urban telecommunications infrastructures of the time fell outside the planners' remit, resting usually with distant, centralised public bureaucracies or virtually autonomous public and private enterprises. Unlike the developers of roads, these tended to roll out their hidden infrastructures according to central government policy or technical factors rather than to any local concept of urban planning. Because these infrastructures were overwhelmingly geared to the basic telephone, which was developed to be universally accessible, there seemed to be no relevance of telecommunications to urban planning.

These general shifts have led an increasing number of urban commentators to argue that older notions of integrated cities, Cartesian space and invariant time are no longer an appropriate basis for urban planning or policy. Back in 1968, Webber suggested that 'the concepts and methods of civil engineering and city planning suited to the design of unitary physical facilities cannot be used to serve the design of social change in a pluralistic and highly mobile society' (Webber, 1968; 1092–3). More recently, Ken Corey has commented that 'urban and regional planning practice throughout many of the world's industrial market economies is in a state of paradigm challenge. In essence, the crisis exists because old planning procedures of how the industrial city functions don't seem to apply for today and tomorrow' (Corey, 1987). To Henry Bakis (1995; 3), tele-communications remain peripheral to urban and regional planning because of the 'persistence of the traditional paradigm whereby the approach to regional development remains, to a large extent, based on the logic of industrial development'.

There are some signs that this is slowly changing, however. The incorporation of advanced telecommunications into urban and regional planning has probably developed furthest in France (see Chapter 9). A recent communications plan for the French city of Lille, for example, comments that:

The traditional concepts of urban and regional planning are today outmoded. The harmonious development of areas towards equilibrium, the correct sharing out of resources, providing support to complementary developments within the city... these ideas have given way to the impression that spaces are fragmented, atomised and strongly competitive.... The insertion of tele-communications into the city makes the development of spaces more complex and introduces today a third dimension into urban and regional planning [after space and time]: this is the factor of *real-time*.

(ADUML, 1991)

But the sense that these old ideas of urban planning are in crisis is about much more than telematics. It stems from a much wider cultural and political–economic shifts from modernism to post-modernism, Fordism to post-Fordism and Keynesianism to post-Keynesianism (Gaffikin and Warf, 1993) to which the discipline and practice of urban planning and policy has yet to fully adjust (Goodchild, 1990).

WAYS FORWARD: POST-MODERNISM, ELECTRONIC SPACES AND THE TELE-MEDIATED CITY

How can we find ways forward in our search for more sophisticated treatments of cities, telecommunications, space and time? Where can useful directions be identified in our search to reconceptualise the city and urban development in ways that pay justice to the diversity of tele-mediated processes now at work? A recent stream of work within cultural studies and debates about the urban effects of post-modernism provides fertile ground for exploration here.

The development of modernity, with its stress on rationality, progress and the search for order, was closely wedded to Cartesian notions of space and universal, absolute ideas of time. 'Modernity is the context of the map', write Lash and Urry (1994; 55), 'it is the era of the "grid", both horizontal grids as in city planning and the "vertical" grids of skyscrapers'.

But the shift away from modernism has led to new notions of space and time as advanced communications networks become infused into the urban world. It has also been associated with a wider set of paradigm shifts within the social sciences and humanities which have shifted urban studies and policy-making away from the scientific orientation of the 1950s. Urban studies have, for example, been revolutionised to incorporate many more critical theoretical approaches which have successfully challenged the dominance of positivism. These see cities and systems of cities as being more complex physical, economic, social and symbolic 'landscapes' which are continually being shaped and reshaped in interaction with the changing social and power relations of capitalism (see Harvey, 1985). Planners, too, have enthusiastically adopted new political–economic perspectives to cities and society, and have, more recently, continually adapted their roles to meet the rapid shifts in the economic, social and environmental nature of cities. Planners are much more suspicious of crude notions of environmental determinism or scientific social theory.

The ways in which space and time are considered have changed too, under the influence of new approaches in physics such as relativity and wider sociological and philosophical debates (see Keith and Pile, 1993; Gregory, 1994; Lash and Urry, 1994). Now they are increasingly seen not just as dead and empty external containers, but as social constructions which are intrinsically produced *within* society (Massey, 1992). Rather than just being separate and external frames of reference within which cities and urban life play out, space and time, like cities, are seen to be continually remade through capitalist societal development in ways that go on to rebound upon capitalist society itself (see Thrift, 1993). Conceptions of space are also becoming more complex and are shifting beyond the physical and locational preoccupations of the past: urban space can be social and psychological, as well as abstract, material and physical (Lefebvre, 1984). Finally, there are new efforts to integrate consideration of time and space. As new networks continue to remake both through their effects on mobility and speed, Nigel Thrift (1993), for example, argues that 'there is little sense to be had from making distinctions between space and time – there is only *space–time*'.

These developments have laid the foundations for much more sophisticated treatments of city–telecommunications relations. Frederick Jameson, in his highly influential essay in 1984, coined the term 'post-modern hyperspace' to signal the break that telematics bring with the familiar, linear and orderly 'modern' sense of space and time in cities when material movement, the friction of distance and transportation dominated. Interestingly, he focuses first on the confusing influence of the changing physical effects that cities have, through post-modern architecture, on their inhabitants. He then links this as a parallel to a wider bewilderment and confusion brought by the new global era of telematics-mediated urban life. To Jameson:

We are in the presence of something like a mutation in built space itself. . . . We ourselves, the human subjects who happen to inhabit this new space, have not kept pace with that evolution. . . . We do not yet possess the perceptual equipment to match this new hyperspace. . . . [It] has finally succeeded in transcending the capabilities of the individual human body to locate itself, to organize its immediate surroundings perceptually, and cognitively map its position in a mappable external world. . . . This alarming disjunction point between the body and its built environment . . . can itself stand as the symbol and analogue of that even sharper dilemma which is *the incapacity of our minds, at least at present, to map the great global multinational and decentred communicational network in which we find ourselves caught as individual subjects.*

(Jameson, 1984; 81–84. Emphasis added)

TECHNOLOGICAL TIME

Paul Virilio, meanwhile, has argued that the new urban era is marked by the use of telecommunications and telematics to support what he calls an 'urbanization of *real time*'. This follows the previous era of the 'urbanization of *real space*' – that is, the physical building of industrial cities (Virilio, 1993; 3). Because time barriers are being so comprehensively challenged by the global 'instantaneity' of telematics, Virilio argues that the physical boundaries of cities – what he calls the 'urban wall' – 'has given way to an infinity of openings and enclosures' (Virilio, 1987; 20), where telecommunications and telematics networks interpenetrate invisibly and secretly with the physical fabric of cities. In global cities, in particular, the intense 'hubbing' of advanced transport and telecommunications networks means that they 'disrupt the time space coordinates of natural space' (Lash and Urry, 1994; 55). Through faster and faster computers linked to global telematics networks, time frames have collapsed so acutely that humans can no longer grasp the speeds of the flows involved (Lash and Urry, 1994; 242). As Adam notes, 'if telephone, telex and fax machines have reduced the response time from months, weeks and days to seconds, the computer has contracted them down to nanoseconds. The time-frame of a computer relates to event times of a billionth of a second' (Adam, 1990; 140. Quoted in Lash and Urry, 1994; 242). Images, symbols, information, data and communications contacts thus simultaneously and instantly saturate the post-modern urban world.

Telematics networks also support asynchronous as well as instant interaction – that is, they uncouple the need for receiver and sender to be available while communicating as with the telephone. Time can be shifted: people can work when they want, communicate when they want, access electronic services and entertainment when they want. This supports the shift from the highly structured time patterns of the modernist city – with its standard business, leisure, sleep and commuting periods – towards more fluid, asynchronous urban lifestyles. William Mitchell (1995; 16) asks us to 'extrapolate to an entirely asynchronous city' where 'temporal rhythm turns to white noise. The distinction between live events and arbitrarily time-shifted replays becomes difficult or impossible to draw (as it often is now on the television news); anything can happen at any moment'.

Notions of friction of distance, location, distance decay or objective space and time often have little meaning here, because of the instant real-time control capabilities of telematics networks across space at all geographical scales. Within telematics networks these concepts can be simply turned on their heads. Urban life becomes increasingly constituted through electronic mediations and representations flowing within the abstract spaces of computers linked via telecommunications networks. 'Spatial and temporal coordinates end up collapsing', writes Olaquiaga (1992; 3), 'space is no longer defined by depth and volume, but rather by a cinematic (temporal) repetition, while the sequence of time is frozen in an instant of (spatial) immobility'. On computer networks, for example, all sorts of services can flow around between and within cities irrespective of distance, often taking different routes between sender and consumer, who can be strung out on a global basis. Computers and equipment can be controlled at any distance via computer networks. Managers can be 'telepresent', issue 'telecommands' and enjoy 'telesurveillance' of their whole global organisation from a single computer screen in a single place. In effect, they can be in more than one place at once. To Virilio, telematics networks cross-cutting cities changes all accepted notions of travel and movement, space and time both between and within cities. He captures this movement to a world beyond that where the friction of distance totally shapes urban life:

In the past, physical displacement from one point to another presupposed a departure, a voyage and an arrival at a final point that remained, however, a *restricted arrival* by virtue of the very duration of the voyage. Currently, with the revolution of instantaneous transmissions, we are witnessing the beginning of a type of *general arrival* in which everything arrives so quickly that departure becomes unnecessary.

(Virilio, 1993; 8)

This challenges accepted notions of the nature of cities and urban life. What Virilio calls the 'unity of place without unity of time' that results 'makes the city disappear into the heterogeneity of advanced technology's temporal regime. Urban form is no longer designated by a line of demarcation between here and there, but has become synonymous with the programming of a "time schedule". Its gateway is less a door which must be opened but an audio-visual protocol – a protocol which reorganizes the modes of public perception' (Virilio, 1987; 19).

ELECTRONIC SPACES

But what, exactly, are these networks a gateway into? In many of the current debates about the Internet, the new computer networks are seen to be new 'virtual' worlds, 'electronic spaces' or 'domains of information'. This reflects Bruno Latour's point above about spaces and times being constructed within networks through which the world becomes 'recombined'. As we have seen, the crucial point is that the growing range of interactions, transactions and simulations which go on within the 'ungraspable electronic terrain' of these 'worlds' respect none of the constraints of space and time familiar within our physical lives in cities. And, as the technologies become more capable, so the spaces accessible though networks like the Internet become more experientially rich and more like places in the sense that people can become perceptually immersed within the worlds constructed by software. William Mitchell argues that the Internet:

subverts, displaces, and radically redefines our received conceptions of gathering place, community, and urban life. The Net has a fundamentally different physical structure, and it operates under quite different rules from those that organize the action in the public places of traditional cities. . . . The Net negates geometry. While it does have a definite topology of computational nodes and radiating boulevards for bits . . . it is fundamentally and profoundly *antispatial*. . . . You cannot say where it is or describe its memorable shape and proportions or tell a stranger how to get there. But you can find things in it without knowing where they are. The Net is ambient – nowhere in particular but everywhere at once. You do not go *to* it; you log *in* from wherever you physically happen to be.

(Mitchell, 1995; 8. Original emphasis)

References to these new electronic spaces, places and cities are increasingly common in the current discourse surrounding the Internet, as are explorations of the ways they parallel and interact with conventional cities. The broad label 'cyberspace' – a term first coined in William Gibson's dystopian novel *Neuromancer* (Gibson, 1984) – is the most common description of this parallel electronic world. Michael Ogden (1994; 715) defines cyberspace as 'a conceptual "spaceless place" where words, human relationships, data, wealth, status and power are made manifest by people using computer-mediated communications technology'. Doug Rushkoff argues that 'we are learning to inhabit a place, what we've been calling "Cyberville" – a territory which is very different from the territory we've inhabited before' (Rushkoff, 1994; quoted in Channel 4, 1994; 3). Harasim, meanwhile, contends that 'computer networks are not merely tools whereby we network; they have come to be experienced as *places* where we network: a networld' (Harasim 1993; 15). David Lyon contends that 'it is no longer the "high frontier" of space which captures the imagination, but the "inner frontier" of conceptual space. Computer game addicts and hackers perceive the screen as a window opening onto a multidimensional world which can be navigated by keyboard joystick or mouse' (Lyon, 1988; 130). On the connections between cyberspace and the city, a recent British television programme, noted that 'the chaos of America's urban sprawl belongs not just to the city of steel and glass, but also to the other city – the phantom city of media and information. With most resources directed to it, it is the cyber-city which is accelerating faster than the real urban space' (Channel 4, 1994; 4).

This spatial/urban metaphor is often taken further when the actual interface between users and telematics services is set up as an analogy to a real city or place. As an example of this, Figure 2.3 shows the interface for the 'Cyberville' telematics service being developed in Singapore by the National Computer Board for the World Wide Web. Cyberville organises access to all sorts of web services – electronic mail, conferencing, games, stock markets, education and a host of information services – using a model of these services imprinted on to an idealised urban landscape. The use of such metaphors helps in the constant struggle to make the intangible tangible; the abstract comprehensible.

But we would argue that, despite the pervasive application of real-time telematics, the constraints of space, place and time still exist; they are modified rather than abolished by these changes. Modern telecommunications and telematics merely operate to make their patterning and interrelationships much more complex and – usually – more bewildering. For instance, these new technologies 'produce a new and more elaborate temporal order' (Wark, 1988)

Figure 2.3 An example of the use of the urban metaphor as a telematics interface –
'Cyberville' in Singapore

– because the many different time zones of the globe interpenetrate continuously within and between cities. Also, it is easy to forget that telecommunications networks are themselves materially and physically fixed with their own highly uneven geography (Kellerman, 1993). They support these 'new electronic spaces' and 'instantaneous times' but only in the physical spaces where the right infrastructure is built and can be accessed. This can actually reinforce the competitive position of cities because they house high concentrations of demand and so are much more attractive to telecommunications operators than rural or peripheral areas. Rather than simply ending the domination of cities in western civilisation, these networks actually tend to 'erupt within the spatial order of the old city' (Wark, 1988). So space and time, and speed and distance are thrown together in new and often bewildering ways within the physical arenas of old cities.

TOWARDS NEW CONCEPTIONS OF THE CITY

These shifts clearly require a redefinition of what we actually now mean by 'city', the 'urban' or 'urbanity'. The shift to telemediated economic and social networks undermines the old notion of the integrated, unitary city which has an identifiable boundary and is separated from others by Euclidean space and the all-powerful friction of distance. New notions of place and urbanity are clearly demanded. Nigel Thrift considers the crucial question 'What is *place* in this new "in-between" world?'. His response offers pointers for reconceptualising the city. To him, 'place' is now:

compromised: permanently in a state of enunciation, between addresses, always deferred. Places are 'stages of intensity', traces of movement, speed and circulation. One might read this depiction of 'almost places' . . . in Baudrillarean terms as a world of third-order simulacra, where encroaching pseudo-places have finally advanced to eliminate place altogether. Or one might record places, Virilio-like, as strategic installations, fixed addresses that capture traffic. Or, finally, one might read them as Morris (1989) does, as frames for varying practices of space, time and speed. . . . No configuration of space-time can [now] be seen as bounded. Each is constantly compromised by the fact that what is outside can also be inside. This problem of 'almost places' becomes particularly acute in the case of contemporary cities. . . . It has become increasingly difficult to imagine cities as bounded space-times with definite surroundings, wheres and elsewheres.

(Thrift, 1993)

Cities and urban areas, then, must now be seen as the fixed sites and places where the many separate and superimposed social, technological, institutional and economic networks which link them intimately into wider social, economic and cultural dynamics coalesce, cross and interact (see Healey *et al.*, 1995). It is the complex interactions between cities as fixed places and the networks that bring intense mobility (telecommunications, infrastructure, transport, the institutional networks of transnational corporations, media flows etc.) that now shapes urban life and urban development (see Chapter 3). Following this, it is the interactions between cities as urban places and the electronic spaces constructed within telematics networks that need to be understood if our understanding of city–telecommunications relations is to be improved.

These changing notions of place and urbanity are encouraging a growing number of urban commentators who are now starting to develop a new range of concepts which address telecommunications-based changes in contemporary cities. They generally start to do this through developing new categorisations of urban space and development. These concepts, some of which are listed in Figure 2.4, represent state-of-the-art attempts to grapple towards analytical frameworks that successfully capture the often bewildering nature of telecommunications-based development in the modern city. They try and elevate our understanding of 'flows', 'telegeography' and 'cyberspace' to match more familiar and common-sense notions of 'place', 'space', 'geography' and 'distance'. In doing so, they help to redress the imbalance between the treatment of the physical and the electronic aspects of cities.

These commentators are now arguing that telematics systems can be best considered as 'abstract' or 'electronic spaces' which parallel and interweave with the wider worlds of physical, social and cultural space (see Poster, 1990). Such electronic spaces cannot be mapped or grasped in the same way as familiar social or physical spaces because they are constructed in 'real time' inside technical network using computer hardware and software. To quote Robins and Hepworth (1988; 161) 'electronic network geography is mapped on *abstract* space'.

'Space of places'	'Space of flows' (Castells, 1989)
Physical presence	'Telepresence' (CEC, 1992)
Physical mediation	'Telemediation' (Richardson, 1994b)
Geography	'Telegeography' (Staple, 1992)
Distance	Speed and time (Mulgan, 1991)
Closure	Openness and exposure (Virilio, 1987)
Locality	Globality (Knight and Gappert, 1989)
'Modern' space	Post-modern 'hyperspace' (Jameson, 1984)
	'Data spaces' (Murdock, 1993)
	'Electronic spaces' (Robins and Hepworth, 1988)
	'Cyberspace' (Gibson, 1984)
	'Netscape' (Hemrick, 1992)
	'Networld' (Harasim, 1993)

Figure 2.4 New conceptual approaches for the telecommunications-based city: old and new characterisations of urban space and development

THE INERTIA OF OLD CONCEPTS

Despite these significant improvements, we would still argue that the basic physical and locational orientation of much of the mainstream of planning and urban analysis – within which telecommunications became so marginalised – remains largely intact. The treatment of space and time as external containers has not, at least implicitly, significantly shifted, beyond the more critical debates about social theory in urban sociology, geography and cultural studies. As Ed Soja (1989; 79) comments, the 'essentially physical view of space has deeply influenced all forms of spatial analysis'. Henri Lefebvre (1984; 292) argues that 'the illusion of transparent, "pure" and neutral space . . . is being dispelled only very slowly'. This means that, while telecommunications are being treated more often, and are receiving much more recognition, these treatments still tend to be tacked on to ideas that originate in more traditional approaches to cities and urban develop- ment where transport and the physical built form are the prime considerations. We believe that the legacy of the physical and locational approach, and the continued attachment to independent notions of time and space, still severely limits the degree to which telecommunications-based effects can genuinely be incorporated into many approaches to urban analysis and policy-making.

This means that many urban commentators and policy-makers still tend to remain wedded to crude versions of the 'grand metaphors' for explaining the shift to a more telecommunications-based society that we explored in Chapter 1. Simple technological determinism is common. Most have, at best, a crude understanding of the technological and regulatory shifts that are underway in telecommunications. Because urban politicians and planners remain firmly wedded to the tangible and salient aspects of cities, the arcane and intangible world of telecommunications still finds little space. A senior politician within a Northern industrial city in the UK recently admitted that 'within the council, we've got a lot of bright local politicians, but they're not very good at working within the conceptual frameworks [of telecommunications]; they like to touch and feel, to know what's happening. They like to be very practical' (Graham, 1995).

Another consequence of these problems is that urban commentators and policy-makers still fail to incorporate telecommunications seriously within their educational and professional programmes. Even today, telecommunications are still included in only a small proportion of courses on urban policy, management and planning. When talking of 'urban infrastructure', built environment educators, policy documents and urban researchers invariably still tend to be talking about the familiar 'hard' and 'physical' infrastructure that we can all see and experience around us in our daily lives in cities at ground level. It is still common to equate urban infrastructure with the dominant and visible infrastructures of the industrial city: roads, railways, airports, bridges and waterways. As Kevin Morgan suggests, this situation – like all paradigm mismatches – is as much a product of people as anything else. He argues that the current generation of urban policy officers and urban commentators are 'simply the product of their age, an age which predated the new ICT [Information and Communications Technologies] paradigm' (Morgan, 1993).

However, it is equally clear that care needs to be taken to avoid overdramatising the telecommunications-based changes underway in cities. A historical perspective, the experience of many failed utopian forecasts, and the danger of falling into the language of hype teach us to avoid sensationalism when approaching telematics and cities (Thrift, 1993). When current trends towards increased speed and mobility are placed in a historical perspective, they actually appear to be an intensification of processes that have a history as long as the modern, industrial city itself. The current analytical challenges to urban studies echo the ones which existed at the beginning of the twentieth century. As Stephen Kern has so ably demonstrated (1983), at that time, the seemingly fantastical technologies of the bicycle, telegraph, telephone, electric light and power networks and railway totally challenged and remade prevailing paradigms of understanding cities, space, distance and time that were rooted in the pre-industrial past. From the slow, linear experience of a single space and time came a brave new world dominated by a bewildering simultaneity. Many spaces and times became superimposed within the strictures of a single experience or economic event. In many ways, then, the current changes that come with the application of computerised communications in urban society reflect merely the latest intensification of a movement towards a speeded-up 'information society' that has roots in the initial industrialisation process (Beniger, 1986). Anthony Giddens has argued powerfully that these trends are fundamental features of modern, industrial societies. They represent what he calls 'a dislocation of space from place', because transport and telecommunications networks allow modern social systems to 'disembed' from their local, physical contexts to operate at much wider scales of space and time than were previously possible (Giddens, 1990; 24).

CONCLUSIONS

In this chapter we have argued that the neglect of telecommunications in the mainstream of urban studies and policy-making is becoming so serious that it threatens to undermine the prevailing paradigms which underpin them. The different elements of this threatening paradigm challenge have been analysed in detail: the challenge of invisibility, the conceptual challenge and the challenge to urban planning. As a route away from this paradigm challenge, we have explored some of the insights provided by recent cultural commentary on urban change that has emerged from debates about post-modernism. This allowed us to highlight a range of new approaches to the conceptualisation of space, time and cities and to review recent approaches which place a more central emphasis on the increasingly tele-mediated nature of contemporary urban life. Improved understanding of city–telecommunications relations, it seems, can best be achieved through exploring the interactions between cities as fixed places where networks intersect and telecommunications as supports for a myriad of electronic spaces which operate free of time and space constraints.

Detailing the elements of this paradigm challenge and highlighting new approaches to conceptualising cities, space and time has provided an important step in the direction of this book's main goal: to help shift telecommunications from the margins to the centre of urban studies and policy. But the question arises here as to how we can best progress from these reconceptualisations to achieve this wider goal? In particular, once we have taken on board the new nature of urbanism and reconceptualised the city and urban life, we need to address issues of causality and explanation. We need to consider how technical developments in telecommunications might be causally linked to social, economic, political and environmental change in cities and vice versa. How, in other words, are urban places causally related to electronic spaces? How best can we move towards an integrated and sophisticated understanding of the relations between cities and telecommunications which does not over-privilege either the technical or the social as causal factors? In what ways can we best move beyond the linear, homogenising and over-simplified models, the assumed technological determinism and the resort to grand metaphors of futurism that so often still dominate urban commentary on these new technologies? It is to these questions that we turn in the next chapter.

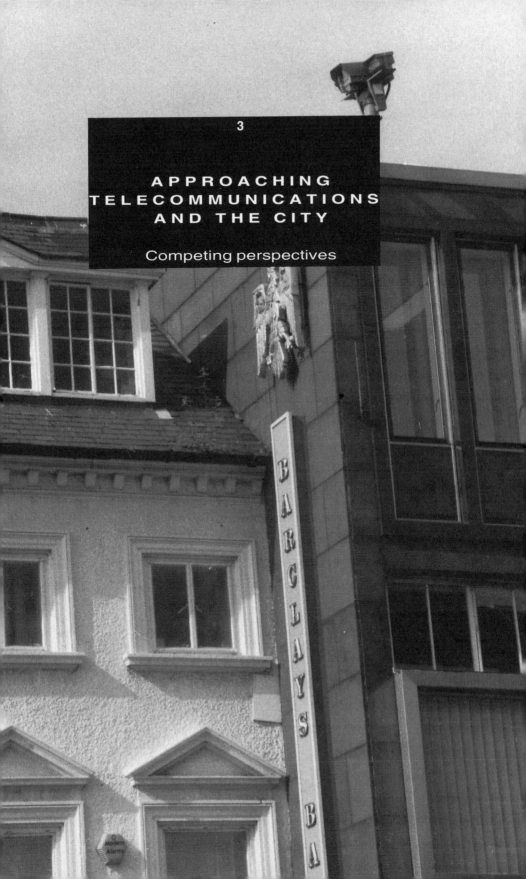

3

APPROACHING TELECOMMUNICATIONS AND THE CITY

Competing perspectives

INTRODUCTION

Looking at city–telecommunications relations is a part of the broader process of analysing the relations between technology and society (Westrum, 1991). As in this wider area, there is not one 'factual' way of studying telecommunications in cities. Rather, there are a range of competing and often contradictory perspectives. Despite this, as with the wider social study of technology, the overwhelming majority of discussions on the subject fail to justify or analyse the approach they take. Instead, commentators tend to adopt one particular perspective without critically weighing up the qualities of competing approaches. Only very rarely do debates arise about cities and telecommunications within which the advocates of different approaches engage.

In this chapter, we critically review the range of analytical perspectives that have so far been adopted – either explicitly or implicitly – in the study of city–telecommunications relations. In other words, following Chapter 2, we explore how theorists have seen cities as urban places and telecommunications networks as supports for electronic spaces to be causally related. This is used to start formulating a new perspective which will provide the basic foundation underpinning the rest of the book.

The competing perspectives reviewed stem from different ideological and theoretical positions within the social and technological sciences. The aim here is to illustrate the variety of contrasting approaches that can be taken to analyse the relationships between urban places and electronic spaces; to evaluate critically their competing qualities; and to use this discussion to develop a new approach which can support a shift towards a more sophisticated and rewarding understanding of city–telecommunications relations.

1. Technological determinism

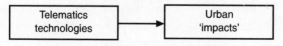

2. Utopianism–futurism

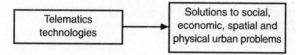

3. Dystopian/urban political economy

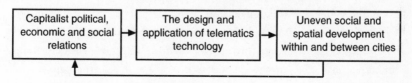

4. Social and political construction of technology

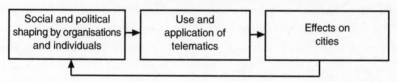

Figure 3.1 The four approaches to studying the relations between cities and telecommunications

In particular, we identify and examine the four dominant perspectives to city–telecommunications relations:

- technological determinism;
- the related approaches of futurism and utopianism;
- a more critical or 'dystopian' approach which draws on urban political economy;
- the social construction of technology ('SCOT') approach.

Crucially for this book, these approaches suggest that the causal relationships that tend to exist between cities and telecommunications are very different. These differences are summarised diagrammatically in Figure 3.1. This chapter also contains four readings as examples of the four approaches. The threads of these debates, and the new perspective we start to develop at the end of this chapter, run implicitly through the remainder of the book, when our consideration turns to specific issues of urban economic development, social and cultural life, the urban environment, the physical form of cities, urban infrastructure and urban planning, policy and governance.

TECHNOLOGICAL DETERMINISM

As with the wider discussion of technology–society relationships, the analysis of the linkages between cities and telecommunications tends to be dominated by a set of approaches which can broadly be termed 'technological determinism'. More often than not, in this 'mainstream' of social research on technology (Mansell, 1994), new telecommunications technologies are seen directly to cause urban change. This is because of their intrinsic qualities or 'logic' as space-transcending communications channels. As Figure 3.1 shows, in their most simple form, technological determinist approaches tend to see the relationships between telecommunications unproblematically as a relatively simple and linear set of technological causes and urban effects or impacts. Electronic spaces are seen to 'impact' on the physical form and socioeconomic development of urban places. The view, to quote Robert Fogel, the 1993 Nobel Prize for Economics, is that 'technological progress is a determinant growth factor' (quoted in Santucci, 1994).

Very often, parallels are drawn here between the current role of telecommunications in determining urban change, and the historic roles of other technological systems. Nicholas Johnson (1970), in one of the first such assertions, argued that 'communications will be to the last third of the twentieth century what the automobile has been to the middle third'. The extract from Anthony Pascal (Box 3.1) is an excellent example of this approach to city–telecommunications relations. He argues that 'technology... shapes destiny. Public actions do modify outcomes; social movements redirect them temporarily. But, ultimately, how we live, where we live and near whom we live depend on the underlying forces inherent in technological evolution and subsequent economic change' (Pascal, 1987; 597). Following this broad technological logic through, Pascal provides a technologically determinist interpretation of the historic centralisation of industrial cities and – determined by the new capabilities of telecommunications – their seemingly inevitable dissolution currently and in the future.

BOX 3.1 THE VANISHING CITY

Human interactions occur across a system of interconnected places, extending from the busiest point in the largest of cities to the furthest spot in the most desolate of boondocks. Technological change ramifies through societies, altering economies and thereby diffusing the inter-actions in space. What once had to happen in the city can now take place anywhere.

Technology, then, shapes destiny. Public actions do modify outcomes; social movements redirect them temporarily. But ultimately, how we live, where we live and near whom we live depend on the underlying forces inherent in technological evolution and subsequent economic change.... Technology-based change breaks out most dramatically in the cities of advanced industrial societies, particularly those that maintain an active market sector. But the collectivist economies and the developing world are not immune....

The era of the computer and the communication satellite is inhospita-ble to the high density city. Clerical and record keeping functions have already begun to deurbanize.... The distant suburbs and small towns of the US are dotted with highly computerized complexes performing bookkeeping, billing and archival tasks for banks and insurance com-panies. The newly emerging technologies will soon begin to provide excellent substitutes for face-to-face contact, the chief remaining *raison d'etre* of the traditional city.

Technological advances in information storage, retrieval and transmis-sion – data, text, pictures and voice through Integrated Service Distribu-tion Networks or ISDN – will soon begin to revolutionize administration and headquarters functions. The falling costs of fiber optic transmission systems make ISDN more practical for more places.

Where it was once necessary to concentrate large staffs of managers who could combine to generate quick solutions for nonstandardized problems, teleconferencing will soon provide an alternative. Teleconferencing even has some advantages over face-to-face meetings: participation need not be sequential and can be anonymous, printout and videotape records are available, subcommittees and caucuses can be accommodated simultaneously. People who grow up in an environment where such things are possible will more easily adapt to and more readily use the innovations. The broad implication? A pension fund with a long time horizon would be ill advised to invest in any fourth tower at New York's World Trade Center.

The ease of face-to-face communication and the concentration of specialized support services also made the city an incubator for businesses.... The city acted as the urban hatchery for new companies, many of which subsequently departed for the suburbs or the countryside. As the cost of disseminating knowledge and information declines and as specialized inputs are more easily assembled in remote locations, incubation too can take place almost anywhere.

Truly revolutionary developments in the assembly of another input, labor, appear in the offing. In telecommuting, workers manipulate and analyze data on individual computers at scattered sites and transmit to central mainframes through modems. And data manipulation and analysis describe what an increasing share of the post-industrial labor force does for a living. Managers who grow up with computers in the classroom, and the family room, will have less difficulty adjusting to the supervision of workers in remote locations.

Although there will always remain functions that require the physical congregation of people and people who prefer tangible proximity to others, substantial new flexibilities will enter the system. Estimates for 1995 suggest that perhaps 15 per cent of the US labor force will telecommute by then. Development of the electronic briefcase and the refinements of the cellular telephone promise even more disengagement between workers and fixed, centralized facilities. In the future then, many will be able to work virtually anywhere. Given a preference for low density residence in amenity-rich areas, which a large fraction of the population expresses, further depopulation of cities seems inevitable....

Consider some additional developments:

.... The addition of robot-staffed warehouses and television catalogs – perhaps holographically enhanced – may one day transform shopping into a largely in-home activity.

Financial institutions are beginning to experiment with home banking through cable television. Customers can deposit, withdraw and shift funds among accounts using personal consoles.

As radio and television reduced the need for access to cities as a source of entertainment, the spread of video cassette recorders implies even more freedom of location. Still to come is two-way television, in which viewers electronically select from a vast library of recorded programs. Somewhat further off is the development of truly interactive programming.

Through such innovations, improved communication substitutes for proximity. Because the city specializes in the advantages of proximity, its attractiveness as a focus for human interactions will continue to decrease. The proper late twentieth century reply to Gertrude Stein's complaint about Oakland, 'There is no there there', is, 'There is less and less there anywhere, anymore. Increasingly, there is everywhere.'

(Pascal, 1987; 597–603)

This approach is based on the linear notion that innovation leads to new technologies which are then applied, used, and go on to have effects and technological impacts upon society. To technological determinists, the urban impacts of telematics and telecommunications are relatively straightforward, linear, direct and easy to pinpoint. Most are virtually inevitable. The decentralisation, or even dissolution, of cities; the free availability of highly capable communications in all locations; the shift towards city economies based on information; the growth of a culture based on tele-interactions; the shift to an 'immaterial' urban life; the growth of telecommuting – all will be shaped by new innovations in telematics in a deterministic and inevitable fashion. The analytical and policy issues suggested by technological determinist therefore centre around how society can adapt to and learn to live with the effects of telecommunications-based change, rather than focusing on the ways in which these effects may be altered or reshaped through policy initiatives.

There are two key points here. First, technological change is of overwhelming importance in directly shaping society. Second, the forces that stem from new telecommunications innovations are seen to have some autonomy from social and political processes (Winner, 1978) – what Stephen Hill calls the 'apparent intrinsic technological inevitability' (Hill, 1988; 2). The social and the technical are cast as two different arenas, the former being shaped by the latter. Machines and technologies are seen to arise and evolve in a separate realm to alter the world (Thrift, 1993). Technological revolutions, such as the current one which many allege to be based on telematics, are seen as virtually unstoppable broad waves of innovation and technological application, which then go on to impact on cities and urban life (Miles and Robins, 1992).

TECHNOLOGICAL 'LONG WAVES' AND TECHNOLOGICAL 'REVOLUTIONS'

Approaches which adopt the notion of 'technological long waves', drawing on the work of economists such as Schumpeter or Kondratiev, tend to adopt versions of this macro-scale 'technology cause' and 'societal effect' approach (see, for example, Hall and Preston, 1988). In this approach, the construction of major new technological systems in society, such as advanced telecommunications today, is seen to stimulate a long wave of innovation through all areas of society, so creating economic growth in new areas and markets and a resulting shift in the makeup and functioning of urban economies (Freeman, 1987). The worldwide diffusion of systems like new telecommunications networks therefore has a 'limited autonomy . . . in determining the scope and direction of major new developments in the economy' (Freeman, 1987; 5). The combination of telecommunications and microprocessors, as so-called 'heartland' technologies which make up telematics (Perez, 1983), is seen to create a technological revolution because of the pervasiveness with which they can be used to revolutionise production processes in offices, factories etc. Chris Freeman argues that 'the combination of radical innovations in computers and telecommunications constitutes a change of "technoeconomic" paradigm. Such a change of paradigm affects all industries and services and brings with it the need for many organisational, managerial and social changes so that the institutional and social framework is well adapted to the potential of the new technologies' (Freeman, 1991; 159). In a similar fashion, Hall and Preston (1988; 285) argue that the right conditions and policies can make 'radical technological clusters' if information technology and telematics act as 'carriers', leading national and international economies out of the current recession and into the next 'long wave' of economic growth based on the 'information age'.

These revolutions are seen to have economic, social, spatial and cultural consequences for cities – either current or predicted (Brotchie et al., 1985; 1). The key task for urban analysis is to map these out or predict them. A classic example from urban history was the separation of offices from factories that the telegraph allowed, the result being the centralisation of offices into central business districts (Tarr et al., 1987). Brotchie et al. (1985; 1), in the introduction to a book of studies on technology and urban form, suggest that the current phase of technological change 'will cause changes in patterns of living and working at least as fundamental and as comprehensive as those induced by the industrial revolution on the eighteenth and nineteenth centuries'. Through its impacts on patterns and processes of work and life, telematics will therefore impact on the form and functioning of cities, primarily through the decentralisation of cities based on 'the uniformly-available high-technology infrastructure' (Hall, 1985; 28).

The broad 'technological cause – urban impact' approach reflects very closely the 'common sense' view of technological change within western culture. As Stephen Hill argues:

the experience of technology is the experience of apparent inevitability... the most influential critics who have sought to understand the experienced 'command' of technological change over twentieth-century life have turned to the machines for explanation, and asserted the 'autonomy' of technology... Each of the technological determinist positions tends to see the technological 'frame' as autonomous, with social and cultural transformations being the consequences of a technologically-inspired trajectory, not the creators of this path.... The technological determinist stance aligns with many people's everyday experience.

(Hill, 1988; 23–24)

FUTURISM AND UTOPIANISM

It is a small step to move from exploring the current impacts of tele-communications on cities to speculating and forecasting into the future. Despite the obvious uncertainties involved, and the risks of being proved completely wrong, a mini-industry currently thrives on predicting the technological future of cities. A central theme of much of such work has been on forecasting the effects of the radical technological changes underway in computing, media technologies and telecommunications upon cities in the future. This tends to be bound up with the many variants of the 'grand metaphor' approach within which western society is seen to be moving en masse to a new and novel stage in its development as some form of 'information society' (Lyon, 1988) or 'information age' (Dizard, 1982).

The speculations of 'futurists' generally tend to take a relatively optimistic view of the future impacts of telecommunications on cities and urban life. The proliferation of electronic spaces and networks are often seen to have very positive effects both for the physical aspects of cities and for urban life more widely. Where there are negative effects, these, too, it is argued, can often be solved through new technologies (Lyon, 1988; 146). Often, these commentaries are breathless and excited, offering tantalising future glimpses of how remarkable advances in new technologies will determine future lifestyles that are incalculably better than those today. Almost always, this future state is offered as a scenario where, as Eubanks (1994) believes, 'potentially huge benefits are to be had by all'.

These views tend to be fuelled by a variety of motivations. Often, direct vested interests exist in the use and application of new computing and telecommunications technologies. Most large media, computing and telecommunications companies, for example, are involved in some way in encouraging the current frenzy of debate and hype over the 'information superhighways', which are often being cast as some sort of technological panacea for all the social, economic and environmental ills of society. Sometimes, futurists have an ideological predisposition to look on technology as a near panacea for the ills of urban society (Slack and Fejes, 1987). Most often, though, futurists simply have faith that the new technologies tend to have positive effects and that the solution to the pressing problems of contemporary urban society are overwhelmingly technological in nature. While the information superhighways are the latest technological project to be the centre of such speculation, videotex, satellite, teleports and cable networks (Strover, 1989) have all had their time at the centre of futurists speculations.

Futurism is often linked strongly with utopianism. Utopianism – the search for radically better and new forms of social life – has attacked the negative aspects of industrial cities – pollution, overcrowding, moral 'degradation', social disintegration – since they first emerged (Gold, 1990). Technological promises have long had a key place in many urban utopias (Gold, 1985; Lyon, 1988; 152). In its most recent guises, treatment of telecommunications as a solution to these perceived problems has been a key theme in utopian thought.

TELECOMMUNICATIONS AS THE 'TECHNICAL FIX' TO URBAN PROBLEMS

Together with the bulk of the futurists, then, utopianists tend to see new telecommunications and telematics technologies as being solutions to the social, environmental, economic and physical problems they associated with the industrial city. With networks apparently replacing the need to commute, travel to shop, enjoy high-quality teaching and museums, this infrastructure is heralded by many as the solution to a disparate range of perceived problems. For example, Santucci, of the European Union's DGXIII, argues that 'it is the task of the information highways to facilitate the transition to the electronic age for example, by helping to bring health care costs under control, solving road and air traffic congestion problems, generating new productivity gains . . . and enabling greater flexibility in the workplace and better skill use' (Santucci, 1994; 22).

The assumption that these trends will be environmentally benign also pervades futuristic and utopian writings; the idea that the clean, demateralised solutions of electronic spaces will substitute for the material ills of commuting and pollution in urban places has also become so common as to exist as an orthodoxy in debates about cities and telecommunications. Telematics networking has even been called the 'alternative fuel' (see Chapter 6).

In this vein, James Martin, in his prediction of the oncoming 'wired society', suggested that the congested and polluted physical spaces of cities would in the future be complemented by 'virtual cities', based on the use of telecommunications to replace physical transport and the need for propinquity (Martin, 1981). These, he argues, 'will allow specialised groups and activities to develop on networks linking a multitude of locations'. He predicted that 'communities, campuses, laboratories or corporate offices [will develop] scattered across the earth but connected electronically so that the chain reaction of human stimulation catches fire as it did in [the physical spaces of] Victorian London' (Martin, 1978; 193). John Gold summarises the new utopian writings based on optimistic predictions of the development of 'wired cities' (Dutton et al., 1987) and 'wired societies' (Martin, 1978) based on advanced telecommunications networks:

Their case may be summarised as being based on an optimistic interpretation of the development of electronically-interconnected or 'wired' societies. In such societies, technology creates opportunities for the restructuring of everyday life. Economic activity is regenerated through reorganisation that reduces costs and makes better use of the total labour force (including those who, for family reasons or disability, could not otherwise enter the employment market). Decentralised working patterns produce opportunities for greater leisure. Ubiquitous, multi-channel cable networks supply the media for cultural enrichment. Electronic banking and shopping bring about increased consumer choice. There is enhancement of democracy through instant electronic referenda. A new urban order arises in which people can move back to the 'natural' surroundings of the countryside while still retaining close connection with their workplaces and with the culturally-enriching aspects of city life.

(Gold, 1990; 22)

These ideas have been extremely influential in shaping debates about technologies and cities. In the 1970s and early 1980s, a wide range of policy agencies began to 'dream of multi-purpose broadband networks as the central technical element of a synthetic conception of urban communication' (Man-dlebaum, 1986; 132). These ideas helped to encourage a wide range of 'wired city' technological experiments in cities which attempt to use telecommunications as policy tools to generate the supposed positive effects (see Dutton *et al.*, 1987). Generally, however, these were unsuccessful because the public demand never matched the anticipation, the technology was too cumbersome or simply because the projects never paid their way (see Chapter 9).

THE 'ANYTHING – ANYTIME – ANYWHERE' DREAM

The almost universal assumption in futurist and utopian writings on cities and telecommunications is that the basic time and space transcending nature of electronic spaces means that we are moving to a world where all information will be available at all times and places to all people. Again, the interactive, integrated and ubiquitous optic fibre network is the technological key here, with its apparently limitless capacities for mediating entertainment, work, culture, administration, health, education, and social interaction. As a result, geography, propinquity and spatial dynamics either cease to matter at all are of much reduced significance. This 'anything, anytime, anywhere' dream and the presumption that it will mean the collapse of the modern city is central to most futurist visions.

Most of these predictions assume that access to future electronic networks will be more democratic and equitable than the familiar social divisions within current cities. The usual assumption is that social access to the new broadband infrastructures and the services on them will become totally universal, so overcoming the familiar patterns of social class division in capitalist society. Jacques Maisonrouge, for example, argues that 'modern information processing capability creates a society where everyone has an equal opportunity to be information literate ... telecommunications and the computer are making information accessible to everyone' (Maisonrouge, 1984; 31). Santucci predicts that a 'truly planetary consciousness will spring from the establishment, at world level, of bi-directional information highways accessible to all individuals' (Santucci, 1994; 16).

ANTI-URBANISM AND THE DE-MASSIFIED
SOCIETY

A strong anti-urban emphasis on the environmentally and socially destructive nature of twentieth-century cities permeates this approach to city–telecommunications analysis. As John Gold argues, underpinning many futuristic visions of the 1960s and 1970s was 'a prevailing hostility towards the metropolitan city and a social philosophy that aims to break up the mass society and recreate smaller communities' (Gold, 1990; 22). Following the long-standing utopianism of people like Ebenezer Howard and Frank Lloyd Wright, industrial cities are seen to be 'sick' (Mason and Jennings, 1982) or 'unnatural' – extreme concentrations that were aberrations created by the industrial revolution. Advances in telematics, as the cause of urban decentralisation or even dissolution, are therefore heralded as solutions to many of the ills of contemporary urban society. These sorts of ideas have a long history. Peter Goldmark, for example, predicted that telecommuting to work would liberate people from the 'conditions of extreme density within the confines of cities and their suburbs' that they were forced to endure as a result of the growth of the great industrial metropolis (Goldmark, 1972). People, he argued, were 'physiologically and psychologically unprepared for' this 'unnatural' life; cities were the areas where 'problems of crime, pollution, poverty, traffic, education etc. are the greatest'.

At their most extreme, these commentators saw telecommunications as a tool for engineering the large, industrial city out of existence through radical decentralisation and the assumption that social transformation would stem from 'wired city' projects. At least, such approaches tended to hold up 'the image of American suburbia as the ideal human environment' (Garnham, 1994; 43).

But by far the most influential futuristic analysis of the effects of tele-communications on cities was developed by Alvin Toffler in his book *The Third Wave* (1981). The second extract (see Box 3.2) presents the essence of his argument. Toffler's approach rests on the argument that advanced western societies are in the midst of a revolution based on their transformation into 'Third Wave' societies – as distinct from 'First Wave' agricultural societies and 'Second Wave' industrial societies. This revolution is based on the new capabilities of computers and telecommunications for supporting a geographical, economic and political decentralisation of urban society. Central to such societies is the idea of the 'electronic cottage' – a household which acts as the locus of employment, production, leisure, and consumption through telecommunications-based interactions with the outside world. Electronic cottages are locationally liberated from the need to concentrate in cities, and so people are freed to locate in the rural areas of their choice. Advanced telematics networks and services support the real-time contact opportunities needed to sustain social, economic and political life in 'Third Wave' societies. The resulting home-centred society therefore escapes the environmental, social and political problems associated with the industrial city. A new convivial, democratic and environmentally sustainable civilisation of small-town or rural life becomes possible. Of course, the other effect of this shift is the virtual dissolution of the great western cities that become mere centrifuges for the 'export' of their contents.

BOX 3.2 THE ELECTRONIC COTTAGE

Hidden inside our advance to a new production system is a potential for social change so breathtaking in scope that few among us have been willing to face its meaning. For we are about to revolutionize our homes as well.

Apart from encouraging smaller work units, apart from permitting a decentralization and de-urbanization of production, apart from altering the actual character of work, the new production system could shift literally millions of jobs out of the factories and offices into which the Second Wave swept them and right back where they came from originally: the home. If this were to happen, every institution we know, from the family to the school and the corporation, would be transformed.

Watching masses of peasants scything a field three hundred years ago, only a madman would have dreamed that the time would soon come when the fields would be depopulated, when people would crowd into urban factories to earn their daily bread. And only a madman would have been right. Today it takes an act of courage to suggest that our biggest factories and office towers may, within our lifetimes, stand half empty reduced to use as ghostly warehouses or converted into living space. Yet this is precisely what the new mode of production makes possible: a return to cottage industry on a new, higher, electronic basis, and with it a new emphasis on the home as the center of society....

Even in the manufacturing sector an increasing amount of work is being done that given the right configuration of telecommunications and other equipment could be accomplished anywhere, including one's own living room. Nor is this just a science fiction fantasy.... If significant numbers of employees in the manufacturing sector could be shifted to the home even now, then it is safe to say that a considerable slice of the white-collar sector where there are no materials to handle could also make that transition....

These are, moreover, among the most rapidly expanding work classifications, and when we suddenly make available technologies that can place a low-cost 'work station' in any home, providing it with a 'smart' typewriter, perhaps, along with a facsimile machine or computer console and teleconferencing equipment, the possibilities for home work are radically extended.

Given such equipment, who might be the first to make the transition from centralized work to the 'electronic cottage'? While it would be a mistake to underestimate the need for direct face-to-face contact in business, and all the subliminal and nonverbal communication that accompanies that contact, it is also true that certain tasks do not require much outside contact at all or need it only intermittently....

In short, as the Third Wave sweeps across society, we find more and more companies that can be described, in the words of one researcher, as nothing but 'people huddled around a computer'. Put the computer in people's homes, and they no longer need to huddle. Third Wave white-collar work, like Third Wave manufacturing, will not require 100 percent of the work force to be concentrated in the workshop....

THE TELECOMMUTERS

Powerful forces are converging to promote the electronic cottage. The most immediately apparent is the economic trade-off between transportation and telecommunication. Most high-technology nations are now experiencing a transportation crisis, with mass transit systems strained to the breaking point, roads and highways clogged, parking spaces rare, pollution a serious problem, strikes and breakdowns almost routine, and costs skyrocketing....

The key question is: When will the cost of installing and operating telecommunications equipment fall below the present cost of commuting? While gasoline and other transport costs (including, the costs of mass-transit alternatives to the auto) are soaring everywhere, the price of telecommunications is shrinking spectacularly. At some point the curves must cross....

Social factors, too, support the move to the electronic cottage. The shorter the workday becomes, the longer the commuting time in relationship to it. The employee who hates to spend an hour getting to and from the job in order to spend eight hours working may very well refuse to invest the same commuting time if the hours spent on the job are cut....

THE HOME-CENTRED SOCIETY

If the electronic cottage were to spread, a chain of consequences of great importance would flow through society. Many of these consequences would please the most ardent environmentalist or techno-rebel, while at the same time opening new options for business entrepreneurship....

Work at home involving any sizeable fraction of the population could mean greater community stability – a goal that now seems beyond our reach in many high-change regions. If employees can perform some or all of their work tasks at home, they do not have to move every time they change jobs, as many are compelled to do today. They can simply plug into a different computer....

At a deeper level, if individuals came to own their own electronic terminals and equipment, purchased perhaps on credit, they would become, in effect, independent entrepreneurs rather than classical employees meaning, as it were, increased ownership of the 'means of production' by the worker. We might also see groups of home-workers organize themselves into small companies to contract for their services or, for that matter, unite in cooperatives that jointly own the machines. All sorts of new relationships and organizational forms become possible.

(Toffler, 1981; 210–223)

TOWARDS THE 'ELECTRONIC COTTAGE'

In Toffler's and Martin's visions, people spend more and more time at home working and consuming services via telematics networks. So the 'electronic cottage', too, is seen to emerge as a centre for computerisation and automation to emerge as the 'computer home' (Mason and Jennings, 1982), the 'electronic house' (Mason, 1983) or the 'smart home' (Moran, 1993). Here, as with the Tofflerian vision, homes will be reformulated to be physical access nodes for the 'electronic spaces' within telematics networks.

This, too, will help address many of the social problems caused by the separation of work and home life in the industrial city. The computer home, linked to advanced telecommunications networks, will revolutionise home life. Nicholas Johnson argued that, ultimately, the home would emerge as a 'home communication center where a person works, learns, and is entertained, and contributes to society by way of communications techniques we have not yet imagined – incidentally solving commuter traffic jams and much of their air pollution problems in the process' (Johnson, 1967; quoted in Abler, 1977). Pelton, more prosaically, argues that 'telepower... will bring hope for high-quality schooling, nutritional advice and medical services through tele-education and telehealth techniques' (Pelton, 1989; 10). George Heilmeier predicts the emergence of 'virtual universities', 'virtual libraries' and 'information malls' where consumers will enter multimedia and virtual reality-based electronic 'shops' and services over electronic networks from the home (Heilmeier, 1992;

52). 'Whereas, in the past, life at home was often confining and oppressive', write Roy Mason and Lane Jennings, 'the home centred life of the future may be exhilarating and mind-expanding, thanks to worldwide networks of electronic communication'. Computer-based access to electronic spaces

will bring into the home environment many of the facilities and services we now often travel many miles to obtain at schools, libraries, offices, theatres etc. . . . . It will provide a new focus for human life – a twenty first century version of the hearth that was so long an essential feature of the 'home' in every age and civilisation.

(Mason and Jennings, 1982; 35)

FROM THE 'ELECTRONIC COTTAGE' TO THE 'INTELLIGENT CITY'

By the linking together of electronic homes with city-wide systems of 'intelligent' networks, many futurists promise radical new ways of managing and organising urban life, as artificial intelligence (AI) and robotics encroach into transport, logistics, manufacturing, retailing, leisure, public administration and home life (Toth, 1990). Thus we move from predictions of the 'smart home' to the 'smart highway' and the 'intelligent city' where, once again, electronic spaces are seen to offer the clear, simple and uncontentious answer to any problems that are currently identified (see Chapter 8).

In extreme cases, futurists go on to predict the complete 'cybernation' or 'automation' of urban society – the emergence of cities as vast 'control systems' geared to maximum environmental and economic performance and social benefit. Kalman Toth, for example, predicts that the pervasive diffusion of 'Silicon-Magnetic Intelligence' (SMI) units, linked via telecommunications, into all areas of life will bring a workless society with 'ease and abundance for all'. He predicts that 'the new American social order will consist of 2 billion SMIs working, doing business, and generating taxes; 5 million people paid as government workers; and the rest of the population getting a salary to enjoy life' (Toth, 1990; 37).

DYSTOPIANISM AND POLITICAL ECONOMY

A third range of analytical approaches to city–telecommunications relations can be grouped under the broad heading dystopianism and urban political economy. Dystopianists and political economists stress the ways in which the development and application of telematics technologies are not somehow separate from society. Rather, they are fully inscribed into the political, economic and social relations of capitalism. Following this, telecommunications and electronic spaces are not seen as simple determinants of urban change. Nor are they cast as panaceas or 'quick-fix' technical solutions to urban problems. According to this approach, city–telecommunications relations cannot be understood without considering the broader political, economic, social and cultural relations of advanced industrial society and how they are changing.

Dystopianists have been influenced by several strands of work. Famous scenarios where advanced telecommunications breed centralised state power and complete social surveillance – such as in Orwell's *Nineteen Eighty-Four* – have played a role. More recent post-modern dystopias of a fragmented and globalised modern urban society in some near future – from Ridley Scott's portrayal of Los Angeles in *Blade Runner* to the 'cyberspace' science fiction of William Gibson's 1984 *Neuromancer* – have also been influential.

More analytical strands of work have developed as a result of criticism of the failings of technological determinism and utopianism/futurism. Much of this draws on recent neo-Marxist work on urban political economy, the political economy of telecommunications, and analysis of the cultural changes involved in the shift to post modernism that we touched on in the last chapter. These approaches offer disturbing anti-utopian or 'dystopian' visions and analyses of telecommunications-based urban life today and in the future.

TECHNOLOGICAL RELATIONS AS SOCIAL RELATIONS

Here, the effects of telecommunications on cities are defined by the ways in which they are used to support wider processes of economic, political and spatial restructuring. These processes are not simply driven by technological change.

Rather, the driving forces shaping the application and development of tele-communications are the political, economic, social and cultural dynamics of capitalism itself. Above all, especially in the neo-Marxist accounts, the develop-ment and application of telematics are seen to be driven by the imperative of maintaining capital accumulation for firms, and the need to overcome crises that reflect capitalism's inherent contradictions. City–telecommunications relations are, in this approach, seen to be driven by the economic forces of capitalism itself and to reflect and perpetuate capitalism's highly unequal social relations.

In this vein, Stephen Hill argues that 'to attribute ... intrusive power to technology *per se* is inherently wrong. Social, political, and economic negotiations are involved in bringing particular technological systems into existence' (Hill, 1988; 6). Robins and Gillespie (1992; 150) stress that 'geographical transforma-tion is not determined by technological innovations, but, rather, it is through the possibilities they offer that new spatial configurations might be elaborated'. The ways in which telecommunications technologies and services are applied in cities are seen to be infused with the wider conflicts between the social and economic class interests that characterise capitalist society and to be set within the wider cultural orientation of modern societies (see Figure 3.1, page 79). As David Lyon puts it, in this approach '"social relations" represented by information technology are seen mainly as class relations' (Lyon, 1988; 151). Like all technologies, a telecommunications or telematics network, and the uses to which it is put are seen to 'encapsulate and embody social relations' (Nowotny, 1982; 97). This is because they are designed and applied to reflect prevailing power relations and social and geographical inequalities. To Eric Swyngedouw, 'changes in mobility and communications infrastructure and patterns ... are not neutral processes in the light of given and changing technological ... conditions and capabilities, but necessary elements in the struggle for maintaining, changing or consolidating social power' (Swyngedouw, 1993; 305).

THE 'BIAS' OF TELECOMMUNICATIONS

The social, economic and spatial effects of telecommunications within and between cities therefore derives from this embeddedness of telecommunications within capitalism. This undermines the common argument that telecommunica-tions, and technologies in general, are somehow neutral in their social and spatial effects and can be shaped to benefit different interests accordingly. Andy Gillespie, drawing on the influential work of Harold Innis (1950), argues that 'tele-communications are not neutral technologies. They are not equally amenable to all users which can be envisaged; an inherent bias is already "locked in" to them, through the network design process' (Gillespie, 1991; 225). Maureen McNeil also attacks the notion that Information Technology (IT) is somehow socially unbiased and the vehicle of 'benefits for all'. Instead, it is loaded with values and intensely biased, coming 'not as a neutral force for change but as the carrier of (minimally) entrepreneurialism, nationalism, versions of masculinity, notions of citizenship and images of meritocracy (with the concomitant denial of class, gender and racial divisions)' (McNeil, 1991; quoted in Robins and Cornford, 1990).

In this view, dominant institutions – primarily transnational corporations (TNCs) – are seen to use the space- and time-adjusting capabilities of telematics to their own benefit while purposefully excluding possible benefits from others. Telecommunications networks, in other words, are seen to be involved as arenas of struggle within a much broader process of societal reproduction which also impinges upon the physical, cultural and social landscapes of cities.

THE FUNCTIONS OF TELECOMMUNICATIONS: MOBILITY AND CONTROL

A key focus of the political economic approach is to highlight the functions that new innovations in telecommunications and telematics are playing in terms of facilitating new ways of organising more mobile modes of social life which can be more easily and flexibly controlled. This is in response to crises in the Fordist era of capitalist development that stemmed from the growth of large, complex firms and organisations beyond the control capabilities of capitalist élites and decision-makers. 'From this perspective', writes Andy Gillespie, 'the development and deployment of advanced telecommunications networks can be interpreted as part of a process by which "Fordist" enterprises are attempting to regain control over their own organisations and labour forces, their suppliers and their markets' (Gillespie, 1991; 214).

The economic and geographical restructuring which is resulting from these changes is seen to be complete reshaping of capitalism on a global basis. In particular, the new capabilities of telematics are underpinning what is called 'time–space compression' – the overcoming or reduction of time and space barriers (Harvey, 1989). This is allowing new solutions to the tensions inherent within capitalism between what Harvey calls 'fixity' and the need for 'motion', mobility and the global circulation of information, money, capital, services, labour and commodities. Because it is driven by the search for new profits and 'capital accumulation', capitalism is seen to be inevitably expansionary. This means that widely dispersed areas of production, consumption and exchange need to be integrated and coordinated. Space needs to be 'commanded' and controlled. New transport and telecommunications infrastructures need to be built which accelerate the mobility of capital and money – so overcoming the spatial and temporal barriers that inhibit this expansion and coordination. The 'goal' of capitalism, and the key to maximum profits with minimum risk is therefore perfect mobility for labour, goods, capital and informational products.

But in a spatial and geographical world, this goal is impossible. Both the new spatial structures that are built up for production and consumption – cities, industrial areas etc. – and the new telecommunications and transport networks, are inevitably fixed and embedded in space (Swyngedouw, 1993). New telecommunications networks 'have to be immobilised in space, in order to facilitate greater movement for the remainder' (Harvey, 1985; 149). This makes them expensive, uncertain and risky to develop – especially for the profit-seeking firms that now tend to control them. This inflexibility means that they then go on to present problems later to further 'rounds' of restructuring, as the continuous dynamism of capitalism plays out. As Swyngedouw (1993; 306) argues, 'liberation from spatial barriers can only take place through the creation of new communication networks, which, in turn, necessitates the construction of new (relatively) fixed and confining structures'. To quote David Harvey:

The tension between fixity and mobility erupts into generalized crises, when the landscape shaped in relation to a certain phase of development . . . becomes a barrier to further accumulation. The landscape must then be reshaped around new transport and communications systems and physical infrastructures, new centres and styles of production and consumption, new agglomerations of labour power and modified social infrastructures.

(Harvey, 1993; 7)

Using this approach, political economists point out how telecommunications are being used to underpin global capitalist restructuring. This is based on the search for flexibility and the control of volatile and global corporations and markets, and on the use of telecommunications profitably to utilise the many different regions of the world through real time coordination and integration of distant production sites. The key relationships between cities and telecommunications is therefore a global–local one in which a city is integrated silently and invisibly into the new global electronic networks through which the many different regions and areas of the global economy are tied together to support profitable enterprise.

As with the design of telecommunications, these processes are biased asymmetrically in favour of the growing transnational corporations. They control the networks privately and exclude others from having access; they also control the roles that cities play in the global networks (Sussman and Lent, 1991). Andy Gillespie argues that 'the instrumental role that advanced communications networks are currently playing in opening up national, regional and local economies to global forces of competition and control seems undeniable' (Gillespie, 1991; 223).

By allowing the erosion of the traditional space and time limits that acted to constrain capitalist development in the past, telematics are seen to support the ever more efficient exploration of new sources of capital accumulation. Central here is the exploitation, control and surveillance of widely separated and varying groups of workers and consumers and the management of all information as private property with availability restricted strictly by ability to pay (Slack, 1987; 10). Schiller and Fregaso, for example, argue that 'the concerted drive to privatize telecommunications should be viewed as an effort to create a thoroughgoing global foundation for systematic corporate exploitation of information' (1991; 202). The result is the so-called 'commodification of information' – the movement of public and free public information services into private ones that are delivered 'bit by bit' on-line but only to those with the ability to pay (Openshaw and Goddard, 1987; Mosco, 1988).

Thus, the landscape of western (and non-western) cities is being re-wrought as a new global search for trade-offs between fixity and motion results in new investments in infrastructures and built environments. To Manuel Castells, this linkage means that a new era of capitalism is upon us based on the construction by transnational corporations of a 'space of flows' in the 'electronic space' of global telematics systems. These are being used to add power and flexibility to corporations in their dealing with the physical spaces of cities. He writes:

The supersession of places by a network of information flows is a fundamental goal of the restructuring process. . . . This is because the ultimate logic of restructuring is based on the avoidance of historically established mechanisms of social, economic, and political control by the power holding organizations. Since most of these mechanisms of control depend upon territorially-based institutions of society, escaping from the social logic embedded in any particular locale becomes a means of achieving freedom in a space of flows connected only to other power holders, who share the social logic, the values, and the criteria form performance institutionalized in the programs of the information systems that constitute the architecture of the space of flows.

(Castells, 1989; 349)

THE REWORKING OF GEOGRAPHICAL DIFFERENCES

These changes do not lead to an ironing out of geographical differences, as the 'anything, anytime, anywhere' dream of the futurists implies. Political economists argue that, just because spatial barriers *can* be overcome in certain circumstances, this does not mean that they *are* being overcome. Rather, they are reworked in new ways to match the new styles of production and consumption and to meet the needs of powerful organisations. New telecommunications and transport infrastructures are seen to allow capitalists to construct ever more sophisticated locational strategies, taking advantage of the fine-grained differences between places in terms of labour costs and quality, public sector subsidies etc., while still integrating widely dispersed sites together into flexible and coherent corporate structures. This, though, is a global process tying together the advanced industrial North with the South – the vast swathes of the Third World now opened up for 'development' and 'modernisation' as cheap production centres and raw material sources to fuel the 'North' – all integrated together using these new technologies (Sussman and Lent, 1991). Advances in telecommunications are seen to help reproduce the political and social relations of capitalism and 'ever more rapid systems of communications are at the basis of an accelerated circulation [of goods, people and money] and, hence, accumulation of capital' (Martin, 1991; 307).

TELECOMMUNICATIONS AND SOCIAL POLARISATION IN CITIES

This perspective means that cities can not be seen merely as artifacts that are shaped by technology or as areas where problems require technical solutions. Instead, cities and telecommunications are bound up together within the same broad scheme of political and economic change. The relationship between cities and telecommunications is therefore more recursive than that suggested by determinists or futurists. As Gillespie and Robins argue, 'new communications technologies do not just impact upon places; places and the social processes and social relationships they embody also affect how such technological systems are designed, implemented and used' (Gillespie and Robins, 1989).

Not surprisingly, these critical approaches cast doubt on the utopian scenarios of an environmentally benign, decentralised and home-centred urban life. They also suggest that the assumptions that telecommunications will somehow iron out social and geographical inequalities within and between cities are little more than absurd. Political economists point to the ways in which telecommunications and telematics are being used to support the restructuring of cities and the exacerbation and intensification of uneven social and geographical development at all levels (Gillespie and Robins, 1989). They stress the key role of telematics in reshaping the time and space limits that confine capitalist economic development, *but in ways that directly favour those economic and political interests who already dominate society.* The result is the globalisation of the capitalist economy, which tends to fragment urban economies, reduce the power of cities over their own destinies and marginalise many disadvantaged social groups.

Rather than simply disinventing cities, political economists stress that the telematics-based decentralisation of certain activities from cities and the likely concentration of urban life on to the home implies the destruction of the civic, public dimensions of cities. Cities are still necessary as the sites of massive sunk investment, as the controlling hubs of global networks, as places with unique concentrations of infrastructure, labour markets, services, information and skills, and as the arenas of social surveillance and control. Cities are the sites where the intensely mobile flows of information, capital, money, people and services become fixed and interact.

Rather than being universal, critical commentators stress the profound unevenness and bias of telematics-based developments within cities. While (usually white and male) affluent corporate élites become saturated with new technologies and the expensive services that are based on them both at home and work, many groups (including women, the poor, ethnic minorities and the disabled) are either excluded from the benefits of the new technologies, or are simply exploited or monitored more efficiently by them. While élites control and shape how telematics are used in cities, many less powerful social groups are controlled by them – they are the surveyed rather than the surveyors, the controlled rather than the liberated.

Telematics are therefore seen not to disinvent cities but promote the development of more polarised and fragmented cities – the latest temporary solutions between the demands for mobility and fixity through which profitability can be resuscitated. The result, combined with the broader shifts towards privatisation and liberalisation, is an exacerbation of the economic and social inequalities in urban life. Thus the future and, increasingly, the contemporary city is seen to be a profoundly polarised and unjust one, a place ridden with extreme inequalities, supported by the biased use of telematics by capitalist élites to maintain their power. David Lyon, for example, argues that 'the ways in which IT is developed and adopted frequently widens the gap between already divided social groups and nations, extends the capacity of the state and other agencies to monitor and control people's lives and augments the power of ever-growing economic interests' (Lyon, 1988; 146). The vision of the electronic home, too, is also attacked as being driven by the need to exploit and survey (low-paid, part-time and usually female workers) in their homes without the costs associated with traditional office work (see Huws, 1985; Zimmerman, 1986).

THE CRITIQUE OF TECHNOLOGICAL DETERMINISM, FUTURISM AND UTOPIANISM

Dystopianists do not only take issue with the explanatory and analytical failings of technological determinism and futurism. Many of these processes, they argue, are actually camouflaged by the construction of glossy and benign ideology for the information age by futurists and telematics industries, where all are heralded to benefit from the telecommunications-based future (Slack and Fejes, 1987). Futurists and utopianists are attacked by this approach for reducing political and moral issues to purely technical ones in an attempt to reduce counter movements against the technological changes underway (Lyon, 1988; 17). This, dystopianists argue, acts to camouflage the real processes taking place in ways that directly support the already dominant position of the multinational corporations who benefit most from the restructuring of cities in the information age. Sussman and Lent, for example, argue that:

In the zealous imagination of 'information age' ideologues, once political impediments are removed from communication technology's revolutionary thrust, its ability to bypass traditional time and space boundaries promises a cornucopia of comforts and conveniences. This would-be 'depolitiza-tion' of technology and economics is, rather transparently, itself a politically charged construction of reality that attempts to disassociate communication/information institutions and industries from the human actors and beneficiaries involved in their design and development. In the technocratic model, it is not the *who* that matters, it is the *how to*, the technical processes downplaying the explicit social applications.

(Sussman and Lent, 1991; x. Original emphasis)

Futurist and determinist visions are therefore lambasted for their 'simplistic, technicist and often technocratic' (Robins and Hepworth, 1988; 155) approach to analysing telecommunications in cities. They are attacked for the way that they 'often obscure the vested interests involved in IT, . . . deflect attention from some embarrassing contradictions, while at the same time giving the coming of the information society the appearance of an entirely natural and logical social progression' (Lyon, 1988; 147).

The extract from Kevin Robins and Mark Hepworth (see Box 3.3) provides an excellent example of this critical stance to analysing telematics in cities. In it they attack the 'infantile utopia' of the home-centred vision developed by the futurists. They then go on to link telematics to the wider political, economic, spatial and social changes underway in the contemporary city – an interlinked set of processes that derive from the global political–economic reorientation of capitalism. They stress how telematics are being shaped and applied to help overcome the crisis tendencies of capitalist accumulation, so supporting new time–space configurations, new control capabilities and new relations of production and consumption, as capitalism restructures from the Fordist arrangements of the post-war era to a new period of 'Neo-Fordism'.

BOX 3.3 ELECTRONIC SPACES: NEW TECHNOLOGIES AND THE FUTURE OF CITIES

INFANTILE UTOPIA

The tendency within futuristic writings on the 'information society' is to see the new information and communications technologies as abolishing problems of space and distance. It is suggested that geographical and locational limits can be overcome through the instantaneous channels of the information and communications networks. All social activities can be electronically transacted from any place, city or country, the next room or the other side of the Earth. The spatial dynamics of the whole world collapse to those of a pinhead. One potent symbol for this meta- or post-spatial future is the idea of the 'electronic cottage'. When the household is plugged into an information and communication grid, then the coordinates of domestic space explode to coincide with the electronic hyperspace of the global village. Or, in reverse, it is as if the outside world had become imploded and condensed within the walls of the electronic household. For futurists, space has become an entirely relative, and empty, category....

HOME AUTOMATION

The computer home scenarios have a narrow and instrumental fixation on technique – the 'evolution' of the household is seen as an expression of some autonomous technological 'progress'. The dream is a domestic machine-utopia cocooned from the outside world in which human agents are passive and infantilized. In such technocratic scripts the household is severed from its surrounding (economic, social and political) contexts. This restrictive and exclusive focus on the (electronic) household tends to foreground and privilege its role in the so-called information revolution. The household symbolizes the human and benevolent promise of the revolution. What we can look to is 'the home's re-emergence as a central unit in the society of tomorrow – a unit with enhanced rather than diminished economical, medical, educational, and social functions' (Toffler, 1981; 364). 'The burden of change', according to another commentator, 'must be towards home-centring our lives' (Aldrich, 1982; 100).

THE INFORMATION CITY

The information household can only be understood within the framework of political economy. In this context it becomes possible to displace technological determinism in favour of the analysis of political and social forces; to shift from putative futures to the here and now; to put the question of agency (political and policy intervention) back on the agenda. We also argue that, in the context of the neo-Fordist information revolution, the spatial aspects of this political economy are particularly important. Furthermore, we argue that a particular level of analysis is appropriate here, the transformation of household space is particularly significant in the context of changing patterns of urban space. While the national and international political economy is, of course, the crucial foundation on which we are building, we believe that important aspects of the restructuring process associated with the new information and communications technologies can be illuminated by focusing on the city and the urbanization process.

This perspective allows us to consider the emerging information grid as a total system to explore the relation between (electronic) households and, more importantly, the relation between households and other points on the network (consumer outlets, workplaces, offices, government institutions and agencies). It also allows us to explore economic shifts as they shape and transform the infrastructure, fabric and ecology of the physical environment, and as they weave the textures and patterns of everyday life. By focusing on the information city we have a street-level appreciation of what the information revolution amounts to; we can explore not just the economic and technological issues, but also the recomposition of everyday experience and culture, the way of life (Robins and Webster, 1986).

(Robins and Hepworth, 1988; 155-176)

THE SOCIAL CONSTRUCTION OF TECHNOLOGY (SCOT) APPROACH

The final approach to city–telecommunications relations – the Social Construction of Technology (SCOT) approach – has also arisen from a rejection of technological determinism. As with the political economy approach, SCOT researchers reject the notion that telecommunications have some autonomous 'logic' which 'impacts' on cities as an external force. Equally, they reject the use of bold metaphors to describe the social 'impacts' of telecommunications as simplistic, unidimensional and overly deterministic (Gökalp, 1988). Instead of seeing technology as somehow autonomous from society, social constructivists argue that 'the compelling nature of much technological change is best explained by seeing technology not as outside of society, as technological determinism would have it, but as inextricably part of society' (Mackenzie and Wajcman, 1985; 14). However, SCOT also tends to reject many of the arguments of political economy, with its stress on the central importance of the structures of capitalism and the overwhelming power of political–economic forces in shaping how telecommunications develop in cities. To John Law and Wiebe Bijker (1992; 290), for example, 'both social determinism and its mirror image, technological determinism, are flawed'.

HUMAN AGENCY AND THE 'SHAPING' OF TELECOMMUNICATIONS

SCOT researchers, rather, stress the degree to which space exists through which human agency and social and political processes can shape the ways in which telecommunications are developed and applied within cities and within wider society (see Figure 3.1). Rather than there being a single technological or political–economic 'logic', or a single 'best way' to develop and adopt a technology, this approach argues that at all levels choices exist in the ways to design, develop or adopt technologies. Thus: 'technology does not spring, *ab initio* from some disinterested fount of innovation. Rather, it is born of the social, the economic, and technical relations that are already in place' (Bijker and Law, 1992; 11). As with the political economy approach, then, the objective of SCOT research is to demonstrate how society influences technology.

Micro-level social processes of human agency are the focus of SCOT research. Individuals, social groups and institutions are seen to have some degree of choice in shaping the design, development and application of technologies in specific cases. Different assumptions are built into the shaping of technology in different places and the development of telecommunications in cities would therefore be expected to be extremely diverse and contingent on the micro-level processes of technological shaping in each case. Kendall Guthrie argues that:

Technology development should be viewed as a series of choices, which begins with whether or not to explore a certain technological area, continues through the design process, and concludes with a decision on whether or not to adopt. Potential design options at various stages can maximise different sets of goals and values. . . . The strategy which designers choose, and therefore whose goals get optimized, is usually more dictated by the distribution of economic and political power than by technological necessities or opportunities. Debates over design choice are, therefore political debates.

(Guthrie, 1991; 46)

The SCOT approach aims to identify, analyse and explain *causal* relationships between social, institutional and political factors and the development and applications of technologies – including telecommunications in cities. The purpose of research in the SCOT tradition is, therefore, to understand how technology and its uses are socially and politically 'constructed' through complex processes of institutional and personal interaction, whereby many different actors and agencies interplay over periods of time (Mackenzie and Wajcman, 1985; Bijker *et al.*, 1987). Kendall Guthrie argues that 'complex technological systems do not develop in a value-neutral, social vacuum. The various decisions about how they are designed are made by many different people, embedded in organizations, which are in turn embedded in the larger societal system' (Guthrie, 1991; 45).

ACTORS, ENTREPRENEURS AND THE SOCIAL CONSTRUCTION OF TECHNOLOGIES

To Callon *et al.*, the initiation of technological projects can be seen as a process through which 'each actor builds a universe around him which is a complex and changing network of varied elements that he tries to link together and make dependent on him' (Callon *et al.*, 1983; 193 [*sic*]). Key technological entrepreneurs work to 'enrol' other people and technological artifacts into the network through processes of 'heterogeneous engineering'. This is done by getting them to accept one's point of view and situating them within the wider framework. Key individuals therefore work to fix the 'issues, the problems, and the line of approach in a manner favourable to [their] own agenda' (Westrum, 1991; 72).

Technological development is thus a profoundly social and political process, not a predefined one. The urban effects of telecommunications are therefore indeterminate – the aggregate results of countless individual examples of the social construction of technology. This means also that it is impossible to define single, all-encompassing 'impacts' of telecommunications on cities in some deterministic fashion. It also means that suggesting that there is some single 'trajectory' of telecommunications-based urban development which is driven by the political economy of capitalism is equally false. The SCOT approach implies that the ways in which telecommunications relate to urban change is likely to vary in time and space in complex ways.

THE BLURRING OF THE 'SOCIETY–TECHNOLOGY' BOUNDARY

Analysing the process whereby telecommunications are socially constructed, however, is difficult to achieve. This is because identifying simple chains of cause and effects is made inappropriate by the complexity of these society–technology relationships. To David Edge 'it is a basic assumption of this approach that the relationship between technology and society is genuinely an *interaction*, a *recursive* process; "causes" and "effects" stand in a complex relationship' (Edge, 1988; 1). Thus the distinction between, technology and society, telecommunications and cities, should be blurred and their interactions considered in an integrated way.

Nigel Thrift (1993) argues that machines and humans are so woven together into broader systems that 'no longer is it possible to see the human subject and machine as aligned but separate entities, each with their own specific functions. No longer is it possible to conceive of the human subject as simply "alive" and the machine just as "dead labour"'. Often, telematics applications are now so closely interwoven with humans and social life that many talk of a new 'cyborg' culture where human subjects are intimately linked into webs of digital technology at all levels (see Bender and Druckrey, 1994; Gregory, 1994, 162–164; Mitchell, 1995). Within this people interact continuously with simulated environments based on artificial intelligence, massive arrays of micro-computerised power, and virtual reality technologies which literally immerse people in multi-sensory worlds constructed by software and hardware (Virilio, 1988; Levidow and Robins, 1989). Humans and machines become fused in ways that make the old separations between technology and society, the real and the simulated, meaningless. Rather, the

human subject is being modernised so that it is able to cope with this new order of reality in which machines are as active as human subjects, in which there is a very high degree of interactivity between machines and humans, and even in which, in certain cases, machines become parts of the human subject's body.

(Thrift, 1993. See Haraway, 1991)

The great difficulties involved in studying these shifts and exploring their implications for cities is one reason why SCOT studies of telecommunications–city relationships remain extremely rare. The SCOT approach leads inevitably to an emphasis on detailed, descriptive case studies of the ways in which specific technologies have been 'constructed' in specific social contexts (there being no real methodological alternative to such approaches). The extract by Kendall Guthrie and William Dutton (see Box 3.4) provides a rare example of a detailed case study of the development of telecommunications and telematics within western cities. Guthrie and Kendall focus on the development on a range of Public Information Utilities (PIUs) – local telematics systems set up for local government and citizen use within cities – within three Californian cities: the Public Electronic Network in Santa Monica, the PALS/PARIS system in Pasadena, the Infonet system in Glendale. In addition, they study a similar city – Irvine – which has decided not to develop a PIU. They demonstrate how very similar technologies have actually been used in very different ways in the different cases, rather than being shaped by some predefined technological or economic logic, the varying capacities, configuration, information content and orientation of the three systems were shaped by processes of decision-making that reflected the political culture and ideas of key individuals. Of central importance were the varying 'technological paradigms' of the officials developing each system – that is, the accepted notions of how the technologies should be applied and shaped in each place.

BOX 3.4 THE POLITICS OF CITIZEN ACCESS TECHNOLOGY:
THE DEVELOPMENT OF PUBLIC INFORMATION UTILITIES
IN FOUR CITIES

During the late 1980s, American local governments began to apply information technology not only to speed up data processing but also to improve delivery of political participation.... This technology allows citizens to access a common computer database and often electronic mail facilities via terminals in public places as well as with private personal computers. The images of their developers closely approached visions of the 'public information utility', first promoted in the 1960s (Sackman and Boehm, 1972; Sackman and Nie, 1973).

.... One can view the adoption and design of a community information system as a process comparable to legislating public policy on citizen participation. In this case policy is embedded in the technology – the arrangement of people, equipment, and technique – rather than in law or regulation (Winner, 1986, p. 29). Like policy, technology is a social construction – the outcome of social and political choice (Danziger, Dutton, Kling and Kraemer, 1982; MacKenzie and Wajcman, 1985; Bijker, Hughes and Pinch, 1987). However, in the case of technology, these policy choices too often are obscured or overlooked because people focus only on decisions about the adoption or nonadoption of a technology rather than also attending to decisions about design and implementation of the technology that influence its use and impact.... Therefore, it is important to study the political shaping of these systems by early adopters....

DESIGN AND POLICY CHOICES

Seldom is there a single best way to do something. There are numerous points of flexibility in the design of technologies, including public information utilities. Some design alternatives can be key policy choices in that they may bias a technology towards particular social uses and outcomes. As noted above, for example, a public information utility can be designed to block or facilitate interactive versus one-way communication. The policy implications of such alternatives are constant, even if policy makers fail to recognize them.

The following are key policy choices about the design of the public information utility which emerged from the case studies: (a) system capacity (both in terms of computer memory and the number of simultaneous users it can support); (b) architecture of the communications channels, which can include broadcasting from one to many, also vertical channels between officials and citizens, horizontal channels between citizens, and many-to-many channels whereby any one user can broadcast to all others; (c) accessibility, based on the dispersion of terminals and the monetary cost of using them; (d) information content, which can be more or less oriented to public affairs depending on the balance of commercial versus non-commercial messages; (e) model of editorial control, which includes viewing the public utility operator as a publisher with complete control over content versus a common carrier system which carries many voices; (f) ownership, which could be public, private, nonprofit ownership or a combination and; (g) financing, which could be public, commercial, and/or subscription based.

FACTORS SHAPING TECHNOLOGY POLICY DECISIONS

The [four] cases [of Public Information Utilities (PIUs) in Santa Monica, Irvine, Glendale and Pasadena] demonstrated a range of technically feasible options. The designers based their decisions not simply on what was technically feasible but also on social and political grounds not determined by the technology....

TECHNOLOGY SHAPING TECHNOLOGY:
THE UNDERLYING PARADIGM

Existing technology presents a variety of opportunities, problems and constraints that shape the future of technology (MacKenzie and Wajc-man, 1985). One way in which existing technology shapes new technol-ogy is by providing mental models of solutions. The most general pattern emerging from the comparative case study was the central role played by what might be called the 'technological paradigm' of the designer (Guthrie, 1991). Technological paradigms are the widely accepted exem-plar puzzle solutions, employed as models or examples, which replace explicit rules as a basis for the solution of the remaining puzzles of technology development....

Across the case study cities, developers employed four distinct paradigms: an electronic mail [Santa Monica], a broadcasting [Pasa-dena], a database [Glendale], and a management information system (MIS) paradigm [Irvine]. That is, technology shaped technology by providing alternative paradigms from which designers could choose to model their public information utilities. Moreover, once designers began working within one paradigm, they overlooked other potential design strategies which did not fit into that model....

In our case cities, the choice of paradigm appeared to be based primarily on the background and professional training of the developers. The technology designers in this study came from different professional communities, which... each had their own training program and set of widely accepted exemplar problem-solutions. The information utility designers' professional training taught them to see a new technological problem in terms of the exemplars from within their own field....

The picture of technology development which emerged from this study of four cities was not of some technologically driven, value-neutral chain of events. It was a series of choices where humans decided at various critical junctures what structure the public information system would take. From this vantage point, technology design is clearly similar to a public policy process, where the outcome is determined more by the authority, influence, and goals of the actors involved and the environments in which they work than by technological necessities.

(Guthrie and Dutton, 1992; 574-597)

A CRITICAL EVALUATION

In this chapter we have critically reviewed the four most influential and common approaches to analysing the relations between cities and telecommunications, between urban places and electronic spaces. We have seen how technological determinism, futurism and utopianism, dystopianism and political economy, and the social construction of technology approach all view city–telecommunications relations in very different ways. We have also begun to look at the implications that these views have for wider questions about the ways in which social structures and telecommunications networks come together with human agency in an urban context. The remaining two tasks of this chapter are to reflect upon the merits and differing qualities of these approaches and to develop a new perspective which builds on this and integrates treatment of cities and telecommunications.

We reject the first two approaches, technological determinism and utopianism/futurism. This is for four reasons. First, while such approaches can be useful in describing the broad, macro-scale historical shifts in the technological orientation of urban society (see Hall and Preston, 1988), they tend to reduce extremely complex interactions between cities and telecommunications to crude and homogeneous models of technologies and their urban impacts. For the sake of the simplicity of argument, telecommunications are portrayed as an active and autonomous agent transforming urban society. This, as we saw in the last chapter, serves to confuse and oversimplify debates about cities and telecommunications rather than to enlighten and reveal.

Second, and related to this, both tend to ignore the crucial social and political processes through which technologies are actually developed and applied within cities and society. Human agency and broader social, political and cultural structures are rarely part of the equation. As David Edge argues, the key point about technological determinism is that 'technological change is [seen as] a prime *cause* of social change' (Edge, 1988; 1). The problem here is that 'technical innovations are themselves "uncaused" – in the sense that they arise out of an

intrinsic, disembodied impersonal "logic" and not from any "*social*" influence'
(Edge, 1988; 1). Technological determinism, he continues, relies on a 'simplified
linear model of the innovation process [which] tends to treat the technology as a
"black box", and is preoccupied with the "social impacts" of a largely pre-
determined technical "*trajectory*"' (Edge, 1988; 1). These comments could be
applied equally to futurism and utopianism. The complex interactions between
telecommunications and cities that we began to analyse in the last chapter cannot
be understood in this way. In reality, as David Lyon argues, 'technological
development does not have pre-set social effects which are predictable, universal,
or, for that matter, just or beneficial' (Lyon, 1988; 157).

Third, it is clear that much futuristic and utopian material on cities and
telecommunications must be discounted simply because it is fuelled by the vested
interests of technology and service companies or the policy-makers who stand to
gain most from telecommunications development in cities. Much of the optimism
which pervades this approach can simply be understood as elaborate marketing
ploys aimed at deflecting criticism or encouraging new markets and new public
subsidies. By developing optimistic and lustrous promises of social transformation
based on telecommunications, and by imparting these to decision-makers and
consumers, very powerful industrial and political groups are obviously set to
benefit. In other words, as Jennifer Slack argues, to a large extent 'descriptions
of the information age, are, in fact, *constitutive aspects* of the information age'
(Slack, 1987; 2. Emphasis added).

Such faith that telematics will automatically have such benign and positive
effects has attracted a good deal of criticism. Many commentators argue that the
conceptions of city–telecommunications relationships employed by both technological determinists and futurists are a dangerous oversimplification of reality.
John Gold, for example, urges that:

A more sophisticated view is necessary of the relationship between technology and urban society . . .
a large proportion of future city schemes have been essentially exercises in technological forecasting.
Their authors are deeply impressed with the potential of technology and adopt a facile, implicitly
deterministic view of its powers to transform . . . without a more balanced view of the complexities
involved, forecasting becomes superficial and flatulent.

(Gold, 1985)

Finally, there is an implicit assumption in determinist and futuristic work on
cities and telecommunications that local social and political actors in contemporary cities have little or no scope to shape telecommunications developments
within cities. Little space is left within these approaches for forces of human

agency or urban and telecommunications policy making at the local, national or supranational level with which to alter the apparent destiny embodied in the telecommunications-based development of a city. The emphasis in this literature tends to stunt the development of critical policy debates on telematics issues in contemporary cities. Views that see telecommunications impacts on cities as inevitable, or herald these new technologies as automatically positive, suggest that current policy-makers need not worry unduly about telecommunications. As Robert Warren argues, 'benign projections give little indication that there are significant policy issues which should be on the public agenda' (Warren, 1989; 345). Thus, crucially from the point of view of this book, technological determinism and futurism work to undermine the very concept that local telecommunications initiatives will develop, or have any impact at all on the ways in which telecommunications develop in cities. We argued in the last chapter that this is one reason why the current policy debates about cities and tele-communications remain so poorly developed. Kevin Robins and Mark Hepworth elaborate on this point, by arguing that:

It is this question of agency that is fundamental. Within this futuristic scenario, technology appears to have its own autonomous and inevitable force. . . . It is a force, moreover, that becomes associated with a higher state of human evolution. . . . Insofar as technological development seems inflexible and unquestionable, and the course of progress to be part of quasi-evolutionary destiny, then perhaps the only appropriate response is that of acquiescence and compliance.

(Robins and Hepworth, 1988; 157)

CONCLUSIONS: TOWARDS AN INTEGRATED APPROACH TO CITIES AND TELECOMMUNICATIONS

We argue that our search for a new, more sophisticated and more integrated approach to understanding city–telecommunications relations should start by blending insights from the latter two approaches described in this chapter: urban political economy and social constructivism. Neither is perfect and both can be criticised on a variety of fronts. Political economy often overplays the conservative effects of the structures of capitalism in shaping technology and neglects the degree to which social processes can change telecommunications development. It often ascribes simple and all-encompassing powers to abstract and macro-level capitalist structures while neglecting the ways in which structures are themselves created by innumerable individual and institutional actions over time. SCOT, meanwhile, can be so attentive to the ways in which social élites shape technology at the micro-level, that it can neglect the wider power imbalances in society and ignore those who are excluded from shaping technology because of poverty, unemployment or marginalisation.

But, crucially, both approaches stress that telecommunications arise and are applied within rather than from outside society: telecommunications-based innovation in cities is therefore socially, politically and culturally shaped rather than being purely technical. This helps greatly in our efforts in the themed chapters in the rest of this book to begin integrating our treatment of cities and telecommunications. Before we go on to do this, however, it is necessary here to start building this new synthesis. How can we best use these two approaches as the starting-points for integrating our approach to cities and telecommunications? How can we develop a framework of causal links between urban places and electronic spaces that complements the new conceptualisations of cities, space and time built up in the last chapter? In short, how can we begin to understand conceptually the ways in which both urban places and electronic spaces are together constructed socially within the broader framework of capitalist political economy?

Although the great complexity of the interactions between cities and telecommunications defies easy analysis, we would argue that there are three key levels of analysis confronting any attempt to build up such a genuinely integrated view of cities and telecommunications. These centre on:

- the functional and material tensions between the fixity of urban places and the mobility supported by telecommunications and electronic spaces;
- the social struggles which develop over the shaping of urban places and electronic spaces;
- the issues surrounding social representation, identity and perception in cities and telecommunications.

URBAN PLACES AND ELECTRONIC SPACES:
MATERIAL TENSIONS BETWEEN 'FIXITY' AND
'MOBILITY' IN SPACE–TIME

First, we can draw on the political economy perspective to view the material functions of cities and telecommunications within capitalism in parallel. When this is done within the context of our discussions in the last chapter about technology and 'space–time', it is possible to approach an integrated perspective of the relations between cities and telecommunications.

To start with, we can see urban places to be founded on spatially fixed and embedded accumulations of physical constructions and networks. Cities are supported by a massive physical built fabric – the land parcels, buildings, streets, neighbourhoods and the material transport and infrastructure networks that support physical flows of goods, people and resources. Urbanisation and the construction of urban places has historically been encouraged by the need to overcome time with space. To a large extent, cities have developed to make communication easier through concentrated physical development. As Tarr *et al.* (1987; 38) argue, 'urban scholars generally agree that cities evolved in order to facilitate human communications'. Melvin Webber comments that 'the history of city growth, in essence, is the story of man's eager search for ease of human interaction' (Webber, 1964; 86. *sic*).

Before electronic telecommunications, when all communication necessitated physical movement, physical concentration in cities meant that this became practical. Physical transport suffered the frictional effects of distance and physical space which, as we have seen, so dominated consideration of cities in the post-war period. Action over distance was only possible through physical movement. Concentration and the small distances involved meant that the time taken to walk or take other transport within the city was not onerous. Thus, cities grew cumulatively by providing all the external agglomerations that physical concentration of employment, services, administration and culture enabled (Figure 3.2). By concentrating all activities – homes, marketplaces, government offices, services, offices and factories – within one relatively small area, cities thus allowed *time constraints to be overcome by minimising distance constraints.*

As we saw in the last chapter, cities can no longer be viewed in this bounded way; they have been redefined by globalisation and political economic change. Now cities must be seen as the fixed places where many relational webs between firms, institutions, social groups and individuals become superimposed and

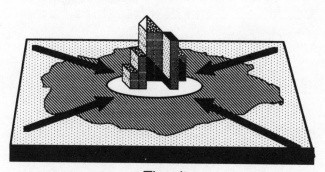

The city
Function: to overcome *time* with space.
Developed to make communications
easier by minimising *space* constraints to
overcome *time* constraints.

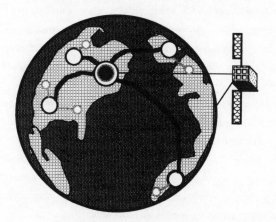

Telecommunications
Function: to overcome *space* with *time*.
Developed to make communications
easier by minimising *time* constraints to
overcome *space* constraints.

Figure 3.2 The relations between cities and telecommunications and time and space
constraints

Urban places (based on buildings, streets, roads, and the physical spaces of cities)	Urban electronic spaces (constructed 'inside' telematics networks using computer software)
Territory	Network
Fixity	Motion/flux
Embedded	Disembedded
Material	Immaterial
Visible	Invisible
Tangible	Intangible
Actual	Virtual/abstract
Euclidean/social space	Logical space

Figure 3.3 Characterising urban places and electronic spaces

'cohabit' within the context of many scales and types of flow and flux – material, people, information, capital, services and media (Healey *et al.*, 1995). But urban places remain highly constrained by their relative fixity and embeddedness in space (Figure 3.3). As Harvey (1985) has shown, the constant expansive drive of capitalism requires new urban places to be created. As new production methods are adopted, new geographies of development emerge and attempts are made to address profit crises. This leads to the continuous remaking of urban places through what Harvey calls the 'restless formation and reformation of geographical landscapes' (1985). But the increasingly dispersed factories, warehouses, offices and shopping malls need to be 'embedded' in space as well as integrated together into coherent economic systems.

Of course, while they are crucial in supporting the mobility and flux, telecommunications are also fixed networks that must be 'embedded' in space. Telecommunications are physical systems made up of links and nodes that are constructed fundamentally as spatial systems linking together only certain places highly unevenly (Hepworth, 1987). The construction and roll out of telecommunications networks is still an expensive and disruptive physical process which has its own complex geography. Network infrastructures need to be rolled out through the congested physical spaces above, within and below cities and in the intervening corridors between them. This shapes the economics of network development within and between cities. Fuelled by more market-based telecommunications regimes, new networks increasingly now tend to concentrate where the best profits can be made – that is, in the 'hot spots' of the largest cities.

Once built, telecommunications support the instantaneous mobility and circulation that, as we have seen, increasingly underpin the integration of all aspects of urban life and urban development across space and time barriers. Harvey, for example, notes that 'the ability of both capital and labour power to move at short order and low cost from place to place depends upon the creation of fixed, secure and largely immobile social and physical infrastructures. The ability to overcome space is predicated on the production of space' (Harvey, 1985; 149, quoted in Goddard, 1994; 279). So the material aspects of both urban places and new telecommunications networks are actually constructed together within the broader processes of capitalist change and restructuring as attempts are made by states and firms to attain new trade offs between fixity in urban place and mobility in electronic space. Of course, this is not some neat and functionalist design process; it is extremely complex, and involves a myriad of individual decisions through which the development of cities and telecommunications are socially shaped.

In terms of their effects on space–time, telecommunications networks *effectively have the opposite effect to concentration in cities in that they help to overcome distance constraints by minimising time constraints* rather than the other way round (see Figure 3.2). To use Marx's maxim, they 'annihilate space with time' – they are 'essentially based on an *indifference to the notion of boundaries*' (Negrier, 1990; 13. Emphasis added).

Where quality telecommunications links exist, distance constraints can effectively collapse altogether (subject, of course, to the cost of using them and the technical and organisational barriers involved). Because they operate at or near to the speed of light, they overcome spatial barriers by minimising – or even eliminating – temporal barriers (Figure 3.2). They help overcome space and time barriers and support the instantaneous or rapid mobility of information, messages, services, capital, images and labour power that are necessary to link widely dispersed sites into fast-moving and integrated economic and social systems.

So the physical telecommunications networks of wires, cables, dishes and antennae support vast and complex electronic spaces within which (near) instantaneous information flow, communication, transaction, storage, processing and interaction becomes possible. These operate within abstract and virtual 'spaces' that are actually 'constructed' by the computer hardware and software that is attached to telecommunications networks (see Figure 3.3). The key difference between urban places and electronic spaces, of course, is that the electronic 'sites' and 'spaces' that these networks support are actually intangible, abstract and virtual (see Figure 3.3, p. 116). Rather than being confined within (relatively) tight, physical boundaries, they are constructed electronically in 'abstract' and 'virtual' space by computers and software linked over space by instantaneous flows of information (Robins and Hepworth, 1988; 161). As Noam argues, in electronic spaces 'territoriality becomes secondary' (Noam, 1992; 413). Here 'constructed space now occurs within an electronic topology' of telecommunications and telematics networks (Virilio, 1987; 18). Such electronic spaces can be developed as systems at all spatial scales, from small systems inside 'intelligent buildings' and 'smart homes', through neighbourhood cable networks, city-wide traffic monitoring networks to national and global telecommunications systems like the Internet.

Here we approach an understanding of the complex material interactions between urban places and electronic spaces. These capabilities mean that telematics or computer networks allow the organisational logic of constructed electronic networks to triumph over the geographical logic of their actual locational arrangements within and between urban places (Mulgan, 1991). This allows 'virtual' organisations to be developed with more flexible relations to the 'outside' world of geography, space, time and place (Gillespie and Williams, 1988). They remark that 'one cannot . . . usefully conceptualise the effect of telecommunications on geographical relationships other than through the intra-organisational and inter-organisational computer networks which bind particular locations together' (Gillespie and Williams, 1988; 1317). 'Relative location' within the abstract topologies of electronic networks and spaces can become more important than 'absolute location' within the physical geography of cities and urban systems (Brunn and Leinbach, 1991; vii). This is helping many organisations to address crises through spatial restructuring by 'doing things in one place that dominate another place' (Goddard, 1994; 282). Through such networks 'telepresence' in electronic spaces can be used to replace or complement presence in physical spaces (CEC, 1992). This supports new types of remote control, coordination, communication, information flows and transactions (Mulgan, 1991; 1). Finally, new forms of social and community life can flourish based on electronic fora and networks which develop as 'virtual communities' or 'telecommunities' operating independently of spatial constraints.

These processes allow new patterns of control and domination to emerge within and between places within which cities still have a key role (both as dominant and subordinate centres, the controllers and the controlled). But it is clear that these merely change rather than destroy the traditional ways in which cities or centres within cities develop as centres of control and influence over other places. New patterns of urban development in fixed urban places become possible through the development of electronic spaces; these urban places, as the main demand centres for communications, in turn go on to influence the further rounds of telecommunications development. In short, this recursive relationship between technology and urban society leads to a new type of urban world rather than a post-urban world.

SOCIAL AND INSTITUTIONAL STRUGGLES OVER URBAN PLACES AND ELECTRONIC SPACES

There are many parallels between the social construction of urban places and electronic spaces; similar social conflicts and struggles shape the production of both. The most powerful social, economic and political groups tend to exercise and maintain their power through the production of interlinked systems of related electronic spaces and urban places (see Harvey, 1993). Power stems from access to, and control over, both the material spaces of cities and the electronic spaces on telecommunications networks. For example, the most powerful institutional groups in the current era of development – transnational corporations (TNCs) – are increasingly dominant in shaping and controlling both the built environments of cities as well as the telematics networks that cross-cut them. Corporate electronic spaces are constructed which allow their many physical sites to be linked together in 'real time' and with great security (Castells, 1989). This gives them great flexibility and power to manage their divisions of labour and relations with the localities through which their networks pass through. Urban policy-makers increasingly fight to attract TNCs; telematics marketplaces are increasingly geared towards meeting their every need. Their dominant power stems from their control over private property (such as land, buildings, and information), telecommunications markets and providers, and all the instruments they have for excluding others from physical and electronic spaces.

But such power does not go uncontested. Each city has a unique terrain within which diverse social, cultural, ethnic and institutional agents struggle to influence the shaping of the city's built environment according to their own material, social and cultural preferences (Zukin, 1991; Ambrose, 1994). Such struggles have recently reached peaks around the massive physical redevelopments in places such as London Docklands. Here, attempts to develop a corporate and exclusionary urban landscape constantly had to deal with the mobilisation of community resistance and the putting forward of a range of alternative community-based plans and visions (Keith and Pile, 1993; 11–13). Thus, politics becomes territorialised around the struggles over the material and social shaping of urban places.

As a result of such social struggles over the built environments of cities, it is now well established that urban places – the gleaming office towers, the shopping malls, transport networks and waterfront developments, the housing areas, suburbs and ghettos – can be read as complex and changing spatial reflections of the dynamics of power in capitalist societies. Thus, social spaces and the 'landscape of cities' are socially constructed through struggles between the many interest groups involved. As Keith and Pile (1993; 8) put it, there are a 'host of competing spatialities' struggling to be imprinted on to the landscapes of cities. The space underpinning urban places is therefore not some objective 'Euclidean' plain within which urban life is enacted. It is 'more like a "melting pot" in which gender, race and class have to be recognized as the basis upon which space is constructed' (Sui, 1994).

Increasingly, strikingly similar struggles go on over the shaping of electronic spaces. Telecommunications networks, too, can be seen as supports for electronic landscapes of power, the results of social struggles over the constructions of electronic spaces accessible through technological networks. Here, too, many unevenly equipped social groups are attempting to shape networks like the Internet and the much-heralded 'Information Superhighway' in very different ways. As in urban places, there are divisions between public and private space and sites which support a massive range of information exchanges, communications and transactions – both social and economic.

The current struggle over the future of the Internet is a good example. On the one hand, a disparate range of social, community and professional groups are advocating the continuation of the 'utter, ungovernable anarchy' of the Internet. They usually want to maintain open access to largely free information along with universal service and publicly funded local community applications for disadvantaged groups. On the other, the giant global TNCs are struggling to turn the Internet or the superhighway into a private commodified 'club', an 'electronic mall' through which global media firms can offer value-added services on a pay-per-use basis using electronic cash systems. The battle is on over the future complexion of electronic spaces. Many of the new telematics networks on the Internet for special interest groups are 'places' that they defend from incursions in similar ways as physical neighbourhoods of cities.

SOCIAL REPRESENTATION, IDENTITY AND PERCEPTION

But social struggles to shape urban places and electronic spaces are not simply about the contest over material benefits within a highly unequal capitalist society. They are also about the meanings that are constructed for places and the ways in which urban place and electronic space become embroiled in the construction of diverse social and cultural identities. Again, this is now well established in terms of urban places (Keith and Pile, 1993). Sharon Zukin, for example, admits that while cities are the 'localization of global economic and social forces . . . space also structures people's perceptions, interactions and sense of well-being or despair, belonging or alienation. This structuring quality is most clearly felt (and most visible) in the built environment, where people can erect homes, react to architectural forms, and create – or destroy – landmarks of individual or collective meaning' (Zukin, 1991; 269).

As daily life becomes more mediated through electronic spaces, and as they take on the characteristics of multi-sensory 'places', they, too, are coming to 'structure people's perceptions, interactions, and sense of well-being or despair and alienation'. While they may only involve small and largely élite social groups, the current struggles over the many electronic spaces accessible over the Internet are struggles over 'landmarks of collective meaning' as much as any neighbour-hood planning conflicts. Many social groups are trying to shape these spaces to represent, reflect and maintain their individual, group, ethnic and/or gender identities. In the United States, disadvantaged groups like the elderly, women, the poor, ethnic minorities and disabled people campaign in parallel for more civilised and equitable cities as well as developing computer networks in 'cyberspace' geared to their own particular needs.

But the ways in which electronic spaces become socially structured is much less legible than urban places. Things are more hidden and fluid, as geography becomes decoupled from social, ethnic and economic status in ways that rarely happen in cities. According to William Mitchell (1995), electronic spaces like the Internet:

eliminate a traditional dimension of civic legibility. In the standard sort of spatial city, where you are frequently told who you are. (And who you are will often determine where you are allowed to be.) Geography is destiny; it constructs representations of crisp and often brutal clarity. You may find yourself situated in gendered space or ungendered, domains of the powerful or margins of the powerless; there are financial districts for the pinstripe set, pretentious yuppie watering holes, places where you need a jacket and tie, golf clubs where you won't see any Jews or Blacks, shopping malls, combat zones, student dives, teenage hangouts, gay bars, redneck bars, biker bars, skid rows, and death rows. But the network's despatialization of interaction destroys the geocode's key. There is no such thing as a better address, and you cannot attempt to define yourself by being seen in the right places in the right company.

(Mitchell, 1995: 10)

Similarities also exist between the perceptual representation of the built environments of cities and 'mental maps' of the abstract topologies of telecommunications networks and electronic spaces. To 'navigate' the Internet, users need to build up perceptual representations not just of the thousands of host computers available but of what activities are going on where and what they are like. As with mental maps of cities, these maps need to have emotional and subjective information about the qualities of the electronic 'places' found as well as basic information designating what is where within the complex and interlocked web of logical spaces. Increasingly, then (some) people have 'mental maps' of both urban place and electronic space. As boundaries blur between the 'real city' and the simulated 'electronic cities' constructed within virtual reality systems, it seems likely that such mental maps will become tightly interlinked.

Considering these three key issues as parallel in this way is the final step in building up the conceptual framework which provides the basis for the more specific chapters in the rest of the book. These begin in the next chapter, when our attention turns to urban economies.

4

URBAN ECONOMIES

INTRODUCTION

We noted in the first two chapters how western urban economies are being radically altered by current processes of economic and spatial restructuring. No longer can we understand cities primarily as centres for the manufacturing and the exchange and production of physical goods and commodities – as most were during the last hundred years. Similarly, the notion that cities are shaped by their positions within single, functional urban hierarchies oriented towards nation states ceases now to have much credence. Unlike the classic Fordist city of the post-war boom, the modern city can no longer be considered as an internally integrated 'locus of mass production, mass consumption, social interaction and institutional representation', as Amin and Thrift (1995; 91) describe. Instead, structural economic change, so-called deindustrialisation and the globalisation of capitalism is transforming the ways in which urban economies function.

Urban economies now tend to be based primarily on consumer services like leisure and retailing and the production, distribution and processing of information and 'symbolic goods' like information services, finance, media, education and advertising. Instead of being integrated on the inside and tied closely into national hierarchies they are now shaped by their linkage into diverse ranges of economic, physical and institutional networks which often operate at international scales. As we saw in Chapter 2, these tie cities and their hinterlands into systems of hubs, spokes and tunnel effects which challenge conventional ways of understanding city economies (see Figure 2.2). These networks mean that patterns of work and economic flow are now much more fluid both spatially and temporally than was the case in the recent past when urban economies were characterised by very regular working hours and geographical routines. Advances in telecommunications are crucial in underpinning these broad shifts. They are basic integrating infrastructures underpinning the shift towards planetary urban networks; inter-urban telecommunications networks themselves comprise a vital set of hubs, spokes and tunnel effects linking urban economies together into real time systems of interaction across space and time barriers.

In this chapter we will explore the radical transformations which are currently underway in the economic makeup and functioning of modern cities. Within these shifts, we will attempt to identify the importance of telecommunications and telematics. Current changes in city economies are complex and interrelated; explanations for them differ considerably between the main theoretical per-spectives available (see Miles and Robins, 1992; Amin, 1994). We will not be concerned here with weighing up the merits of the various theories of urban change available. Our concern, rather, will be to slice through the economic segments of cities so that an overall perspective of the importance of telematics in urban economic change can be built up.

URBAN ECONOMIES AS THE INFORMATION-SWITCHING CENTRES OF THE GLOBAL ECONOMY

A major structural shift away from manufacturing towards producer services, consumer services, and knowledge-based industries is remaking the economic constitution of cities. World trade in services and information is now equal to that in manufactured electronic goods and automobiles combined (Johnston, 1993; see Lanvin, 1993). In general, the jobs which surrounded the old manufacturing heartlands of cities are being swept (or automated) away, often to 'offshore' locations in newly industrialised or less developed countries (NICs and LDCs). Pelton argues that 'most product-oriented industries are discovering that they can exist only in a global market. Their global competitors can out compete them if they scale their production to local markets only' (Pelton, 1992; 10). In their place, the bulk of the high quality jobs created in cities currently are for highly skilled 'information workers' in so-called quarternary or high-order, decision-making functions. These jobs involve the skilled manipulation, processing, adding value to and dissemination of information, knowledge and symbols (Locksley, 1992) – what Robert Reich calls 'symbolic analysts' (Reich, 1992). Much of this information is electronically encoded in computerised form and transmittable via telecommunications and telematics. In Europe, for example, 50 per cent of all jobs and 80 per cent of all new jobs now come from information-based services (Johnston, 1993). By the year 2000, the European Union have predicted that telecommunications themselves will account for 7 per cent of all Gross Domestic Product (from 4 per cent in 1984), and will indirectly support 60 per cent of all employment (Mulgan, 1991; 14).

Jobs for information workers in cities arise for several reasons. As the global economy becomes more complex and volatile, the 'costs of coordination and control tend to rise faster than the material capacities' of the economy (Mulgan, 1991; 2). 'Information jobs' are geared, first, towards reducing uncertainties for firms and institutions as business environments become more complex and volatile. Often, these jobs arise as a result of organisational attempts to control and monitor the vast global systems that increasingly characterise the global economy and shape their individual sector(s) (see Beniger, 1986). Second, they are needed to service the increasingly complex educational, administrative, financial, legal and technological underpinnings of modern economies. Finally, they supply the burgeoning demands for 'commodified' information goods such as computing, media and on-line information services offered via telecommunications in the 'network marketplace' (Dordick et al., 1988).

Because of these economic shifts, urban areas can now be considered fundamentally to be 'information cities' (Hepworth, 1987), 'transactional cities' (Gottmann, 1983), or the 'information switching centres' of the world economy (Mulgan, 1989). But this restructuring towards higher levels of information intensity is not merely a post-industrial force. In fact, as Geoff Mulgan points out, societies are becoming 'super-industrial' rather than 'post industrial' (Mulgan, 1991; 13). Increasing speed, complexity and volatility right across the economy means that employment for information workers is growing across all economic sectors – from mining, agriculture and manufacturing through to retailing, services and government. As we shall outline in the rest of this chapter, large, globally oriented and dynamic cities are particularly strong centres for information workers. London, for example, shows higher proportions of 'quarternary' of information workers in all economic sectors than does the UK as a whole (see Figure 4.1).

The opening up of national space economies to global economic forces, driven by such supranational institutions at the European Union and the General Agreement of Tariffs and Trade (GATT), means that these processes are increasingly global in scope. Particularly important here are the increased freedoms in moving around capital, money, goods and services across the globe

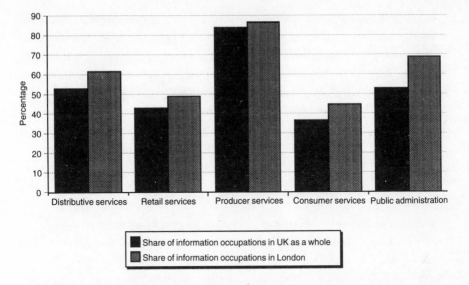

Figure 4.1 Share of information occupations in the labour force by sector in London
and the UK, 1981

Source: Hepworth *et al.*, 1987

on a rapid and flexible basis (Shachar, 1994). Telecommunications and telematics networks are providing a crucial technological support to this globalisation process.

This globalisation is closely related to an increasing dominance in all economic sectors of transnational corporations (TNCs). The recent era has seen a dramatic rise in the size, power and influence of multinationals (Dicken, 1992); they are increasingly dominant in shaping urban economic fortunes (Dabinett and Graham, 1994). At the start of the 1990s, there were 37,000 TNCs globally, covering all economic sectors; their stock of foreign direct investment reached $2 trillion dollars per year (Thompson, 1993). The top 500 TNCs now generate 30 per cent of gross global product, 70 per cent of global trade and 80 per cent of international investment flows (Myers, 1994). This is helping to shift both urban economies and national telecommunications regimes from a local or national orientation to a global or international one (Knight and Gappert, 1989). As we saw in Chapter 3, TNCs straddle many local and national economies and weave their operations together using real-time telematics networks crossing many time zones. Telematics, along with rapid transportation networks, provide key enabling technologies for the extension of TNC activity.

Interwoven with this globalisation and the increasing power of TNCs over urban economies, there is a speeding-up of economic processes and an increasing uncertainty and volatility of markets. Ironically, telematics networks, developed and applied to bring control and improve predictability to global firms, often further encourage this volatility because they support the real-time flow of finance, capital and services around the world (Mulgan, 1991). This, along with increasingly segmented consumer tastes and competition from newly industrialised countries (NICs), means that there has been a crisis in the inflexible and hierarchical Fordist forms of industrial organisation that dominated between 1920 and around 1980 (Capello and Gillespie, 1993). The result has been a breakdown in the vertically integrated Fordist firm, as epitomised by the routinised and standardised production systems in the first automobile production lines. Instead, there is now a stress on continuous innovation and a constant search for flexibility, efficiency and responsiveness. This is leading to what the business and management literature refers to as 'business process re-engineering' – the development of flatter and more 'customer-centred' organisational structures, a greater degree of sub-contracting, and a switch towards new industrial spaces and processes organised along 'post Fordist' lines (Hammer and Champy, 1993). As Smith argues, 'the corporate structure that we have known is simply incapable of managing in the lightning-fast competitive environment of the "information age"' (Smith, 1994). Thus, new technology based 'batch production' is replacing mass production (Piore and Sabel, 1984) and firms are restructuring themselves as 'network firms', using telematics to integrate all stages in their operation in a seamless and efficient manner (Capello and Gillespie, 1993; 49). Increasingly, for example, links between firms are based on Electronic Data Interchange (EDI) and Electronic Fund Transfer (EFT) operating through telematics networks.

There is even talk, particularly amongst futurists and utopianists, of a shift towards the 'virtual corporation' (Smith, 1994) or 'virtual office' (Bleeker, 1994). In this scenario, leading-edge firms shift to flexible working patterns, increase short-term contracts, and set up telematics and information systems that support much more fluid movements of staff, while carefully controlling the delivery of products and services to customers. Bleeker, for example, argues that using integrated computing and communications technologies, corporations will increasingly be defined 'not by concrete walls or physical space, but by collaborative networks linking hundreds, thousands, and even tens of thousands of people together' (Bleeker, 1994; 9). Here, using the approach we developed in the last chapter, we can consider new innovations in telematics to provide electronic spaces through which business processes can be reorganised in ways which are likely to affect radically urban economies. Suzuki (1993) predicts that, in the near future, 'the biggest change will most likely result from the concept of a virtual electronic space created by electronic tools' including 'virtual rooms, virtual buildings' for meetings, databases linked up into 'virtual information warehouses' and networks supporting transactions. Through these, he argues, seamless formal and informal communications will be supported within and between firms to radically influence business processes (Kishimoto and Suzuki, 1993). If significant, these trends obviously have major implications for both property markets and the physical form of cities (see Chapter 8), and urban transport and infrastructure (see Chapter 7).

These trends lead to an increased reliance on the use of LAN and WAN telematics networks to underpin production and intra-firm linkages. Crucially, all of these shifts are leading to an increased importance for advanced tele-communications and telematics as key mediating links underpinning economic processes both within and between city economies. Telecommunications and telematics are now involved as key 'generic technologies' in attempts to innovate and compete in all stages of the production process and all economic sectors (Dicken, 1992; 101). Key results are new telematics-based ways of organising and covering space and territory, new patterns of 'teleworking', and new diversified prospects for cities as centres of economic activity. As Jean Gottmann suggests, 'in the modern world, with its expanding and multiplying networks of relations and a snowballing mass of bits of information produced and exchanged along these networks, the information services are fast becoming an essential component, indeed a cornerstone, of transactional decision making and of urban centrality' (Gottmann, 1990; 197).

THE METROPOLITAN DOMINANCE OF TELECOMMUNICATIONS INVESTMENT AND USE

Telecommunications provide technical systems which allow 'action at a distance' without physical movement. They tie places together but not on the same terms or in neutral ways (Samarajiva and Shield, 1990). Through them asymmetric power relations and control are exercised between the powerful and the less powerful or powerless (Castells, 1989). From single nodes, modern telematics networks mean that multi-site organisations or distant markets can be controlled over whole territories and hinterlands (Goddard, 1994; 282). This does little to even out patterns of uneven fortunes between cities and regions. In fact, it may even exacerbate this unevenness, because it becomes possible to centralise power further on to key global control centres at the hubs of telematics networks.

This feature of telecommunications networks has, historically, actually served to enhance the power of the largest and most dynamic cities as control and transaction centres (Gottmann, 1983). Telecommunications innovations made the huge industrial city possible just as much as transport innovations. Use of the telephone, in particular, has become so woven into every aspect of the functioning of modern cities that its use is often ignored or taken for granted (Pool, 1977, 1983). The increased penetration and use of the telephone was closely woven with the continued growth of more complex cities as the twentieth century progressed. As well as supporting urban sprawl and suburbanisation since World War II, telecommunications were crucial in allowing the large industrial metropolis to grow in the first place. Without telephones, the extreme concentration of office and control functions within central business districts in the first half of the century, and, indeed, the development of the skyscraper itself, would have been impossible (Pool, 1977).

The telephone made action at a distance a cheap, immediate and effective possibility without leaving the office. Thus it enhanced the degree of power and control that centralised business élites could maintain over widely dispersed routine production functions – a pattern that continues today with telematics. As Jean Gottmann suggests,

the telephone provides, when needed, quasi-immediate verbal communication between all the interdependent units at minimum costs. . . . It would have been very difficult for all these complex and integrated networks [in cities] to work in unison without the telephone, which made possible the constant and efficient coordination of all the systems of the large modern city. . . . The telephone helped to make the city bigger, better, more exciting.

(Gottmann, 1990; 198)

Large, international cities therefore sustain remarkable concentrations of telephone connection and use. Tokyo, for example, has three times more telephones use by for its 23 million population than the whole of Africa has for five hundred million (Harrison, 1995).

Despite all the predictions that telecommunications will lead to the end of cities, this metropolitan dominance in telecommunications investment and use remains strong today (Hepworth, 1989; 183). The enduring dominance of at least some of the largest and most privileged cities as centres of control, coordination and transaction in a global, information-dominated economy is clearly shown in what little empirical is available on the urban/rural composition of tele-communications investment and use (Pred, 1977). In general, the largest, most dynamic and most globally oriented urban economies tend to be most advanced in terms of the level of development, reliability, cheapness and sophistication of their telecommunications infrastructures and the use of advanced services by the economic and social groups within the city.

This is shown extremely well by the map in Figure 4.2, which ranks European cities according to their importance as centres for telecommunications investment and use. In general, it is the largest cities which act as global centres for control and high-order corporate functions that are most dominant here. Indeed, this dominance seems to be growing. This is because of the growing centralisation of key control functions and company headquarters on the largest cities at Europe's core – primarily London, Frankfurt and Paris between 1973 and 1988 (Sassen, 1991). American evidence from cities in Indiana suggests that telecommunications are 'creating a gap between [the] largest cities and medium and smallest cities' because information-based 'quarternary' functions remain centralised in the larger cities (Alles and Esparza, 1994).

But the metropolitan dominance of telecommunications extends to social uses too. A recent survey on the propensity of social groups to use 'POTS-plus' telecommunications services like answering machines, cordless phones and call forwarding, for example, found a 'linear relationship between urbanity and technology ownership' (Shields et al., 1993). This relates to greater exposure to new telecommunications innovations, the more capable telecommunications infrastructures in large cities, and the faster and more movement-dominated lifestyles that tend to exist in urban areas.

A further factor supporting the metropolitan dominance of telecommunications investment and use is the current shift away from national monopolies towards the liberalisation and globalisation of telecommunications regimes (Graham and Dominy, 1991; Graham and Marvin, 1994). This process inevitably leads to a concentration of telecommunications providers in the most profitable areas and markets while others tend to become neglected. Cities, and particularly large, dynamic and globally oriented cities, provide concentrations of information-intensive firms and organisations that are tempting markets for profit-driven providers (Gillespie and Robins, 1991). These cities thus benefit from the roll out of new infrastructures and services well ahead of their rural hinterlands, which are much more expensive and much less lucrative to serve.

This is particularly the case within the largest 'global cities'. In the UK, for example, the second operator, Mercury, initially concentrated overwhelmingly on

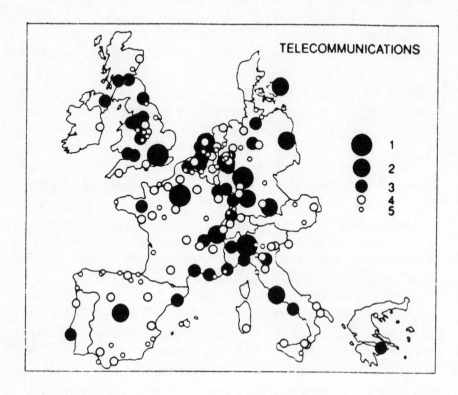

Figure 4.2 European cities ranked by their importance as centres for telecommunications
Source: Hall, 1992; 164

the City of London – which, until recently, gave it 90 per cent of its profits (Morgan, 1993) – and the other main business services cities in the UK (Manchester, Birmingham, Glasgow, Edinburgh). Its services remain inaccessible to much of the country, especially the periphery and rural areas, unless BT is used as a link (Graham, 1992a). A similar bias exists in the development of cable networks, which is overwhelmingly focused on the most prosperous and information-rich areas centred on London and the South East of England. This process can be considered one whereby operators engage in the spatial and social 'cherry picking' or 'cream skimming' of a highly differentiated marketplace (Graham and Marvin, 1994).

The simplest way to track the dominance of telecommunications investment and use by (particularly large) cities, is to compare the proportion of the national population they take up with their proportion of national telecommunications investment, facilities or flows. In the case of broadband cable in the UK, for example, while London and the South East have only 24 per cent of UK households, they have 39 per cent of all cable connections (Cornford and Gillespie, 1993). Other examples of this metropolitan dominance in use of, and investment in telecommunications are shown in Figure 4.3, which draws examples from the UK, France, the USA and Japan. Similarly, Figure 4.4 shows the effect

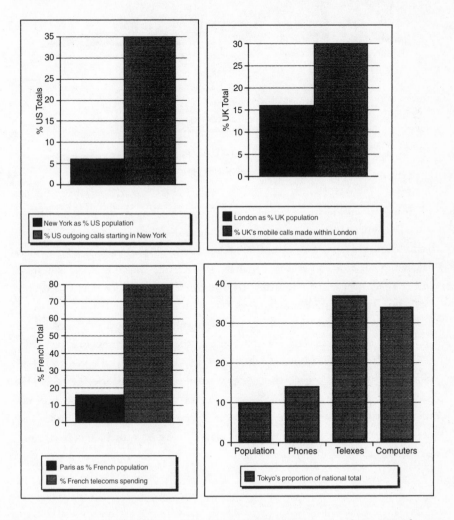

Figure 4.3 The urban dominance of telecommunications investment and use: examples from the USA, UK, France and Japan

Source: Financial Times, 1994

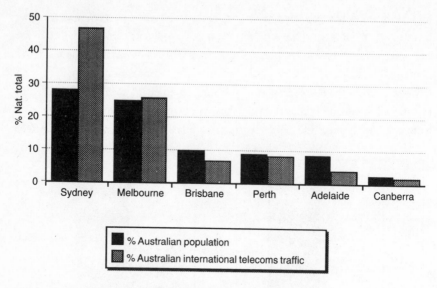

Figure 4.4 Australian cities and their share of international telecoms traffic
Source: Newton 1993

of Sydney's world city status in Australia by breaking down of international telecommunications traffic for the whole Australian urban system. In all cases, investment in, and use of telecommunications – particularly international telecommunications – is dominated by the largest, most dynamic and most internationally oriented city within the nation. This dominance reaches extreme situations, such as with Paris (Figure 4.3).

These factors mean that it is easy to overestimate the degree to which high order services which are concentrated in the largest cities can be easily decentralised through advanced telecommunications. We would agree with Jean Gottmann when he argues that 'despite all the propaganda by the technologists ... there may be a deconcentration of some activities, but basically transactional activities are not likely to be scattered through rural territory just because technology is becoming available to overcome distance' (Gottmann, 1990; 198).

CITIES AS NODES ON CORPORATE TELEMATICS NETWORKS

A crucial aspect of the urban dominance of telecommunications in the modern, global economy, is the roles that large, metropolitan cities play as the nodes and hubs of private, corporate telematics networks (Valovic, 1993). Private corporate investment in telecommunications rose 400 per cent between 1980 to 1987, from $15 to $75 billion (Mulgan, 1991; 224). The number of such networks doubled in only three years between 1979 and 1982, from 500 to 1,000 (Howells, 1988; 130).

The growing influence of multinational corporations is also encouraging the trend towards liberalisation and privatisation in national telecommunications policies (Irwin and Merenda, 1989). Corporate consumers are at the vanguard of criticism of the high costs and obsolete services delivered by national PTT monopolies. Their threats not to invest at all in a location or to use their new telematics-based flexibility to move elsewhere are pushing nation states to liberalise and/or privatise telecommunications ownership (Irwin and Merenda, 1989).

TNCs are therefore beginning to benefit from a shift from national public telecommunications monopolies to an integrated and globalised telecommunications 'marketplace' which is directly oriented to meet their needs (Schiller and Fregaso, 1991; *Financial Times*, 1992). Morgan (1992) estimated that 3 per cent of customers (the largest corporations and government agencies) accounted for 50 per cent of all telephone revenues. Not surprisingly, a wide range of service innovations and cost packages are continually targeted at these customers to help attract their accounts in the face of spiralling competition from other providers. The drive now amongst the largest suppliers is to develop global service coverage so that they can offer corporate customers a 'one-stop' telematics service for 'Intelligent Networks' on a global basis (Iwama and Kano, 1993). This in turn reinforces the urban dominance of telecommunications noted above, because the largest cities also act as the main nodes for corporate networks.

New telematics technologies allow TNCs to construct single, real-time networks through which they can control their production, research and development and the flows of goods and services on a global scale, across many sectors and locations. In many sectors, these networks are also the delivery vehicles for products and services to the market. As Peter Dicken argues, 'it is the possession of such instantaneous global communications systems that enables the TNC to operate globally, whether it is in manufacturing, resource exploitation or business services' (Dicken, 1992; 108). James Davis (1994) considers that telematics:

have made global production serving a global market possible, the nature of which the we have never before seen. It is feasible and economic to have design done in Silicon Valley, manufacturing done in Singapore or Ireland, and have the resulting products air-shipped to markets again thousands of miles away. Along with global production and global consumption, we also have a new global labor market. U.S. workers compete against Mexican or Thai or Russian workers for all kinds of jobs – not just traditional manufacturing and agriculture jobs, but also software design and data analysis – and capital enjoys remarkable fluidity as it seeks out the lowest costs and the highest returns.

(Davis, 1994)

This increases the locational flexibility of TNCs, because they can use corporate telematics networks to increase the effectiveness of producing goods and services using inputs from many locations (Hepworth, 1989; Chapter 5). To Kevin Robins and Andy Gillespie, 'network configurations afford a new mobility, a hypermobility, to global corporations. Within the global system, it becomes possible to shift activities around the network topology and across territories' (Robins and Gillespie, 1992; 159). As Geoff Mulgan argues, in this process of globalisation and the growth of TNCs:

communications technologies have played a decisive role. The great power of capital has always been its ability to choose, to decide where to locate and when to withdraw. The satellite, cable and microwave greatly enhance this mobility and the leverage it confers, allowing much more sophisticated location policies: mental labour can be separated from production, skilled labour from unskilled, service tasks from manufacturing ones.

(Mulgan, 1989; 19)

In the cases of corporate computing facilities forming the hubs of corporate telematics networks, this mobility of capital is made infinitely fast for practical purposes (see Figure 4.5). Here, centralised computing facilities (based usually in large cities) actually allow productive capacity to be distributed in real time throughout the corporate network, on a planetary basis if necessary (Hepworth, 1989). These massive but hidden intra-corporate information flows, which also carry finance, labour and services, are a crucial yet virtually invisible dynamic underpinning modern city economies and raise the question: 'how do we define the limits, contours and workings of the contemporary economy?' (Hepworth, 1989).

The uses of these corporate telematics networks is helping to support the complex 'spatial divisions of labour' of TNCs within which cities compete for the best position possible (Massey, 1984; Irwin and Merenda, 1989; Janelle, 1991). For example, while routine manufacturing and data preparation activities typically now take place in less developed or newly industrialised countries (LDCs and

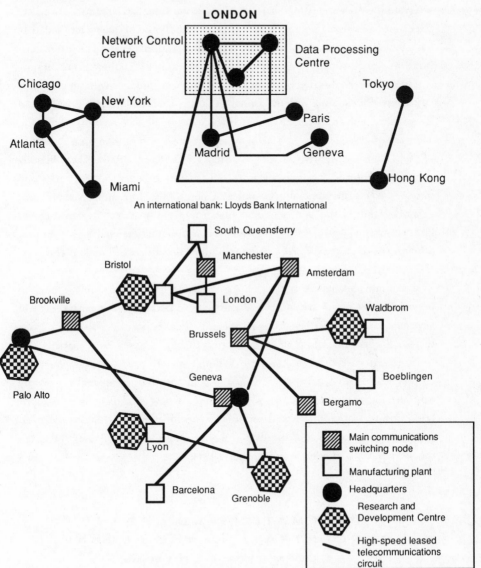

Figure 4.5 Examples of the linkage of cities into corporate telematics networks

Sources: Howells and Green, 1986; 170; Howells and Wood, 1993

NICs), global control and managerial functions are rooted in the world cities of London, Tokyo and New York and research and development tends to locate within high-tech, innovation-rich areas around new industrial spaces. The two networks in Figure 4.5 illustrate how corporate telematics networks are used to integrate these elements together across space and time barriers. Such corporate telematics networks are the basis for real-time flows of the voice and data communications, services, electronic financial flows, image and video transmissions and – increasingly – remote working networks between these many sites.

Such networks also support transactional electronic linkages with suppliers and customers, for example with the growth of Electronic Data Interchange (EDI) and Electronic Fund Transfer (EFT) within the automobile and retailing industries. They are now seen as the key factor in supporting speed of innovation, mobility of capital and competitive advantage within TNCs (Keen, 1986). These literally glue multinationals together across the diverse time zones, spaces, cities and localities where they operate (Gonzalez-Manet, 1988). The importance of such networks, and the global economic dominance of multinationals is shown by the fact that internal data and voice transmissions within multinationals now comprise 50 per cent of all cross-border communications flows between nations (Mulgan, 1991; 220).

Corporate telematics networks are designed strictly to limit access to information flows so that a business can achieve maximum responsiveness and advantage. Rather than benefiting peripheral cities, political economists suggest that such networks allow effective control to be exercised over markets and territories from central locations. To Gillespie and Robins, for example, these corporate telematics networks 'favour the penetration of peripheral regions by centrally located organizations', so instigating 'a new system of inter-regional trade in information goods and services . . . , with dominant regions the main exporters and weaker economically peripheral regions [and cities] relegated to the role of importers' (Gillespie and Robins, 1991).

TELEMATICS AND URBAN CONCENTRATION: THE GLOBAL COMMAND CENTRES

We saw above how certain larger, globally oriented and dynamic cities such as New York, London, Paris and Tokyo tend to dominate telecommunications investment and use within their respective nations. We also stressed how the positions of these cities as centres of control, and concentrationes of high order decision-making and 'quarternary' functions are being strengthened by techno- logical developments in telematics and the liberalisation of telecommunications. It is clear that, against the prophets of urban dissolution, developments in telematics compound the many existing advantages of cities. These stem from their nodal positions on global transport (especially airline) networks, their concentrations of a wide range of services, their social and cultural milieux for global business, their locational prestige, their size as centres of property investment, the flexibility that derives from their highly skilled and large labour markets, and the versatility that comes from the many possible suppliers and clients within these cities.

GLOBALISATION AND THE GLOBAL COMMAND CENTRES

The emergence of this closely interconnected group of 'global command centres' is a direct result of the major shifts in the global economy discussed above. A range of interconnected factors are involved (Moss, 1987). These interact through cumulative links which encourage the continued centralisation of key producer services, high-order financial services and corporate control functions on to the three legs of the global financial marketplace – London, Tokyo and New York – and, to a lesser extent, second-tier cities like Paris and Hong Kong (Sassen, 1991). These cities have emerged together to provide the apex of the 'planetary metropolitan network' (Dematteis, 1988).

Three preliminary factors can be identified. First, the globalisation of markets, the increased dominance of TNCs and and the dispersal of TNC manufacturing and production activities across the globe has led to growing demands for centralised TNC control functions. These are directed to large cities because such locations reduce risks, provide the maximum range of opportunities and limit uncertainties. Second, the greater ease and freedom of moving capital and money across national boundaries has led to demands for financial centres through which these global flows can occur and be coordinated. Third, large transnational firms have restructured and now subcontract a greater degree of their demands for high- order producer services like accounting, legal services, insurance, consulting, and advertising. This has fuelled demands for centralised centres to produce and distribute these services, often on a global basis. These factors weave together and combine with the need to maintain face-to-face contacts in complex business negotiations to enhance the dominance of these cities. Critical to the success of these global command centres is the concentration of specialised information. To Mitchelson and Wheeler (1994; 87), for example, New York has 'the greatest concentration of non routine information ever assembled in one place'.

Most importantly, though, are the intense uncertainties thrown up by the volatility, velocity and unpredictabilities of the global economy. As Mitchelson and Wheeler argue, 'in times of great uncertainty, select cities acquire strategic importance as command centres and as centralized producers of the highest order economic information' (Mitchelson and Wheeler, 1994; 88). In the extract in Box 4.1, Mitchell Moss shows how, in a volatile and global economy, the development by corporations of a wide range of decentralised functions scattered across the globe demands a parallel centralisation of corporate control and coordination functions on to these global command centres. Thus, as Moss argues, the decentralisation of routine functions within TNCs actually *requires* the central-isation of control and headquarters functions on to the global command centres. To quote Saskia Sassen (1991; 5), it is 'precisely because of the territorial dispersal facilitated by telecommunications that agglomeration of certain centralizing activities has sharply increased'. These in turn fuel the growth of high-order accountancy, banking, legal and other services, who service the concentration of corporate headquarters and can also use telematics to link themselves to their global markets. In these services, as Simmons argues 'improved technology broadens the hinterland which the existing centres can service' (Simmons, 1994).

BOX 4.1 INTERNATIONAL INFORMATION CAPITALS

While advances in transportation and communications technologies have long made it possible to disperse both the headquarters and production of manufacturing activities to suburban locations, cities that are centers for information-intensive services' (e.g. accounting, advertising, banking, law, management consulting, publishing) are likely to benefit from the greater use of sophisticated information and telecommunications technologies. The opportunities that communications technologies present for multinational firms to engage in the provision of global services has also heightened the value of information to co-ordinate international activities.... In fact the greater the extent of geographic decentralisation, the greater the need for centralisation of key control activities.

Telecommunications has eliminated the dichotomy between central-

isation and decentralisation and allowed decentralisation with centralisation (Keen, 1986). Not all cities will benefit from telecommunications
technologies. Rather those cities whose economic life is based on the
exchange of information, both face-to-face and electronically, will be
strengthened by the capacity to participate in the increased global
marketplace for business services through communications technologies....

New telecommunications technologies, in conjunction with the inter-
nationalisation of services and finance, are strengthening a handful of
world cities such as New York, London, Tokyo, Los Angeles and Hong
Kong. These cities, at one time centers for the manufacture of goods, are
now the centers for the production of information that is distributed
electronically around the globe.... Communications technology, by
extending the global reach of cities that are centers for information-based
services, also affects the relationship of a city to its home nation. The
world information capitals increasingly resemble the city-states of
ancient Greece, for their destiny is remarkably independent of their own
domestic national economies. Such cities are intricately linked to each
other through sophisticated telecommunications networks that operate
on an around-the-clock basis. The face-to-face activities that occur in
these cities have not been made obsolete by new technology rather,
technology has extended the geographic reach of the individuals and
firms that transact business in these world capitals. The operational
boundaries of a city are no longer defined by geography or law, but by the
reach of phone lines and computer networks....

The deregulation of the telecommunications industry in the United
States and the privatisation of telecommunications in other advanced
industrial nations in leading to the creation of a new telecommunications
infrastructure designed to serve the information-intensive activities of
large metropolitan regions.... The new communications technologies,
whether they be teleports, optical fiber systems, mobile telephones,
'smart buildings', or radio paging systems are initially being built to serve
the major information users located in large cities and metropolitan areas.
Moreover, the very cities that are the centers for face-to-face communication are also the ones which will benefit most from the spread of
advanced telecommunications systems. For, as electronic means of
communications are used to control and co-ordinate geographically
dispersed activities, there will be less face-to-face decision making in
outlying areas and the city's role as a hub for the interpersonal exchange
of information will be even more important.

(Moss, 1987; 536)

Of course, this concentration of headquarters, producer service and finance
firms in turn generates spiralling demands for a wide range of lower order
consumer services such as restaurants, retailing, cleaning, leisure etc. These often

casualised, part-time and low-wage jobs generally outnumber the high-skill professional jobs in corporate and business services by a factor of two or three to one. This contributes to the development of a highly unequal and polarised class structure in these cities (Sassen, 1991; Chapter 5).

The privileged position of these cities as hubs for the world's best, cheapest and most competitive sets of telecommunications infrastructures, and the beneficial effects of liberalisation, also supports this centralisation. It is no accident that the USA, Japan and the UK were the first nations to liberalise their telecommunications regimes during the explosion of service growth of the 1980s. In all cases, this was geared heavily towards enhancing the positions of their global command centre through allowing competitive innovation in telecommunications for these sectors. It is also no accident that it is the global command centres which often provide the most lucid illustrations of metropolitan dominance of telecommunications investment and use, as outlined above. In Britain, for example, the Corporation of London argue that the UK's liberalisation of telecommunications since 1981 has contributed directly to London's continued competitiveness as a global information capital: 'product innovation and service provision in telecommunications . . . have been directed towards the financial services. Competition between telecom providers appears to have had a considerable effect on customers. Services have been redefined and costs reduced' (Simmons, 1994).

But there are also important social and cultural reasons for the continued dominance of the global command centres, even when much of the formal information they traditionally dominated may now be accessed on-line from almost anywhere. In the highly uncertain world of the global command centres, tacit, informal and clandestine information based on social networks and trust is extremely valuable. This information is tightly confined to those on the inside of the social networks within the key institutions of these cities. It will never be possible to substitute the tacit information exchange that is possible over lunchtime or after-work bar meetings through telecommunications. In short, the people involved at the apex of corporate power, and also those at the apex of financial markets and professional services, need to be 'in the thick of it' in ways that cannot be substituted by telematics (Pryke and Lee, 1994). Rob Atkinson (1995; 15) recounts the true story of an attorney in a Washington DC law firm who was moved to the other side of the building from her colleagues. After two weeks of feeling 'isolated' she asked to move back to be next to the rest of her colleagues.

These concentrating tendencies mean that the social and cultural production of urban places – gleaming financial services, offices, the 'together' culture of the corporate and service professional – therefore underpin these global command centres as much as telecommunications (Allen and Pryke, 1994). Power is effective through global telematics networks as well as the transformation of physical and social spaces within the cities themselves. To quote Mitchelson and Wheeler:

> To know the language of power requires intimacy, to be an insider mandates presence within, and to be effective therein requires face-to-face communication. Being within an environment enriched by first-hand information, however informal, is a matter of survival and not of discretion. These people and their knowledge, the glamorous segment of the so-called quarternary sector, are concentrated within select cities that are dense with scarce information and with the power that it enables. These places provide information that other places need. And through the economies of localization, they sustain specialities that are demanded across continents if not the globe.
>
> (Mitchelson and Wheeler, 1994; 87)

TELECOMMUNICATIONS AND THE INTEGRATION OF FINANCIAL MARKETS

On top of their roles as corporate headquarters centres, and centres for high-order producer services, London, Tokyo and New York also together form an effective global financial marketplace which is intimately integrated via telematics (Budd and Whinster, 1992). This again is driven by the emergence of truly global markets and TNCs, which require financial systems and flows to match their operations. The liberalisation of national trade, finance and investment regulations also directly encourages demand for international financial services. To quote Sassen, these cities function as one 'transterritorial marketplace' (Sassen, 1991; 327) for finance, with Tokyo the main exporter of capital (based on yen), London the main centre for the processing of international capital (based on Deutschmarks and Eurodollars), and New York the main receiver of capital (based on US Dollars). Together, they provide a set of integrated twenty-four-hour global financial marketplaces which dominates financial flows and services within the capitalist world. The trading hours of the stock markets in these cities are coordinated so that one is always open, the possibilities that they will all open twenty-four hours per day is being explored (Hepworth, 1991; 135). 'As London is going to bed, Tokyo is beginning to trade, and the Wall Street bars start to become populated at the end of a day's trading' (Budd, 1994; 11).

Crucially, these global financial transactions are increasingly mediated within complex systems of advanced telecommunications networks that are filling the corridors between these cities – especially on global satellite and optic fibre systems (Warf, 1989). This is driven by the goal of eliminating delays in financial trading, so improving the rate of flow and circulation of capital and the degree to which investors can take advantage of tiny fluctuations in stock values and exchange rates. Current digital dealer boards – computerised trading systems attached to global telecommunications access – give traders virtually instantaneous (sub-100-millisecond) connections to customers. ISDN is currently being used to support multi-media connections between traders and customers. The result of this increase in velocity, though, is an increasing volatility of markets, instantaneous adjustment to changing exchange rates, and ever greater switched circulations of global funds in search of the highest rates of return. The scale of these electronic flows of finance are astonishing: the average value of cross-frontier stock trading, for example, now reaches $10 trillion per day (Myers, 1994). These flows are also major stimulants of other communications and information flows between the three centres, for example with on-line financial information centres and international phone traffic. Daily telephone calls emanating from Wall Street, for example, grew from 900,000 in 1967 to 3 million in 1987 (Warf, 1989).

There is little doubt that the electronic integration of these financial and capital markets represents the single most important application of telematics within the global economy. To quote Mark Hepworth (1991; 132), telecommunications 'have effectively removed the spatial and temporal constraints on twenty-four hour global securities trading and created pressures for [financial and tele-communications] "deregulation" in all countries across the world'. This is perhaps the best example of the development of telematics networks creating tunnel effects that link some widely separated spaces closely in terms of abstract electronic space while pushing physically near places further away. In other words, while some cities converge in electronic space, the physical and electronic separation of these cities from the rest of their hinterland economies can become more pronounced. As Geoff Mulgan suggests, in effect, 'speed replaces distance, so that the centres of two cities are often for practical purposes closer to each other than to their own peripheries' (Mulgan, 1991; 3).

ELECTRONIC FINANCIAL MARKETS

Interestingly, a new range of electronic trading systems for national and international stock exchanges are also emerging which are not physically confined to the trading floor and which take place within electronic spaces. These are accessible nationally and internationally in real time, and erode further the spatial and temporal limitations on stock market activity. To Salsbury (1992; 27) such systems 'threaten almost every aspect of world security markets'. The best developed of these, the National Association of Securities Dealers Automated Quotation System (NASDAQ) is actually now the third largest stock market in the world, after New York and Tokyo. London has also developed its SEAQ system and Chicago has set up a system known as GLOBEX, which allows electronic trading of stocks outside normal trading hours from anywhere in the world (Budd, 1994). Because these networks do not have a physical presence and are accessible irrespective of locations, there has been some speculation that they may even out the financial landscapes or even 'end the geography' of financial transactions altogether (Budd, 1994). While they may undermine the competitive position of the big financial centres, it remains doubtful, however, whether they will really challenge the territorial concentrations of global financial markets within the global command centres. The complex and self-reinforcing advantages of these centres seem too strong and the ultimate logic of electronic trading networks seems to be one of extending the reach of these cities rather than undermining their prominence. To quote Leslie Budd, 'even where electronic trading systems appear to overcome the constraint of locality, one finds that their development and that of global alliances between exchanges is informed by notions of territory and territoriality' (Budd, 1994; 24).

TELEMATICS AND URBAN DECENTRALISATION: TELE-MEDIATED PRODUCER AND CONSUMER SERVICES

The reverse side of the centralisation of corporate control functions, high-order producer services and global financial markets is a broad-scale decentralisation process which is underway for more routine producer and consumer service functions. In fact, these two processes are symbiotically linked. The initial decentralisation of corporate productive activity drove the centralisation of corporate control functions; this now drives further decentralisation of routine services in the search to cut costs, revive profits, and maintain competitiveness. As Mark Hepworth argues, 'the more information-intensive organisations become, the more they will seek to minimise their information costs' through the so-called 'back-officing' of routine functions (Hepworth, 1992; 35).

Routinised back office functions that provide services and support to the headquarters and top offices in the world cities are now beginning to utilise telematics to try and spread out to lower costs locations in the areas around the major global cities – and, increasingly, farther afield. They are therefore using telecommunications directly to access cheaper labour, services and property outside the high-cost global command centres at a rapidly increasing rate. This is associated with the re-engineering and downsizing of centrally located firms. Many more peripheral cities are pinning major hopes on their chances of emerging as back office centres, a crucial source of new jobs in the urban economies most damaged by the collapse of manufacturing employment. This decentralisation is occurring at four levels: between the North and South in global terms towards newly industrialising and less developed countries (NICs and LDCs); between core and peripheral regions within advanced western nations; from metropolitan to non-metropolitan areas and smaller cities; and from city centres to suburbs. The result, to quote Judy Hillman (1993; 10) is that 'it is becoming increasingly difficult to know who is doing what and where. The invisible economy can be as elusive as the black'.

'ELECTRONIC IMMIGRANTS': BACK-OFFICING PRODUCER AND CONSUMER SERVICE FUNCTIONS

Businesses vary considerably in the degree to which they can offer services to firms or consumers remotely via telecommunications. It is possible to split them into three types (Sociomics, 1992). First, 'tele-defined' businesses, such as consumer service centres, telephone operators, telemarketing and market research, actually deliver their services via telecommunications. This makes them particularly suitable for 'back officing'. Second, 'tele transacted' businesses, such as airline reservations, banking, insurance agency and administration, are using telecommunications more and more for on-line transactions between offices and sites. It is therefore increasingly possible to deliver these services remotely. Finally, 'tele-reliant' businesses rely on telecommunications, but only for a small portion of their transactions. They find it difficult, if not impossible to deliver their services remotely because of the importance of face-to-face contact. Two types of decentralisation process can be identified: one for high-level corporate élites and another for low-order routine services.

'Flexecutives': teleworking and flexible work processes for corporate professionals

The first type of decentralisation is geared towards the use of telematics to allow more flexible work processes for élite corporate executives and professionals. This process is also linked in with trends towards teleworking by professionals from small, rural locations. The proliferation of 'telecottages' – small centres offering support for rural teleworkers – is also a potentially important response to this process (Qvortrup et al., 1987).

Here, the flexible mixing of city centre work with home teleworking, the use of teleworking centres and mobile telecommunications is much more important than the Tofflerian image of full-time work from 'electronic cottages'. To quote Judy Hillman (1993; 1), 'the vision of teleworkers hunched over equipment in the home has receded. Today's and tomorrow's teleworkers, particularly those employed by firms large and small, may well work at home part-time, at a neighbourhood office part-time, from their cars or hotels when convenient'. The key, she argues, is a flexible combination of physical and electronic movements and spaces, not a total substitution of the physical by the electronic. This means that most of these people need to stay within reasonably easy access of the main business cities by air, road or rail. She predicts that 'individual decisions, combined with those of organisations and government, will gradually shape the future, a more flexible lifestyle and city, if not flexicity' (Hillman, 1993; 1).

The development strategies for environmentally attractive rural and tourist areas are now being geared up to attract these working corporate executives and professionals for portions of the year. In the Balearic Islands in Spain, for example, a network of resort offices is being constructed linked into advanced telecoms networks to Northern Europe. In the United States, many resort towns in the South and West are growing fast based on the attraction of corporate teleworkers, [people that Leinberger calls 'Flexecutives'] because of the flexibility that telecommunications and fast transport networks lend to their patterns of work (Leinberger, 1994). Such small, élite groups of self-employed professionals as consultants, head hunters, and writers, he claims, can now 'live where they choose and still remain plugged into the economic mainstream. They are redefining the American dream of living in suburbia on a one-acre plot with a spouse, two children, a dog, two cars – and a two-hour round trip commute to work' (Leinberger, 1994).

Urban and national back officing

The second type of decentralisation is geared towards back officing routine functions and services, with telematics being used to access cheap, non-professional staff in decentralised locations. The first generation of back-office growth for producer and consumer services operated within or between western nations, allowing decentralisation within cities and national space economies. In 1989, for example, Merrill Lynch, a key financial institution on Wall Street, shifted 2,500 routine jobs over the Hudson River to New Jersey in order to cut costs, reduce staff turnover and access qualified staff (Brasier, 1989). A large portion of BT's directory and operator services for the South East of England are now operated in Scotland. Routine data-processing activities in all economic sectors can similarly be shifted to lower cost locations. In cheque and mail processing, for example, image communications are being used in the USA so that workers doing the data processing are actually separated from the physical plants where the cheques and letters are being sorted. The technology is manipulated remotely, allowing workers to be used in places with poor postal or overnight air services (Atkinson, 1995).

The economies of a few strategically centred cities have been transformed by an influx of back-officing jobs. In the last 10 years, for example, Omaha, Nebraska, claims to have created 100,000 tele-related jobs – a result of its labour surplus, low costs and its position at the crossroads of the the USA's national optic fibre infrastructures (Richardson, 1994a). Many business 'outsourcing centres' are now being set up, usually by rural development authorities, to take advantage of the possibility of attracting tele-mediated back-office work to their areas. Ireland's whole national development strategy centres on exploiting its high educational standards, English language, and high-quality telecommunications infrastructure, to attract back offices from the USA and UK in financial services, publishing, data processing and software development.

Tele-mediated consumer services

A particularly rapid area of growth currently is in the tele-mediation of consumer services such as banking, insurance and airline reservations (Richardson, 1994b). Here, freephone services, computerised office systems and new technologies for managing huge flows of telephone calls are leading to the development of new centralised back offices for dealing with whole national and international markets. As the example of the First Direct telephone bank shows (see Box 4.2), these tele-mediated links between the supply and consumption of services are starting to challenge the accepted needs for physical propinquity between the suppliers and consumers of services within cities. For example, a national network of bank branches can be radically reduced while a single back-office centre is set up on one location to support telephone banking. At the same time, employment in the remaining branches can be cut through the application of automation and telematics technologies. This is particularly the case of Automatic Teller Machines (ATMs), which are themselves linked up to the private corporate networks of banks. These are already being substituted for human-mediated routine banking services such as cash dispensing. Ranald Richardson notes that 'as these [ATM] machines become more sophisticated, they are likely to be at the heart of new self-service branches in the near future' (Richardson, 1994c).

BOX 4.2 THE BACK OFFICING OF CONSUMER SERVICES:
THE CASE OF THE FIRST DIRECT BANK

The development of First Direct, the UK's most successful telephone banking operation, presents a good example of how telematics-based reorganisation can radically change the social and spatial aspects of consumer service delivery (Richardson, 1994b). Increasingly, this makes the traditional assumption that consumer services must rely on networks of service outlets distributed in ways that reflect the distribution of population an unhelpful one.

Keen to target the most profitable and affluent segments of the consumer banking market, the Midland Bank set up the First Direct telephone banking operation in November 1988. It now employs over 1,300 staff who use telematics and telecommunications to deliver a

24-hour national banking service from a single site – a 10,000 sq.m office in Leeds – at the geographic centre of the UK. First Direct has no physical branches at all; its 400,000 customers do all their banking transactions over the telephone – including arranging loans and transfers. Cash withdrawals are handled by the physical branches and Automatic Teller machines (ATMs) of other banks. Customers only pay the cost of a local telephone call to access First Direct from anywhere in the UK. Computerised information and call management systems allow First Direct's employees to provide personal service to customers, who are specifically picked from the incoming applicants to be from higher socioeconomic groups (with incomes over £15–20,000). The fine detail of consumer profiles built up by this approach allows a specific targeting of follow-on products which boosts profitability.

At the same time, a physical reorganisation of Midland's branch network has taken place, with over 750 branches shutting during the past four years. Over 15,000 jobs will be lost nationally during this process. This physical withdrawal of services has tended to occur in areas of low socioeconomic status such as inner cities (see Chapter 5). Thus, the jobs created in Leeds have been dwarfed by national losses from branch networks – an inevitable result of the overall cost-cutting drive. Many of the remaining staffed branches are also being turned into automated, self-service outlets, with the use of advanced ATM and videoconferencing technologies. Thus, in both physical and electronic senses, there is a process of polarisation. Affluent groups and areas are being 'cherry picked' by competing banks while poor groups and areas are being socially dumped from physical branch networks and the growing range of tele-mediated bank services. There are also increasing dangers, as telecommunications improvements and further globalisation come into effect, that these tele-mediated services will prove extremely mobile, adding to the volatility and insecurity of the urban economies that manage to attract them. There are no technological reasons, for example, why First Direct could not relocate to a less developed country to cut costs even further.

Wider trends towards cashless consumption, with the proliferation of credit cards, Electronic Fund Transfer at Point of Sale (EFTPOS) systems, and 'smart cards' – which can be 'refuelled' with cash from bank accounts and then used to buy goods and services – are also emerging. One system, Mondex, is being developed by a consortium of NatWest bank and the telecommunications company BT. Following a trial of 10,000 people in Swindon, England, the aim is to develop Mondex as a 'truly global payment system' based ultimately around a totally electronic cycle of drawing money by electronic transfer from bank accounts to 'recharge' smart cards, often over the domestic telephone, so that they can be used to purchase a wide range of goods and services (Bannister, 1993).

Such technologies support a further shift towards electronic trading for goods and services via telephone and telematics systems. They also heighten the possibility of the closure of many physical financial services outlets in cities and their substitution with locationally free back-office-based networks. Between 1989 and 1991, for example, 1,526 bank branches closed in the UK. Between 1991 and 1995 25 per cent of all Barclays Bank – the UK's largest clearing bank – will close (Richardson, 1994c). The branches to remain are being carefully selected using the analytical capabilities of Geographic Information Systems (GISs), based on their profitability and geared towards more affluent spatial and social groups within cities (Christopherson, 1992; Chapter 5). This trend has major implications for the type and distribution of employment – jobs in British clearing banks declined from 355,000 in 1989 to 300,000 in 1993 (Bannister and Atkinson, 1995). It is also reshaping the geography of physical service outlets and leading to a heightened unevenness of social access to services – whether they are mediated by physical space or electronic space.

Another good example of the use of telemediation for customer service is the case of British Airways (BA) (Richardson, 1994b). Through an integrated corporate telematics network, which straddles the whole world, BA have recently set up a new strategy for dealing with the millions of calls it receives from consumers and travel agents about flight availability and reservations. Within the UK, 90 per cent of BA sales staff were traditionally based in expensive offices next to Heathrow airport in London, geared towards the concentration of markets in the London region. Now, however, a new national back-office strategy has been set up, using automatic call distribution technologies to integrate a range of offices that have been dispersed across the peripheral cities of the UK into a so-called 'virtual single office' (Richardson, 1994b). (See Figure 4.6.)

This strategy has cut costs per employee by between £3,000 and £4,000 per year. A private, 'intelligent' telecommunications network automatically routes calls from various parts of the UK to these various centres based on their opening hours and availability for business. This network ensures the best use of resources, maximum flexibility and the fastest response: the aim in all centres is to answer 80 per cent of all calls within 20 seconds – the so-called '80:20' target (Richardson, 1994b). This provides an excellent example of 'electronic integration' of services, where telematics are used to integrate widely dispersed sites in many cities into single functioning centres (Dabinett and Graham, 1994). Richardson comments that 'BA's telesales operation, although located at five sites,

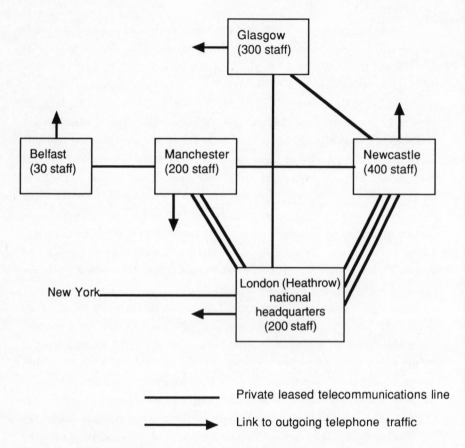

Figure 4.6 British Airways' 'virtual single office' for dealing with customer reservations in the UK

Source: Richardson, 1994a

can be seen as a "virtual single office". Central control extends to preparation of a "national roster" – the number of staff required at any one site at any time – to allow the national operation to reach its "80:20" [call-answering] target as efficiently as possible' (Richardson, 1994b).

Towards global back-office and teleworking
networks

Plummeting international telecommunications costs, the rapid growth of high-capacity global telecommunications networks, and the increasing locational freedom of TNCs is fuelling a global decentralisation of routine tele-transacted and tele-defined services at all spatial scales. The 'off-shoring' of a wide range of service support services on a global basis is burgeoning. Pelton predicts a rapid growth of what he calls 'electronic immigration' – the on-line 'importing' of (often female) labour and skills from cheap 'telecolony' locations around the globe (Pelton, 1992). He writes, 'the ability to recruit and electronically-import cheap professional services into the United States, Europe and Japan could become the top international trade issue of the 21st century' (Pelton, 1992). Mark Hepworth, argues, from the point of view of London, that:

transferring 'back offices' to cheaper cities is one established option: the superior on-line control capabilities of computer networks and strengthening 'internal markets' in big companies and government will speed up this exodus. In fact, rather than going to cheap labour markets like Newcastle and Sheffield, routine office work may be increasingly transferred by satellite to the Third World (New York's financial companies are already doing this).

(Hepworth, 1992; 35–36)

There are already many examples of back office centres in LDCs and NICs which are integrated in real time into supporting higher order functions in the global information capitals. Robert Reich, in *The Work of Nations* speaks of 'the foot soldiers of the information economy' being 'the hordes of data processors stationed in back offices at computer terminals linked to worldwide data banks' (Reich, 1992; 210). Phillip Money notes that 'a trend that was widely forecast in the early 1980s – the transfer of significant numbers of "back office" white-collar jobs from developed countries to offshore locations where costs are lower, working regulations more relaxed and unions non-existent' (Money, 1992).

Back office centres are now booming in NICs and LDCs which have the skills, apparent political stability and telecommunications and transport infrastructures to be employed as 'off-shore' service centres. In the Philippines, for example, single centres now generate over 700 million keystrokes per month. Their advantages for corporate customers are considerable: they offer wages of 20 per cent of those typical in the west; staff turnover levels are 1–2 per cent per year compared to 35 per cent in North America; and discipline often reaches military levels to ensure accuracy (Money, 1992). Moreover, many LDC and NIC governments are directly encouraging the growth of these trends as attempts to diversify fragile economies through financial incentives, 'free port' zones, and subsidised teleport and 'digiport' infrastructures (developed in partnership with western telecommunications companies) (Wilson, 1994). In Jamaica, for example, the government has invested heavily in transport, telecommunications, property and tax inducements for its 'digiport' initiative, which is booming as a back office centre for US companies, primarily for catalogue retailing, telemarketing, information and reservation services (Wilson, 1994).

The use of such new work practices, which are also blossoming in low-cost western nations such as Ireland, was initially particularly strong in such routine and information-intensive jobs as the processing of insurance claims, telemarketing, airline reservations, secretarial and translation services and the processing of government records. All eight bids for the UK contract to computerise national criminal records, for example, came from various LDCs (Money, 1992).

Increasingly, however, high skilled and higher value added services are being restructured in this way. Services such as international financial analysis, software programming, film animation, image processing, computer programming, academic editing and publishing are now involved. Here, complex chains of 'value added' and complex geographies of computing equipment are developing. Raw data for information services companies is usually collected from western nations, shifted to LDCs for data preparation, inputted on to computers in western cities, and sold back to western customers (Goddard, 1992).

Software development industries in India, for example, are growing at a rate of 30–40 per cent per year, as the Indian government lures in large software companies like Dell and Microsoft with telecom connections, training support and tax breaks (Bannister, 1994). Between 15–20,000 people now develop software in India, and many telephone 'hot lines' for software support to users in the west are now based in India, where highly qualified labour can be recruited at very low cost. Other global services based in India include a 60-strong team who coordinate the catering aspects of British Airways' global reservation system.

As the capabilities of transglobal optic fibre and satellite systems increase, and as costs plummet, the prospect is for increasingly diverse networks supporting 'electronic immigration'. These risk to undermine service employment in advanced western nations – the main economic motor of western cities – as many services follow manufacturing in search of lower costs in NICs and LDCs. It will increasingly be possible to switch visual and video services across global distances as cheaply as data and voice. Already, the World Bank has suggested the idea of using labour in Africa to monitor the Closed Circuit TV (CCTV) camera systems in American shopping malls. A spokesman said recently 'there is potential for African countries to come into the global economy through these types of technologies' (Bannister, 1994). CCTV monitoring for the computer group Tektronix is already carried out over 2,000 miles away from their Portland headquarters in Atlanta, Georgia. They are already exploring the possibility of shifting this to lower cost centres in Africa (Bannister, 1994).

TELESHOPPING: THE PROSPECT FOR TELE-MEDIATED RETAILING

Many utopianist and determinist treatments of the city assume that the era of teleshopping – that is, shopping mediated in various ways by telecommunications – is upon us. Tomkins, for example, anticipates a world where telematics, linked to virtual reality technology, creates a set of virtual electronic spaces within which retailing can take place. This substitutes for the physical 'difficulties' involved in conventional shopping:

the trudge round the stores has been replaced by the speed and convenience of computer-aided selection in the home, car or office. [When] virtual reality enters the world of retailing… consumers will be able to try on clothes by watching computer-generated images of themselves (or, indeed, someone else) wearing them. Later, instead of watching a video screen, people will probably be able to don a virtual reality helmet and gloves, then transport themselves into the stores of their choice. They will roam the virtual aisles, examining virtual goods and quizzing virtual sales assistants for more information if required.

(Tomkins, 1994)

In fact, teleshopping – while growing – is still a small industry, covering just 0.2 per cent of all retailing sales in the USA (Tomkins, 1994). The main teleshopping at the moment – such as the QVC cable channel in America and Europe – is technologically crude, offering constant product demonstrations, 'infomercials' and advertising along with a free telephone number for orders.

Telematics have been far more important so far in underpinning new ways of organising and managing physical flows of goods and physical retailing spaces in city centres and out-of-town malls than it has for totally substituting these with 'virtual' types of shopping based totally on electronic mediation. The aim here is to match supply as closely as possible with demand. This is done by developing more flexible forms of retailing, with larger, centralised stores, just-in-time delivery of stocks, the intense surveillance of individual consumers (see Chapter 5), and highly responsive systems for moving as near as possible to real-time coordination with fast-changing customer preferences. Telematics-based systems are being used to capture the flows of goods in real time, creating a picture of the products sold and the socioeconomic profile of customers. In Sainsbury's, a major UK supermarket, for example, 'the use of scanning technology at the checkout in 95 per cent of Sainsbury's stores, together with in-store inventory checks using hand-held data terminals, provides information which is used to calculate the replenishment requirements of each store on a daily basis' (Ducatel, 1990). This is linked into wider telematics-based systems of logistics management, and electronic fund transfer with suppliers, creating a kind of 'extended conveyor belt' between production, distribution and consumption that is integrated via telematics (Ducatel, 1990).

Full teleshopping, where consumers shop from home from displays about products, place their orders via telematics systems, and arrange for electronic forms of payment, seems unlikely to match the utopian predictions. The prospects for tele-based retailing are much more problematic than the prospects for tele-mediated banking services. Its main focus is likely to be on several specialised markets, such as for the housebound or the potentially lucrative markets of the busy corporate professional or technology enthusiast. But while the Internet currently only supports catalogue-like displays of specialised goods such as books or music, many firms see it as a foundation for full teleshopping based on electronic money flows and interactive links into people's homes. Microsoft, for example, are investing heavily in systems that support such transactions on a 24-hour-a-day basis, linked into interactive television (see Chapter 5).

However, in the short term, such teleshopping seems likely to complement rather than replace the vast majority of face-to-face shopping in malls and city centres. This is because shopping has emerged as a key leisure activity, where the *whole point* is to leave the confines of the home to explore physically new consumer spaces in cities (see Chapter 5). As Judy Hillman argues 'the telestore may be too remote, except for the housebound, the isolated and the hyperactive. For most people, shopping is likely to remain a social, visual, tactile, stimulating, if sometimes exhausting, acquisitive experience, which brings particular pleasure in the unexpected bargain or encounter' (Hillman, 1993; 2). Even in France, where 6 million freely distributed Minitel videotex terminals now hook up 30 per cent of all households, only a small proportion of the 20,000 consumer services available are teleshopping based.

MANUFACTURING, INNOVATION AND NEW INDUSTRIAL SPACES

A combination of the deindustrialisation of western economies, a shift of manufacturing capacity to rural areas, less developed and newly industrialising countries and in-situ automation has a meant a continuing collapse of manufacturing employment in western cities. Currently standing at around 20–35 per cent in most western nations, some have predicted that manufacturing employment will fall to only 5–10 per cent of employment by the early 21st century (Hall, 1987).

TELEMATICS AND MANUFACTURING

While manufacturing is therefore of declining importance as a source of jobs, this does not mean that it can be ignored as an important factor shaping urban economies. For the urban or local economies that manage to attain nodal status as manufacturing and research and development locations on the corporate networks of TNCs, the rewards can be substantial (see the Hewlett and Packard example, Figure 4.5). It is in this sense, where corporate telematics networks are used to integrate manufacturing and research and development functions across many local economies, that telematics are most important in manufacturing.

At the plant level, however, telematics are the basis for other key innovations in changing the organisation of production processes. Local Area Networks (LANs) in factories are being used to link automated or robotic manufacturing machinery to try and integrate together the elements of manufacturing processes (through Computer Integrated Manufacturing (CIM) technologies). Here, the key is to use telematics to direct the production process (Hepworth, 1989; 129). Ideally, CIM leads, ultimately, to higher levels of automation and a fully integrated, extremely flexible, telematics-based production system. To quote Antonelli, 'telematics can particularly affect the coordination costs of all those processes that require real-time monitoring and feedback. In this way, telematics can completely change the reorganisation of management or production; the design of logistics plans; the reorganization of production cycles among factories and firms; the management of finance – that is to say in all the fields where real-time control allows a further degree of automation' (Antonelli, 1988).

In reality, though, the many problems involved in implementing CIM mean that it has tended to be limited thus far to 'islands of automation'. Nonetheless, CIM helps to support new degrees of control and efficiency because it allows the continuous reprogramming of production, real-time troubleshooting, reduced waste, and design and production to be tailored to the increasingly diverse segments of the marketplace because of the 'economies of scope' involved with reprogrammable machinery (Malecki, 1991).

The application of similar principles between plants, either within or between firms, is also having major implications. Prime amongst these are changing patterns of organising the supply chain, with the growing use of telematics to underpin just-in-time (JIT) systems right through from raw materials to final consumption (see Chapter 7). Evidence suggests that, because the infrastructure demands of JIT and flexible manufacturing are great, the transport and telecommunications advantages of cities means that they are reasserting themselves as dominant centres of innovation. Along with the existing advantages that dynamic urban areas have for information gathering and an 'innovative milieu' and high-quality university institutions, it seems that 'the essence of economic development, the process of technological innovation, is being centralized in urban agglomerations, and in the wealthy countries of the world, to a greater extent than we have seen for some time' (Malecki, 1991; 255).

NEW INDUSTRIAL SPACES

In some privileged cities and regions, new processes of innovation and manufacturing growth are having an important impact on the urban economy. As part of the broader shift away from Fordism and towards more flexible forms of industrial organisation, self-sustaining innovation and concentration have created a range of new industrial spaces, linked into global markets (Antonelli, 1988). The key sectors involved include electronics and telematics, biotechnology, aerospace, nuclear technology, medical technologies, environmental technologies and space. Well-publicised examples of such 'localised production complexes' (Scott, 1988) include Silicon Valley and Route 128 in the United States, Toulouse in France, certain small towns in Tuscany (Amin and Thrift, 1992), and Bäden Württemberg in Germany.

Telecommunications and telematics play a paradoxical role here (Robins and Gillespie, 1992). They tend not to be used to link innovators together because the ongoing innovation requires intense face-to-face contact and ongoing trust-based relationships. Rather, the growth of telematics actually 'encourage the appearance of new specialised production activities, which themselves then frequently cluster together in geographical space' (Storper and Scott, 1989; 26).

Within these spaces, continuous intellectual and knowledge inputs are far more important than in previous eras of production (Florida and Kenney, 1993). This makes research and development an ongoing activity, as short product cycles call for constant improvements to products. It also means that links with academic research institutes and universities tend to be very important, as are their (generally) suburban environments and good global transport and telecommunications infrastructure.

THE 'TECHNOPOLE' PHENOMENON

To Castells and Hall, the 'campus-like atmosphere' of these so-called 'technopoles' is now to be found 'on the periphery of virtually every dynamic urban area in the world' (1994; 1). Many city and regional development agencies have attempted to stimulate the 'take off' of such industrial complexes within their areas through a wide range of 'technopolis' initiatives (Malecki, 1991; Gibson et al., 1993; see Chapter 9). But the importance and effects of high-tech innovation is contested. To some, technopolises, linked via telecommunications, represent the emergence of a new urban era of the decentralised 'smart city', heralding the onset of the demise of the large industrial metropolis. Naisbitt and Aburdene, for example, see 'a new electronic heartland of linked small towns and cities as laying the groundwork for the decline of cities . . . in many ways, if cities did not exist, it now would not be necessary to invent them . . . truly global cities will not be the largest, they will be the smartest' (Naisbitt and Aburdene, 1991; 329).

Such districts rely on face-to-face and informal links between key entrepreneurs and innovators, and the wider 'innovative milieu' within which continuous innovation is able to flourish (Malecki, 1991). Because of the need for collaborating firms and entrepreneurs to engage in 'permanent innovation' based on intense trust, 'spatial proximity between customers and suppliers may be of increasing value' (Cooke and Morgan, 1991; 22). Nevertheless, many of these new industrial spaces are not only intensely information rich, they are also becoming highly dependent on advanced telematics infrastructures and services (Hall, 1992; Castells and Hall, 1994). The importance of telecommunications and telematics within these new manufacturing spaces, and, indeed, within the broader modernisation of manufacturing should not, therefore, be overlooked.

BUSINESS SUPPORT AND RESEARCH AND DEVELOPMENT NETWORKS

Despite apparent problems in trying to mediate innovation via telecommunications, many attempts are being made by a wide range of public development agencies to use telematics to stimulate innovation. This centres on two main areas. First, there are many initiatives aimed at supporting firms within industrial districts or local economies through the use of telematics to access business support services remotely and 'on line'. The assumption amongst policy-makers here is that 'telematics devices allow real time transfer of complex information and, without limits of space, tend to transform communications and decision-making processes', so improving the competitiveness amongst small firms (Rullani and Zanfei, 1988; 62). In Prato, Italy, for example, a telematics system called SPRINT was established in the early 1980s as an on-line link between local authorities, trade associations, firms and chambers of commerce, as an electronic space within which information could be exchanged between them. This, however, failed because the traditional cooperation of local entrepreneurs was already embedded into the local social structure and because of its 'premature appearance in an area without a developed telematics culture' (Capello and Gillespie, 1993; 47). The lesson was that 'only a few of the [information flows in the industrial district] can in reality be transported on telematic support, replacing interpersonal relationships' (Capello, 1989; 19).

Probably more important is a second trend towards the development of high-capacity telematics networks supporting collaborative research and development across national and international boundaries (Williams and Brackenridge, 1990). While, as we have seen, the global research and development networks of the largest TNCs are already extremely sophisticated (Miller, 1994), a range of initiatives are now attempting to link smaller research and development centres within small firms and universities as a tool of development policy. This allows distributed access to the most sophisticated computers and the transfer of high quality data, image and video signals between researchers and entrepreneurs. It also supports more rudimentary services like electronic mail, bulletin boards and news to encourage the sense of a shared social space for those accessing the networks. The first of these networks – such as BITNET and NSFNET – developed in the USA linking major research universities and military research establishments. These, in fact, were the seeds of the Internet, which itself is a key global support network for academic and research and development

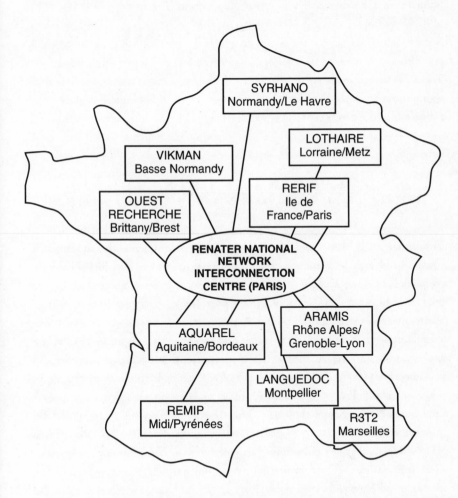

Figure 4.7 Renater: an example of a national telematics network aimed at supporting collaborative research and development

Source: Parfait, 1994

collaboration. More recently, national policy efforts to encourage innovation have led to much more capable networks such as the French Renater system (Academic Network for Universities and Research Laboratories), which claims to have the highest capacity (at 34 million bits of information per second) of any network in the world (see Figure 4.7). Renater links 280 major research centres, technopoles and major cities in France, as well as providing gateways to the international networks like the Internet (Parfait, 1994).

It is difficult at this stage to evaluate the success of such projects, or their influence on the urban economies through which such networks pass. As with the SPRINT example, there is a constant risk that sophisticated networks such as Renater fail to attract the use expected and under-estimate the degree to which it is difficult to mediate innovation and research and development via telematics. The constant risk is therefore that network projects 'might find themselves with a grandiose infrastructure which is not used, the technological equivalent of cathedrals in the desert' (Morgan, 1992; 21).

CONCLUSIONS: AN ELECTRONIC REQUIEM FOR URBAN ECONOMIES?

This chapter has developed a sweeping perspective tracing the dynamics of economic change in contemporary western cities. It has also highlighted the complex and multifaceted roles being played by telecommunications and telematics networks and services within these fast-moving processes of change. We have seen how, in all the economic sectors that dominate urban economies, telematics are closely involved in wider and extremely rapid processes of spatial and organisational restructuring. It would appear that telematics are currently providing broad support to a new round of global urban economic restructuring which is leading directly to a latest set of crises in urban economies. These seem fundamentally to change the relationships between economic activity and the location within which it takes place. Telecommunications are providing a new 'glue' which integrates cities into the wider 'networked economy' of instantaneous information, service, capital and labour flows which increasingly shape their fortunes (Gillespie and Hepworth, 1988; 113). They herald a new more flexible, volatile and unpredictable set of economic dynamics which raise critical questions for the economic future of western cities. Two questions, in particular, can be highlighted. First, can city economies be sustained against the shift towards tele-mediated decentralisation, business re-engineering and globalisation? Second, what are the implications for urban economies of the search for telematics-based flexibility in business organisation and the apparent logic of polarisation that is embedded in this process?

AN 'ELECTRONIC REQUIEM FOR URBAN ECONOMIES'?

It is clear that the use of telematics to change the organisational logic of previously urban services is fundamentally shifting the patterns and processes of urban economic development in many sectors. There are some signs that the functions that actually require face-to-face contact, chance meetings and the 'soft' cultural, environmental and social advantages offered by the centres of large cities are either declining or – at least – shedding staff through telematics. Back offices are shifting to take advantage of lower cost cities and locations with reduced rates of staff turnover – often the deindustralised cities with high rates of unemployment that were hit hardest by the collapse of manufacturing. In addition, many of the sophisticated information services previously monopolised by large cities are now available on-line in all cities and many non-urban locations. A new set of electronic relations between cities and their hinterlands is being constructed, founded on telematics-based flows of work and services, increasingly to meet the global dynamics of TNCs. A report by the European Commission argues that new forms of 'telepresence' are widely being employed to replace or complement the 'physical presence' that was necessary before what they term the 'new telecommunications paradigm' (CEC, 1992a) Mark Hepworth talks of the emergence of an 'electronic hinterland' around the great cities with 'people travelling to work by "wire" and firms doing business by "wire"' (Hepworth, 1992; 36). These trends, as we have seen, are increasingly global in nature.

The shifts towards back-office decentralisation, telemediation, tele-based outsourcing and flexible teleworking amongst corporate employees have led some commentators to declare an 'electronic requiem' for urban economies (Kellner, 1989). The routine service employment that offered hope for resuscitating urban economies in the wake of the collapse of manufacturing seems increasingly under threat. Peter Kellner, for example, argues that:

Many organisations ... find that modern technology makes it possible to separate policy work (which often still needs to be done in a major city, but which employs few people, from administrative clerical work (which can be done almost anywhere and employs many people). This is opening the way to some remarkable costs savings. . . . It used to be the case that goods and services could be provided more effectively if the people who supplied them worked closely together. Cities offered their residents a bargain: we shall provide the work and the wealth if you put up with the dirt and the danger . . . Suddenly the centuries-old rationale for cities is vanishing. Residents depart, their retreat being sounded not by trumpets but by the warble of electronic pulses.

(Kellner, 1989)

As high rates of structural unemployment continue in western cities – with attendant social and political crises – these processes of change raise major concern about the future of urban economies. How sustainable is an urban economy when it has been so fragmented, exposed to global telematics-based networks, and pitched into the global battle for investment? Is there an employment future in telematics-based services for the western cities so devastated by the continuing collapse of manufacturing employment?

Despite the efforts of many from the telecommunications and corporate worlds to suggest that the urban economic future is rosy, it is hard not to conclude that the answers to these questions – from the current perspective at least – look rather bleak. The current restructuring processes are driven by the imperative of reducing costs, employing fewer permanent staff, replacing labour with capital, increasing automation, and thus supporting greater responsiveness, flexibility and competitiveness. Much heralded innovations such as 'back offices' and 'tele-working' – while bringing undoubted benefits to those cities and groups that gain access to new employment through them – often only serve to redirect the jobs that remain following restructuring as they are shifted to cheaper locations or cheaper and less secure employees. Only rarely do they create genuinely new employment of high quality. As Bannister argues, with teleworking innovations, 'although often dressed up as altruistic, the motives for the company are usually self-interest' (Bannister, 1994).

But these processes, although worrying, represent the restructuring rather than the dissolution of cities. They are also very uneven. Larger cities – particularly the global command centres – are maintaining their central roles as service, control and command centres. They remain the economic powerhouses of western society. Centralisation is occurring here as well as decentralisation. In particular, headquarter and control functions seem to be centralising further on to the élite group of global command centres and other world cities, while at the same time reorganising so that they employ fewer people in these places. Michael

Parkinson (1994) argues that 'major metropolitan areas, situated at the centre of communications networks and offering easy access to national and international institutions, the arts, cultural and media industries, are, if anything, becoming more attractive to international finance houses, corporate headquarters and producer service companies'.

So continued urbanisation is occurring as well as a global scattering of previously urban functions. These processes are occurring at different levels, leading to a complex picture of economic change. For example, the decentralisation of routine service functions from some larger cities is leading to new processes of urbanisation in other, more peripheral cities. Many cities are also strengthening their roles as centres for consumption, leisure and tourism services both for their regional and national hinterlands, and as nodes within the booming markets for global tourism.

The main problem seems to be that telematics are associated with a reduction in the capability of many advanced industrial cities to deliver employment of the quality or quantity required to sustain their socioeconomic fabric. While this applies to all cities, it is especially the case in the old manufacturing centres which have a weak service base such as Detroit and Liverpool. Here, talk of an 'electronic requiem' for urban economies may, in some extreme cases, be almost justified.

'RE-ENGINEERING', THE COSTS OF FLEXIBILITY AND THE LOGIC OF POLARISATION

The pervasive shift towards the 're-engineering' of businesses based on telematics is changing the very nature of firms. As Bob Kuttner argues, these processes can mean that 'in this conception, the company is no longer a physical entity with a stable mission or location but a shifting set of temporary relations connected by computer network, phone or fax. Only a small set of corporate managers have permanent jobs' (Kuttner, 1993; 2).

Increased competitiveness in many of these 'information' sectors often comes at the expense of closed operations, automation, radical reductions in managerial staff, less secure working conditions and a geographical shifting round of the remaining jobs to lower cost locations. These are then intimately linked together via corporate telematics networks. Increasingly, also, such locations are non-metropolitan, peripheral and even in less developed countries. Further advances in computing and telematics, global movements towards liberalisation in service markets, and the continued drive to strengthening competitiveness seem set to make such re-engineering a virtually continuous process.

Even the proponents of such re-engineering stress that it represents the obliteration of large parts of existing business and workforce structures rather than simply the automation of existing business processes (Hammer, 1990; 104). A recent survey showed that 69 per cent of all US firms and 75 per cent of all European firms were currently engaging in some form of 'business process re-engineering'; on average, these projects resulted directly in job losses of 336 in North America and 760 in Europe (Caulkin, 1994b). In their efforts to reduce the core of stable jobs, downsize and sub-contract the largest 500 companies in the United States lost 4.4 million jobs between 1980 and 1993 (Davis, 1993). City economies and urban economic policy-makers are faced with the paradoxical situation of needing to re-engineer themselves in order to compete, while being fearful of the social consequences of the processes they encourage. This paradox of greater telematics-based economic competitiveness leading to worsening social conditions in cities is captured perfectly by Bob Kuttner:

When one company becomes more competitive by doing more with less, the result is greater productivity. But when the entire economy competes in this way, the income lost to displaced workers may outweigh the gains in productivity. Unless we devise complementary strategies to yield jobs for the displaced, the 'competitiveness' craze is likely to worsen the problem.

(Kuttner, 1993)

In this context, Joseph Pelton argues that 'it is clear that both unemployment and underemployment due to technological advances are likely to be very big problems in the future. . . . The key element in charting corporate and economic activity . . . is that no important trend can or will be "local" . . . technological unemployment and the 168 hour working week will be experienced around the world' (Pelton, 1992; 11). The focus of this process are the service industries that now dominate city economies. Many of the growing service sectors that offered hope to deindustralised cities during the 1980s – insurance, finance, legal services, media, government and education – are currently using telematics to cut jobs and restructure their organisations in ways which seem to erode – or at least reorientate – their dependence on metropolitan locations. Such sectors are now shedding hundreds of thousands of jobs as they re-engineer themselves to take advantage of telematics-based operations on an increasingly global basis. In London, one of the world information capitals, for example, all of the jobs created in these sectors during the 1980s boom that Moss discusses have now been shed (Caulkin, 1994a). Thomas Hirschl argues that these shifts represent a shift away from 'post-industrial' to 'post-service' economies, as telematics are used to replace service labour across a whole range of sectors (Davis, 1993).

Ironically, such business agendas are often being supported by local, national and supranational economic policies, as apparent solutions to current economic crises and new urban economic futures are sought. Teleworking, homeworking, sub-contracting services to distant companies, the need for a flexible work force operating increasingly over telematics networks has, for example, been placed at the top of the economic agenda in the European Union (Johnston, 1993). Telematics-based working is seen to be a solution to 'inefficient' labour markets which cut off the urban unemployed from sources of new jobs. The hope is to move jobs from the areas of creation to the areas of 'surplus labour' – i.e. the inner cities. Such initiatives represent positive readings of the potential for telematics-based revitalisation of urban economies. To the European Commission (1992), for example, telecommuting heralds a 'new perspective toward full employment' which can 'create or expand job opportunities for persons with disabilities, elderly people, and residents of rural areas' and 'enable companies to reach untapped pools of highly qualified workers' (CEC, 1992b).

But, in reality, these processes seem to be driving a fundamental logic of polarisation between privileged 'cores' of particular social groups and geo-graphical areas and the remaining 'peripheries'. This logic of polarisation has two inter-related sides – social and spatial. First, there is a process of social polarisation. On the one hand, there is a diminishing group of élite corporate decision-makers – the full-time, secure and permanent staff at the helm of the global command centres and the TNCs. On the other, there are the wider groups of back-office workers, sub-contracted and outsourced workers, the casualised employees in the low-order retailing and leisure services, and the widening groups of unemployed and underemployed. Robert Warren argues that 'telecommunica-tions and information processing are central to this spatial restructuring and the resulting reduction in jobs in high-waged unionized sectors and growth of low-paid, largely non-unionized jobs in personal and consumers services, manufactur-ing and finance and business services' (Warren, 1989; 342).

In the immediate future, as consumption becomes more dominant in shaping urban economies, the bulk of genuinely new jobs in most urban economies looks set to be in the lower order consumer services – in such sectors as retailing, leisure, cleaning and catering. These are generated by the burgeoning growth of business and consumer tourism and are also necessary to support the presence of multinational corporations, government, administration and education within cities. The bulk of the urban labour force is left to 'compete for dead-end jobs as fast food workers, amusement park employees, janitors, [and] security guards' (Davis, 1993).

These trends prompt Sally Lerner to argue that the United States, at least, is polarising 'into an increasingly poor, "redundant", deskilled underclass and a small affluent technical–professional elite' (Lerner, 1994; 187). Hirschl notes 'increases in the section of the economically marginalized population obtaining poverty or near-poverty incomes', but also a growth of a 'destitute, economically inactive population' (Hirschl, 1993). These shifts, he argues, seem to be generating a qualitatively new form of poverty. People caught by this poverty may 'ultimately form a new social class' (ibid.). Related to this, is a parallel polarisation in access to basic consumer services such as banking and retailing, as electronic and physical reorganisation takes place to adjust to these changing markets, as shown by the First Direct tele-banking case. This is leading to the physical withdrawal of services from marginal areas and a concentration on 'cherry picking' from affluent and lucrative markets through both physical and electronic means (Christopherson, 1992).

In a context where many western inner cities already suffer the multiple disadvantages deriving from high localised unemployment rates, the effects of these labour market changes threaten further to polarise the social fabric of cities. The corporate élites who remain in permanent employment seem likely to become more and more detached, both socially and spatially, from the casualised and flexible work forces and the structural unemployment that surround them (see Chapter 5). The signs of this polarisation and fragmentation are easy to spot in most western cities. Even the corporate élites working at the hearts of 're-engineered' network corporations in the world information capitals face every day the social fallout from the deindustrialisation and restructuring of their cities. The danger is that these processes will further intensify as new leaps in the technological capabilities of telematics combine with stronger global shifts towards a market-based, liberalised and corporate-oriented economy.

Second, and related to this, processes of spatial polarisation are underway. It is clear that those cities who are still able to benefit from corporate centralisation and the growth of new industrial spaces are widening their advantages over the remainder – the old industrial and peripheral cities – who are increasingly locked into a war with less developed countries and newly industrialising countries over the spoils of decentralisation. The related shift towards market-based and globalised telecommunications regimes also means much greater unevenness in

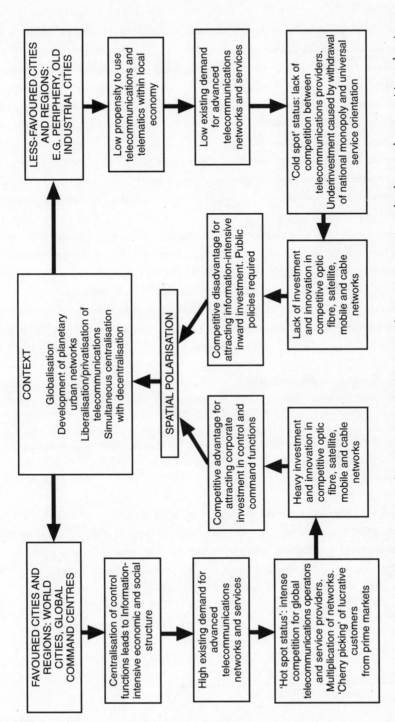

Figure 4.8 Vicious circles and virtuous circles: how telecommunications are contributing to more uneven development between cities and regions

the development of advanced telecommunications infrastructures. These in turn go on to influence the development prospects of cities and regions at different points in what Mitchell Moss calls the global 'informational' urban hierarchy (Moss, 1987).

As Figure 4.8 demonstrates, these processes can be considered as parallel 'virtuous circles' – where advantaged core cities and high-technology zones increase their attractiveness – and 'vicious circles' – where the disadvantages of peripheral and deindustrialised cities are magnified by their lack of competitive telecommunications infrastructures. As Robins and Gillespie argue, the clear danger here is that 'the new global order will be marked by a new segmentation of "on-line" and "off-line" territories, as it were; by a new hierarchy that differentiates those localities that can harness territorial endowments to the network structure and those characterised by both internal fragmentation and external disarticulation from the network' (Robins and Gillespie, 1992; 161).

For these reasons, we would argue that the issues thrown up by these global changes are moral and ethical rather than technical or economic in a narrow econometric sense. Many dilemmas arise here. Bannister, for example, suggests that:

The movement [to global back-office networks] poses a threat and a moral dilemma. The threat is to existing jobs [in the advanced industrial world]. Companies faced with falling telecom costs and increasing dependence on computer-based information systems are being encouraged to separate out routine processing functions. It matters little whether the computers, or people, carrying out these tasks are in Manchester or Bombay; what matters is the cost. Handling information can amount to 25% of a company's costs, so the savings are potentially enormous.

(Bannister, 1994)

These telematics-mediated changes in urban economies are clearly a vitally important force driving the complex range of social, cultural, infrastructural, environmental, physical and governmental changes that are underway in contemporary urban life. It is these that we turn to in the next chapters, starting with the social and cultural dynamics of cities.

5

THE SOCIAL AND CULTURAL
LIFE OF THE CITY

Any metropolis can be thought of as a huge engine of communication, a device to enlarge the range and reduce the cost of individual and social choices

(Deutsch, 1966; 386)

All information in all spaces at all times. The impossible ideal. But the marriage of computers with existing communications links will take us far closer to that goal than we have ever been

(Godfrey and Parhills, 1979; 10)

INTRODUCTION

Modern culture is, as Thompson puts it, profoundly 'mediatized' (Thompson, 1990). Technical media support a large and growing proportion of modern cultural flows and social interactions. Current innovations in telecommunications and telematics support new types of rapid social interaction or cultural transmission across previously impossible distances. This extends long-established processes whereby the 'spaces' of social and cultural interaction have separated from the particularities of social and geographical 'place' (Giddens, 1990). He argues that these technologies act as 'disembedding mechanisms' which 'lift our social activity from localised contexts, reorganizing social relations across large time and space distances' (Giddens, 1990; 53).

In these cases, physical propinquity becomes substituted by the mediating effects of technological networks for linking people across time and space (Meyrowitz, 1985). A widening range of telecommunications technologies and services are diffusing into the domestic sphere which can support this process in a variety of ways (see Figures 1.3 and 5.1). Basic person-to-person telephone networks – the 'Plain Old Telephone Service' – are well established as by far the most important social telecommunications networks – having near universal penetration in most western nations. Passive television and radio are also near universal household technologies supporting mass communications. Basic wired phones are now being complemented by a range of new telecommunications systems (see Figure 5.1). In many western cities broadband cable networks reach over half of the population; Direct Broadcasting by Satellite (DBS) beams international mixtures of channels to many people who are not on cable (Figure 5.1). Video recorders are becoming almost universal in some countries. Mobile phones and notebook computers are diffusing rapidly. Sophisticated personal computers (PCs) for the home are emerging as mass market consumer durables. With the explosive recent growth of domestic connection to the Internet, Batty and Barr (1994; 705) even argue that, 'ultimately there is no reason why every computer in the world should not be connected to the Internet directly or indirectly'. Finally, there is much speculation about the convergence of digital telecommunications, media and computing technologies to support multi-media and interactive connections to the home through PCs or interactive television. This feeds into the widespread current debates about the potential social and cultural effects of 'cyberspace' and 'virtual reality' within advanced industrial cities.

These developments are important not simply because such new technical media can establish different patterns of social and cultural life but because they can also act 'as a *potential reorganization of social relations themselves*' (Thompson, 1990; 217. Emphasis added). The potential exists for social and cultural identity and social experience and interaction to become reconstituted with more freedom from time–space constraints within the theoretically unlimited domains of electronic spaces. Already, communications technologies, particularly television, 'have brought individuals and families into the presence of places and events that were previously distant or unknown, enabling them to identify with dispersed yet knowable communities and to imagine themselves as embedded in regional, national and even transnational collectivities' (Moores,1993; 623).

Murdock (1993; 535) discusses some of the social and cultural relations that are emerging based on these new technologies. Telephones help to support the maintenance of intimate links irrespective of distance; television 'rituals' such as major sporting occasions support new social events and gatherings; computer games support new sorts of social isolation and withdrawal; and such networks as electronic mail and bulletin boards generate social meetings in 'virtual space' between people who never actually meet in 'physical space'.

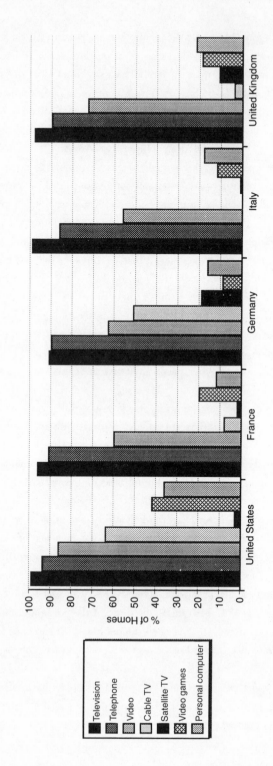

Figure 5.1 Penetration rates of consumer telecommunications into the home in various advanced capitalist nations

Source: Bannister, 1994

The proliferation of truly interactive social communications networks means that, in some cases, social groups are now establishing closer relations through electronic networks than through physical proximity (Storgaard and Jensen, 1991). To adapt Anderson's (1983) terminology, new technical media may be used to establish and maintain 'imagined communities' of widely separated people, based on some common interest of characteristic. As with the economic forces in cities, therefore, it becomes impossible to understand cities socially and culturally without looking at their different social and cultural groups and the various specialised networks through which they establish their relations – often on wider and wider scales. As Calhoun has argued, 'large cities are at once dissolving into a plethora of local communities and being absorbed into a larger-scale and more tightly-knit web of indirect relationships' (Calhoun, 1986; 329). This view, however, must not overlook those groups who are excluded from access to these network technologies or the information that flows over them and who are therefore prevented from entering into these 'indirect' telecommunications-based relationships. The many possibilities that these processes seem to bring for reorganising and individualising social and cultural life, at least for those in control of the technology, are well summarised by Storgaard and Jensen:

The individual can dissociate him/herself from others, and along with this, by applying information technology, he/she can obliterate practically any restraint. By one's lifestyle, one will dissociate oneself from strangers at the very same time as emphasising one's affiliation with a certain group, location, or local community. On the other hand, by applying information technology one will attempt to obliterate the barriers of the given group or location. Thus, it becomes easier to pick and choose one's future associates instead of being tied up with those individuals one just happened to end up with. That is: it will be easier to associate with those sharing one's interests instead of those sharing one's location.

(Storgaard and Jensen, 1991; 136)

But what do these processes actually mean for the social and cultural life of cities? Do these technologies radically change the nature of social interactions and social inequalities within cities? Are we really, as Godfrey and Parhill (1979) argue, approaching truly universal access to all information via telematics – irrespective of any locational, temporal barriers or social inequalities? Does this portend the end of the city as the social container and cultural dynamo of modern western civilisation? Will urban life withdraw into electronic spaces? Or – to carry on the theme of this book – are things more complex, subtle and even contradictory?

This chapter explores the complex relationships between modern telecommunications and the social and cultural life of advanced capitalist cities. Instead of trying to summarise this vast subject, it cuts across five key areas of these relations. These we label: post-modernism and urban culture; social inequality and polarisation; the household as a locus of urban social life; the development of telematics-based social surveillance; and the emergence of 'virtual' communities based on telematics networks.

POST-MODERNISM, TELECOMMUNICATIONS AND URBAN CULTURE

There is little doubt that rapid advances in telecommunications and telematics technologies are currently supporting shifts towards a truly globalised post-modern urban culture. The development of these media in the century between the late nineteenth and late twentieth centuries has been a central component of the shift towards modernity in society – towards a forward-looking, technological and urbanised culture that allowed forms of temporal and spatial experience which were impossible in 'pre-modern' times. With modernism, technological development was equated with progress; there was confidence that single conceptual frameworks could explain all social phenomena; and planned action was seen as the route towards the emancipation of all social groups. As Phil Cooke argues, 'modernity is . . . a mode of spatial and temporal experience which promises adventure and self-transformation while threatening to destroy the familiar. It bisects geographic, ethnic, class, religious and ideological boundaries' (Cooke, 1988). Modernism relied fundamentally on telecommunications systems – primarily the telephone and television – to support what Graham Murdock calls its 'characteristic separation of space and place, together with the displacement of locations by networks and the creation of new forms of social relations combining intimacy with distance' (Murdock, 1993; 536).

Debates currently rage on the degree to which this modern, urban culture is genuinely being transcended by some new form of 'post-modern urban culture' (Savage and Warde, 1993). The use of the word 'post-modernism' is extremely loose and contested, tied in as it is with alleged sea changes in philosophy, literature, media studies, aesthetics, architecture and conceptions of identity and the self. Most interpretations of post-modernist urbanism, however, stress several key attributes: it rejects or questions the notions of modernism derived since the enlightenment; it lambasts the failings of modernist planning and architecture; it is sceptical of scientific and technological knowledge and especially social 'meta-narratives'; it underlines the diversity of cultures and social processes in western cities; and it stresses the increasingly 'reflexive' consideration by individuals about their cultural identity and the places they live (Savage and Warde, 1993; 138; Madani-Pour, 1995). Also common is a stress on new urban aesthetics based on the consumption of highly symbolic goods through which people help to construct their identities. Finally, the celebration of cities as centres of consumption and the 'spectacle' of festivals and events are also widely considered as key elements of post-modern urbanism.

TELECOMMUNICATIONS AND CITIES IN AN INTEGRATED GLOBAL CULTURE

Post-modernism is closely tied in with what David Harvey (1989) calls 'time–space compression' – the collapse of spatial and temporal boundaries. The girdling of the globe with massive arrays of near-instantaneous telecommunications and fast transport networks now means that it is possible to see the world as an integrated global cultural system – as 'one place' (although obviously not a homogeneous one). This incorporates flows of people, money, capital, goods, information and symbols within which economic flows have an important place. Mike Featherstone argues that 'the intensity and rapidity of today's global cultural flows have contributed to the sense that the world is a singular place which entails the proliferation of new cultural forms for encounters' (Featherstone, 1990; 9). This is the culmination of a process that has been underway since the beginning of modernity and since the diffusion of the first telegraphy, telephone, radio, newspaper and television systems.

As we saw in Chapter 1, a major motor of time–space compression is a frenzied process of alliance formation between computer, media and telecommunications companies, as attempts are made to offer the goods and services made possible by technological convergence around digital telematics (Morley and Robins, 1990). To Buckingham *et al.*, for example,

new digital technology is the mechanism driving this global business shakeup. It makes mergers between companies in the computer, entertainment and telecommunications sectors far more rational. And you need the financial muscle which can come from mergers and takeovers if you are to undertake the massive investment needed to establish global super information highways.

(Buckingham *et al.*, 1993)

The result is that 'a business revolution is clearly under way, as fast changing technology offers enormous new markets across the world, and fresh ways of building on their broadcasting base, international media and telecommunications organisations are becoming aware of undreamt of opportunities . . . and terrified at the risks of falling behind' (Buckingham *et al.*, 1993).

But technological changes and digitalisation support rather than directly determine the globalisation of urban culture. Political, regulatory and institutional forces are more important than just technology in opening up national cultures to penetration by these globalising and centralising media conglomerates. This opening up of new markets through liberalisation and privatisation is, however, supported by digital telematics for transcending the confines of space and time and delivering cultural products to these markets. This process is therefore a global business shake-up facilitated by both political and technological shifts. National monopolies are being replaced by an integrated global marketplace for digital media and communications services. This fuels the globalisation of urban culture. Morley and Robins suggest that 'the new media conglomerates are creating a global image space, a space of transmission [that] cuts across the geographies of power, of social life, and of knowledge, which define the space of nationality and culture' (Morley and Robins, 1990; 47).

A good example of this is the way in which Rupert Murdoch's global News International empire is diversifying beyond newspapers and satellite TV into telecommunications, computerised information and entertainment services. Their recent purchase of a stake in the telecommunications operator MCI testifies to their global strategy to position themselves favourably to take account of the synergies now possible between owning the hardware of the telecommunications channels (cable, satellite and long distance telecoms capacity) and the software to distribute over the networks (TV programmes and value added services). In the UK, US cable telephone and cable companies are building the urban cable networks, which offer TV, telecommunications, and, increasingly, value added services. In the USA, the largest merger in corporate history ($60 billion) was attempted recently between Bell Atlantic phone company and Tele Communications Inc – a cable company. Although it failed, dozens of others are in the offing, geared towards the predicted 500 channels of commercial digital and interactive TV that many predict will be delivered to the home in the USA by the end of the century by new broadband cable technologies.

CITIES AS 'PRODUCTS' IN THE GLOBAL IMAGE SPACE

Cultural globalisation and the new relationships between electronic spaces and urban places are also clearly shown in the ways that modern cities are planned and marketed. Increasingly, cities are being designed, packaged and marketed for global consumption by tourists, businesses, media firms and consumers via technical media. Competing in the production of attractive urban imagery is a key element of the new urban entrepreneurialism that we discussed in the introduction. Global media flows are therefore being used to support this competitiveness as well as to justify the physical remaking of city centres along post-modern lines: to improve the 'product' for the global media 'marketplace' through the creation of pedestrian plazas, cosmopolitan cultural facilities, festivals, 'spectacles' and sports and media events.

Cities now go beyond traditional newspaper and magazine-based marketing and scramble to be represented electronically via TV and film productions and place marketing videos. These are often prized more for their symbolic qualities than their direct economic multipliers. Place marketing strategies attempt to lob convivially constructed images of particular cities into what David Harvey calls the global 'image bank', as a direct support to the economic competitiveness of cities on global networks (Harvey, 1988). Images of places and cities therefore join with images of global food, architecture, style, pop culture, and beer in the global media marketplace. A growing range of cities are also setting up hosts and servers on the Internet containing maps, information, photographs and guides to their facilities which real and potential visitors may access. A wide range of partnership bodies in Manchester, including urban development agencies, recently came together to construct 'Virtual Manchester' (http://www.u-net.com/ manchester/) – an electronic analogy of the real city to promote, inform and market their place in the global image space. This city server and many others are even linked up into a specialist World Wide Web network called 'CityNet' through which Internet users can 'visit' a list of hundreds of cities all over the world just by clicking their name (available at http://www.city.net/ on the World Wide Web). Not surprisingly, these are geared to the international travelling élites who dominate use of the Internet.

The proliferation of such electronic urban imagery, however, can lead to a blurring of the reality of the city and its image, and to social conflicts over how a city is marketed. Many actual residents of the city may take issue with the disinfected and idealised urban images often portrayed by the marketing agencies. Such gulfs between the real and the simulated are, once again, a pervasive quality of post-modern urbanism that is encouraged by the instantaneous flows of images across time and space boundaries. To Paul Knox (1993; 17), 'postmodernity, because of its anarchic, impulsive and parodic qualities, makes for particularly fluid and unstable relations between signifier and signified, image and reality'.

LIBERATION OR DISEMPOWERMENT? POSITIVE AND NEGATIVE INTERPRETATIONS OF CHANGE

One result of this new global culture is that the old symbols of the locational rootedness of urban life increasingly seem to be lost or remade, to be replaced by an intangible and accelerating set of global flows of images, data, services, products, people and commodities. But what are the effects of such time–space compression on the social and cultural landscapes of cities and the experience of urban life? How does it impinge on the sense and experience of place of different individuals and groups in cities? How, in short, does the local level of urban places interact with these global forces and globally stretched electronic spaces (Robins, 1989)?

Not surprisingly, interpretations of these complex questions are diverse and highly contested. On the one hand, a wide range of up-beat, optimistic, even utopian commentaries are emerging which stress the positive effects and potential of post-modern urban culture and telematics. Two main strands of work sit uneasily together here. First, there are more positive post-modern commentaries on the changing nature of urban culture. These stress the new recognition of variety, pluralism and diversity in contemporary urban culture, as epitomised by the new emphasis on the ways in which gender, race and sexuality provide the basis for constructing urban identity, experience and place (see Wilson, 1991; Hall and Jacques, 1988). These optimistic views resonate with the work of some critical commentators. Donna Haraway, for example, stresses the positive transformative potential of 'cyborg' technologies – telematics and biotechnologies which are integrated seamlessly into the individual human body. 'A cyborg world', she writes, 'might be about lived social and bodily realities in which people are not afraid of their joint kinship with animals and machines, not afraid of the permanently partial perspectives with contradictory standpoints' (Haraway, 1991; 154). Here, personal telematics systems linked into global networks might allow the construction of new, empowering social worlds that break free from the 'tradition of racist, male-dominated capitalism' (Haraway, 1991; 150). They might help their 'users' to respect difference and 'otherness' and operate free from many traditional social and cultural constraints, particularly those imposed by men on women.

Second, massive hype in current media, press and cultural debates is also emerging which tries to highlight the fundamentally empowering and liberating importance of electronic 'cyberspaces' or the emerging 'information superhighway' based on the Internet. Drawing on simple technological determinism and utopianism, these approaches tend to evangelise and hype up the transformative powers of telematics; these are seen to be virtually inevitable and woven into some technological logic. In particular, they stress the capabilities of telematics for supporting individual, interactive communication on a one-to-one, one-to-many or many-to-many basis across complex electronic webs and networks on a global basis. Through the humble personal computer and modem, it is argued, individuals can now shape, participate in and interact with a vast range of new multimedia spaces rather than merely being passive consumers of a narrow range of TV, newspapers and radio as in the past (Leary, 1994; Negroponte, 1995). The explosion of the Internet makes massive information resources available on all subjects all over the world to be accessed from anyone's desk. This occurs irrespective of their location, disability, gender, race or other characteristics that have led to repression and subjugation in previous eras. Publishing information, through the World Wide Web, for example, is almost as easy as accessing it; no central authority stipulates who can do what, where and why.

Combined with the exponential growth in both the power of computers and the bandwidth capability of telecommunications, these trends mean that the information resources – and therefore powers – available to the individual are also exploding exponentially. In addition, television will soon be interactive rather than passive, as it merges with the personal computer and as digital media services blur with information services. Personally designed media consumption will replace the crude aggregation of narrow broadcasted choice as 'software agents' scour the globe in search of digital information, images and video, briefed to the exact requirements and interests of their individual subject. The widening diffusion of interactive technologies such as e-mail, the World Wide Web, distance learning and virtual reality are, according to Doug Rushkoff, being used to 'enhance the individual . . . , to break down the walls between individuals', so creating a humanistic 'cultural renaissance' of truly historical proportions (Channel 4, 1994; 15). Liberation, freedom and flexibility are the watchwords here. In this view, whole swathes of 'cyberculture' are emerging within which any person can interact with others in an almost infinite variety of electronic spaces; these are less hierarchical, less centralised and more participatory than any media in human history. 'Never before', writes Timothy Leary, 'has the individual been so empowered' (Leary, 1994; viii).

But for each optimistic interpretation or utopian scenario there is a pessimistic, even dystopian, riposte. Many critics stress the disorientating and alienating effects of time–space compression and globalisation, the social polarisation and commercialisation underway and the shift towards intensified surveillance and social control in cities (see Davis, 1990; Christopherson, 1994; Knox, 1993; Olalquiaga, 1992; Jameson, 1984). The image here is of the bewildered individual, passively caught up in a global cacophony of mass and personal communications, images and signs over which there is no control and little comprehension. The explosion of telematics and media is seen increasingly to mediate the interaction between individuals and 'reality'. This produces a disturbing new 'technological culture' which, according to Tim Druckrey, 'complicates and confounds many of our notions of reality. Beneath facades of ownership through consumption and conceptualization, technology subsumes experience. New technologies supplant and disenfranchise precisely when they seem friendly' (Druckrey, 1994; 2–5).

Gregory (1994; 156), following Fred Jameson, argues that these proliferating electronic spaces are 'overwhelmingly a space of fibre optics and video screens, calling up, dissolving, and (re)producing a world of traces in nanoseconds'. In this vein, Lash and Urry (1994; 3) argue that urban subjects are becoming 'overloaded by this bombardment of the signs of the city' through the proliferation of video recorders, multi-channel cable and satellite TV, computer networks and all the other accountrements of world urban culture. Arthur Kroker drawing on recent French theorists such as Foucault, Virilio and Baudrillard, takes this one step further. He talks of the 'possessed individual' as the defining image of post-modern urban culture. Here, the subject is trapped in an increasingly cynical world of 'virtual' aesthetic experiences, drifting aimlessly within the chaos of tele-mediated signs and symbols which together constitute a simulated and hypereal 'world of digital dreams come alive' (Kroker, 1992; 3; see Poster, 1990).

Finally, such critiques also stress that the explosion of information available is not simply synonymous with people being more knowledgeable about 'reality' than in previous eras or more in a position to exert influence over it (see Roszak, 1994). Far from it, in fact. Neil Postman argues that the vast corporations driving today's technological culture are obsessed with providing more and more information without any notion of the problems that this solves. He calls this the 'elevation of information to a metaphysical status: information as both a means and end of human creativity. In Technopoly, we are driven to fill our lives with the quest to "access" information. For what purpose or with what limitations, it is not for us to ask; and we are not accustomed to asking, since the problem is unprecedented' (Postman, 1992; 61). In this vein, Michael Curry (1995; 81) even suggests that, with the growing automation and telemediation of contemporary culture, 'people actually need to learn less and less, but they have access to more and more information'.

But such a polarised and general debate, while interesting and important, is clearly only of limited use because it suggests an either/or future of 'all good' or 'all bad'. As we saw in Chapter 3, the ways in which the uses of telematics are socially constructed, while strongly biased, are likely to have diverse effects between different places, groups and organisations. Positive effects are likely to co-exist with negative ones and there will be complex patterns of winners and losers. This makes all-encompassing generalisations hazardous. In fact, the debate can only continue in this rather simplistic, black–white/utopian–dystopian manner because of the absence of critical empirical research tracking in detail the actual effects of telematics on urban culture, patterns of interaction and the balance between 'empowerment' and 'disempowerment' for the many diverse social and cultural groups that make up cities. Following on from the work of Moyal (1992) on the social uses of the telephone, we would expect great diversity in the social uses of telematics. For example, her research suggests that men are more likely to be interested in the 'instrumental' use of technologies to 'get things done' (such as transactions, obtaining information, etc.) while women are more likely to be interested in the networks as a way of sustaining distant social relationships (through 'intrinsic' communication) (see Silverstone et al., 1992). But without comprehensive research tracing the diverse social construction and uses being made of electronic spaces, set against the broader political economy of telematics, the debate seems likely to remain unhelpfully polarised between such 'visions of heaven and hell' (Harrison, 1995).

TELEMATICS AND THE CULTURAL MEANING OF THE CITY

Whether one stresses the negative or positive aspects of post modern urban culture and telematics, it can be argued that the trends they embody are remaking the social and cultural meaning of the city. A city is now less a physical site for social interaction in public space – as in the modernist vision – and more a fixed place for the intersection of global networks that carry the instant flows of signs and information which currently shape urban social and cultural life. But, once again, we must be careful to avoid the assumption that there is one universal experience of these trends; rather, the meaning of urban places in the context of globalisation is extremely diverse both between places and between the social groups who inhabit them (Massey, 1993).

Box 5.1 presents an extract from the work of Paul Virilio, an influential and dystopian French commentator on the effects of telecommunications on urban culture. To Virilio, the shift to real-time and twenty-four-hour networks of information flow has destroyed the cultural meaning of the city as a place where time and space barriers were overcome, a centre for the creation of special rhythms of social interaction and cultural life. In this vein, Michael Sorkin (1992; xi) believes that 'computers, credit cards, phones, faxes, and other instruments of instant artificial adjacency are rapidly eviscerating the historic politics of propinquity, the very cement of the city'. With an accelerating and broadening range of global cultural flows, the danger is that the ubiquitous feeling emerges, as Olalquiaga (1992; 2) puts it, of 'being in all places while not really being anywhere'.

BOX 5.1 THE OVEREXPOSED CITY

Does a greater metropolis still have a facade? At what moment can the city be said to face us? The popular expression 'to go into the city', which has replaced last century's 'to go to the city', embodies an uncertainty regarding relations of opposites ... as though we were no longer ever in front of the city but always inside it. If the metropolis still occupies a piece of ground, a geographical position, it no longer corresponds to the old division between city and country, nor to the opposition between center and periphery....

On the other hand, with the screen interface (computers, television, teleconferencing) the surface of inscription – until now devoid of depth –

comes into existence as 'distance', as a depth of field of a new representation, a visibility without direct confrontation, without a *face-à-face*, in which the old *vis-à-vis* of streets and avenues is effaced and disappears. Thus, differences between positions blur, resulting in unavoidable fusion and confusion. Deprived of objective limits, the architectonic element begins to drift, to float in an electronic ether devoid of spatial dimensions yet inscribed in the single temporality of an instantaneous diffusion. From this moment on, no one can be considered as separated by physical obstacles or by significant 'time distances'. With the interfacade of monitors and control screens, 'elsewhere' begins here and vice versa....

Solid substance no longer exists; instead, a limitless expanse is revealed in the false perspective of the apparatuses' luminous emission. Constructed space now occurs within an electronic topology, where the framing of the point of view and the scanlines of numerical images give new form to the practice of urban mapping.

In this realm of deceptive appearances, where the populating of transportation and transmission time supplants the populating of space and habitation, inertia revives an old sedentariness (the persistence of urban sites). With the advent of instantaneous communications (satellite, TV, fiber optics, telematics) arrival supplants departure: everything arrives without necessarily having to depart. Only yesterday, metropolitan areas maintained an opposition between an 'intra-mural' population and a population outside the city walls; today, the distinctive oppositions between the city's residents occur only in time: first, long historical time spans which are identified less with the notion of a 'downtown' as a whole than with a few specific monuments; and second, technological time spans which have no relation to a calendar of activities, nor to a collective memory, except to that of the computer. Contributing to the creation of a permanent present – whose intense pace knows no tomorrow, the latter type of time span is destroying the rhythms of a society which has become more and more debased.

(Virilio, 1987; 17–19)

THE BLURRING OF URBAN PLACES AND ELECTRONIC SPACES

This sense of confusion is not just an electronic phenomena – many argue that it also applies to the changing nature of the material and physical landscapes of cities and urban places (Olalquiaga, 1992). These are increasingly being redeveloped and fragmented into diverse 'packaged' or 'themed' and 'simulated' elements

(Knox, 1993; 1). 'Urban regeneration' initiatives are implanting individual post-modern buildings and developments into cities – office blocks, shopping malls, gentrified or new housing estates, heritage and tourist areas (Sorkin, 1992). These tend to be geared to specific social and cultural uses while often excluding others. They are also increasingly 'commodified', meaning that they are engineered as sites for consumption with access restricted strictly according to ability to pay (Zukin, 1991). The best examples here are indoor shopping malls and new post-modern office developments that are often carefully controlled to tailor to consumers and social élites while excluding social 'undesirables' (Boyer, 1993; Christopherson, 1994). Susan Christopherson (1994; 417) notes the emergence of many 'buffered and isolated . . . urban spaces' such as Bunker Hill in Los Angeles, Harbor Place in Baltimore, Battery Park in Manhattan and the Renaissance Centre in Detroit. These, she argues, are 'organized around a set of functions oriented to the business traveller or commercial tenant and intended to separate this "person as a profit centre" from the urban environment'.

At the same time, the old public parks and spaces of cities can be so undermined as to prevent their use for social and cultural interaction. In New York, for example, Christine Boyer notes a trend towards a fragmented and privatised urban landscape where corporate developments, backed by the City authorities, are 'focussing on the provision of luxury spaces within the center of the city' while investment and planning ignores 'the interstitial places' and public parks which have become 'littered with broken glass, trash and abandoned cars, while the employment of maintenance workers has been reduced dramatically' (Boyer, 1993; 115, 117).

While ostensibly public spaces, many of these new packaged and themed developments are actually controlled so that what happens in them and who visits them is extremely predictable and consumption driven. In this 'City as a Theme Park', urban design becomes geared towards the simulation of mythical 'historic' environments, as epitomised by Disneyland itself (Sorkin, 1992). These trends have blurred the distinction between the 'public' and 'private' spaces of cities 'with the multiplying of "private public spaces"; that is, privately owned and managed public spaces offered for public use' (Bianchini, 1989; 5).

While these trends are actually more varied and complex than the usual reliance on American literature suggests, they raise interesting questions about the ways in which electronic spaces may be subject to similar patterns of control, as corporations attempt to 'package' electronic spaces for the purposes of electronic consumption. One (admittedly untypical) example of this is Las Vegas. Here, 20 million visitors a year now spend $4.5 billion in themed hotels, gambling and leisure spaces. An increasing attraction, though, is a range of enormous complexes filled with the latest virtual reality entertainment technologies such as the Virtualand centre. In these, participants explore electronically constructed worlds for pleasure which have many similarities with the themed physical environments within which they are placed (Channel 4, 1994; 5). According to Fred Dewey, such trends represent a blurring of the differences between electronic spaces and urban places, as both succumb to commodification, simulation and packaging for individualised consumption. He argues that:

we are already very much inside a 'virtual' environment, and what's really being impoverished is the world of real experience, and people interacting with each other. . . . What's developing now is an entirely new form of controlled environment. We find malls, theme parks and themed environments. These provide safe, secure environments where people can interact. It looks very much like public life, but in fact really isn't, because the environments are owned and controlled and heavily regulated by, generally, large global corporations. People interact somewhat randomly, but the actual experience is entirely manufactured – all of its terms are defined ahead of time. The experience is very similar to going through virtual reality. While this provides a kind of vitality, at the same time it's based on leaving behind the mess of real urban life. Everyone expects that the Cyberworld is not going to have these kinds of parameters and controls. This is extremely unrealistic.

(Dewey, 1994; quoted in Channel 4, 5–6)

So tensions exist between the public and private nature of both urban place and electronic space. There is a sense of a parallel shift towards a kind of 'universal particular' or 'generic urbanism' (Sorkin, 1992; xiii) of commodified, globalised urban spaces which embraces both physical and electronic elements. Graham Murdock draws the striking parallel between the 'fortress effect' generated by many post-modern buildings, and the development of vast, private 'dataspaces' on corporately controlled networks. He argues that 'here, as in territorial space, a continuous battle is being waged between claims for public access and use, and corporate efforts to extend property rights to wider and wider areas of information and symbolization' (Murdock, 1993; 534).

These trends, we would argue, tend to make it more difficult to 'read' and understand the nature of urban landscapes. Michael Sorkin writes that 'the city has historically mapped social relations with profound clarity, imprinting in its shapes and places vast information about status and power' (Sorkin, 1992; xiii). But the physical and electronic landscapes of the post-modern city increasingly undermine this legibility. It is becoming more and more difficult to 'read' the city as a cultural landscape that reflects the real processes of cultural, social and economic change at work (Dear, 1993). In other words, just as the city is being infused with a multitude of communications devices, it is becoming less effective as a communications device itself. The information necessary to read the city is becoming more hidden, complex and contradictory due to globalisation and the development of invisible telematics-based 'electronic spaces' across all areas of urban life. To Michael Dear, for example, who speaks from the point of view of Los Angeles, 'the phone and the modem have rendered the street irrelevant; social hierarchies, once fixed, have become "despatialized"' (Dear, 1995; 31).

Los Angeles is always quoted as the emblematic example of these trends. As an extreme example of the trends underway, it shows a particularly diverse set of confusing electronic and physical spaces which, according to some commentators, have ceased to relate to each other in any comprehensible way. Edward Soja remarks that Los Angeles is now:

divided into showcases of global village cultures and mimetic American landscapes, all-embracing shopping malls and crafty main streets, corporation-sponsored magic kingdoms, high-technology based experimental prototype communities of tomorrow, attractively packaged places of rest and recreation all cleverly hiding the buzzing workstations and labor processes that keep it together.

(Soja, 1989; 246)

Tension therefore exists here between the physical and electronic spaces of the post-modern city. The new time–space coordinates and fabric of physical urban life can just about be observed. While this may be becoming more difficult due to the effects of post-modern planning and architecture, with their tightly controlled spaces, the physical patterning of land use and the observation of transport system still enable that. But the electronic spaces, as we argued in Chapter 2, are hidden, diverse and complex. Little is actually known about them or the relations they have with urban physical landscapes.

The sheer lack of understanding of the relations between electronic spaces and urban places means that there are clear dangers here of overgeneralising the relationships between telematics, urban landscapes and social processes in cities. Los Angeles and Las Vegas are not typical cities; the experience of individual places within the globalising urban culture is likely to be diverse. It is also clear that, just as the social and cultural construction and experience of places is diverse, so the social and cultural uses made of telecommunications and telematics are also extremely varied (Moyal, 1992). This means it is hazardous to generalise about the ways in which telecommunications relate to places and the social experience of the myriad of social and cultural groups that make up cities. The social construction of both urban landscapes and electronic spaces leads to variety and contingency, even if there are wider political–economic trends towards increased corporate power, the commodification of urban space and growing social polarisation in cities.

SOCIAL EQUITY AND POLARISATION

The world of information technology is a world made for the very fortunate few, maybe 20 per cent of the population; the people who are called these days the 'symbolic analysts', who can work with numbers and ideas, and who live in a little leafy isolated suburb surrounded by high spiked gates and guards; who sit there with their little computers and their telephones and deal with information all over the world. And they don't venture downtown, and they don't use public transport, and when they do travel it's in the front parts of international aeroplanes. . . . And then they'll be the rest, who don't have access to this technology, who don't know how to use it, who don't know how to make products out of it. And they live downtown, and they use public transport, and they'll have a tough time.

(Handy,1995; Quoted in Channel 4, 1995; 20)

We saw in Chapter 3 that there are many predictions of plenty in the 'information society' or the 'computerised age' (see Lyon, 1988). Usually these utopianist treatments suggest that telematics will herald some point of departure from the familiar social inequalities of capitalism. The onset of a benign and utopian society with universal access to technology and an end to geographical barriers and drudgery is the usual prediction (see, for example, Graves, 1986). Current evidence suggests, however, that such predictions fly in the face of contemporary urban reality. As Pedersen (1982; 254) argued, there are dangers that what he calls the 'radical democratic ideal of an information society' popularised by the utopian visions 'may turn out to be a myth'. Urban societies are becoming more unequal not less unequal. There are many related processes underway here. We have already seen in the last chapter how global economic restructuring is underpinning a shift toward a more polarised urban world.

The factors linking economic restructuring to social polarisation in cities that we looked at in the last chapter have been widely discussed (see Mingione, 1991; Healey et al., 1995). Much less common, however, is discussion of the social inequalities in access to telecommunications and telematics infrastructures – and the resulting unevenness in the ability of social and cultural groups to participate in the increasingly telematics-mediated life of cities (Nowotny, 1982; Calhoun, 1986; Robins and Hepworth, 1988; Murdock and Golding, 1989). Many debates about globalisation and the shift towards telematics-based social networks imply some degree of uniformity in these processes. Universal access to technology is often assumed or implied. As Doreen Massey argues, the concept of 'time–space compression' which tends to be invoked in these debates tends to be 'a concept without much social content' (Massey, 1993; 59).

ACCESS TO NETWORKS: THE UNEVEN 'POWER GEOMETRY' OF URBAN ELECTRONIC SPACE

It is increasingly obvious, however, that such inequalities are profound, pervasive and probably growing (Massey, 1993; 60). These inequalities in access to telecommunications networks are important because they influence the ability of people to participate in any meaningful fashion within modern information-based society. They are also important because social élites, who have the best access to networks, are able to use them to reinforce their social privileges and, in many cases, their domination over those denied access to telecommunications. Our perspective developed in Chapter 3 reminds us that these networks and the services available to them are an increasingly important means whereby power is exercised over space, time and people. As Eric Swyngedouw argues, 'the increased liberation and freedom from place as a result of new mobility modes for some may lead to the disempowerment and relative exclusion of others. This in its turn, further accentuates economic and social inequalities' (Swyngedouw, 1993; 322). But these inequalities and differences are also complex, and are not entirely the result of the unevenness of global economic networks and uneven wealth.

It is also clear that issues to do with accessibility of people to jobs, services and amenities in cities are now conditioned in electronic spaces as well as urban places. The relations between time and space within social lives in cities change radically as a result of this. Geographers have traditionally been concerned with documenting the time–space 'rhythms' and 'choreography' of urban residents through their daily lives in physical space (Hägerstrand, 1970). The traditional view was geared entirely to the idea, discussed in Chapter 2, that space and time were the external containers of urban life. For example, it was assumed that 'each individual has a moving pattern of his own, with turning points at his home, his place of work and his shopping centre during the week and his recreation grounds on a holiday or a Sunday' (Hägerstrand, 1970; 44 [sic]).

Those social groups with access to sophisticated telematics can now transcend

*"I can't make my mind up – another pint or go home
and surf the Internet!"*

Figure 5.2 Source: Courtesy *Private Eye*, 24 March 1995

these physical limits and rhythms because services, amenities and jobs can be accessed in electronic space without (necessarily) moving in physical space (through teleworking, teleshopping, telebanking, etc.). What Hägerstrand called the 'time–space choreography' of every day life is no longer confined to urban physical spaces. For many, it also encompasses the use of a multitude of electronic networks and spaces as phones, faxes and electronic mail are used to keep in touch, distant computers support the petty transactions of everyday life, and the mass media themselves begin to take the form of global electronic networks.

Considering how constraints and lifestyles are patterned within the complex social structures of cities therefore now requires a parallel consideration of physical and electronic spaces and their interactions. Inequalities in physical and electronic space tend to be mutually reinforcing. On the one hand, affluent social élites working in transnational corporations tend to have intense physical mobility as well as an intensive use of electronic spaces. This simultaneous choreography of urban places and electronic spaces is shown in Figure 5.3.

The story is very different for those who are spatially trapped in the urban ghettoes and who have very little, if any, access to urban electronic spaces and telematics networks. As Michael Dear argues (1995; 27), with telematics, 'time–space coordinates have been stretched to as yet unknown dimensions' for élite groups while at the same time 'for minorities, the poor, the disabled, and women, the time–space prism closes rapidly to become a time–space "prison"'. The boundaries between these different social areas in cities can now be considered to be social edges both in physical and also electronic spaces. With tele-communication increasingly developing according to the logic of liberalised

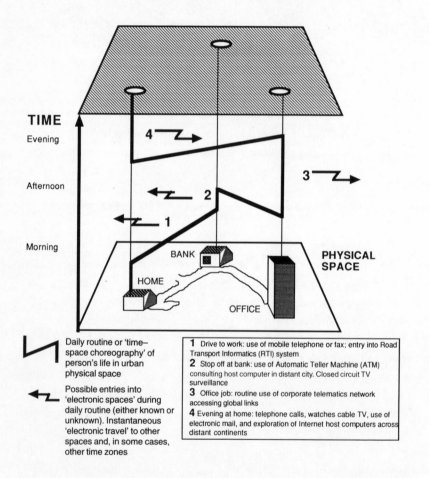

TIME

Evening

Afternoon

Morning

BANK

PHYSICAL SPACE

HOME

OFFICE

Daily routine or 'time–space choreography' of person's life in urban physical space

Possible entries into 'electronic spaces' during daily routine (either known or unknown). Instantaneous 'electronic travel' to other spaces and, in some cases, other time zones

1 Drive to work: use of mobile telephone or fax; entry into Road Transport Informatics (RTI) system
2 Stop off at bank: use of Automatic Teller Machine (ATM) consulting host computer in distant city. Closed circuit TV surveillance
3 Office job: routine use of corporate telematics network accessing global links
4 Evening at home: telephone calls, watches cable TV, use of electronic mail, and exploration of Internet host computers across distant continents

Figure 5.3 The combined use of urban place and electronic space in a hypothesised daily routine for an office worker (adapted from Hägerstrand's time geography)

markets, wide social imbalances in access to networks and the services which run on them are an intrinsic part of their current development. While some see telecommunications to be 'technologies of freedom' (Pool, 1983), we would argue that they tend to offer freedom only to already powerful social groups.

This can be seen with the examples of the basic telephone and personal computer. Despite many assumptions that it is now a 'universal service', the basic telephone – that most fundamental point of access into the information society – shows profound unevenness (see Box 5.2). The phone remains by far the most mature and important interactive communications network – in most western nations it is now considered a right and not a luxury. However, current assumptions that 'many advanced countries guarantee a certain minimal access to networks such as the telephone' (Hall, 1992; 256) are not confirmed when we look closer at who has access to a telephone. This is revealed in Box 5.2. As Figures 5.4 and 5.5 show, household penetration rates for the basic phone vary strikingly in the UK according both to socioeconomic status and urban location.

BOX 5.2 THE MYTH OF UNIVERSAL TELEPHONE SERVICE

It is now widely accepted that access to the basic telephone provides a key support to social and economic participation in modern western society. Modern society and modern economies rely fundamentally on intense telephone-based interactions across distance at every level. Indeed, the use of telephones is now so woven into every aspect of modern life that it is often taken for granted: universal access to telephones is often assumed and telephones themselves are now treated as an 'anonymous object' – a basic part of the everyday environment (Fischer, 1992). The telephone, according to Graham Murdock and Peter Golding,

is the hub of most people's personal information system. Not only does it connect them with the informal networks offered by friends, neighbours and relatives, it also provides a major point of access to the professional information services of organizations.

(Murdock and Golding, 1989: 185)

The social importance of the telephone means that the consequences of not having a telephone for the social, political, cultural and economic participation of groups and individuals are dramatic. As Claire Milne has put it, 'continuing social and economic change will reinforce this [importance], making it ever-more serious disadvantage to be deprived of a telephone' (Milne, 1991). As well as being used to maintain ties in scattered families, telephones also allow participation in more flexible labour markets. The burgeoning population of the old and the tendency to support independent living in the 'community' also make telephone links more and more essential. In addition, welfare services are relying more on cheaper telephone-based services; new innovations in welfare and support services invariably include telephone helplines and information services (see Chapter 9). Finally, access to all information, goods and services is increasingly geared to people with access to telephones in the public, private and voluntary sectors.

But access to the phone is far from universal. Within all western nations, national phone penetration rates hide substantial variations in spatial levels of access to telephones. In the United States, in 1987, five million households were without telephones because they could not afford them. The 'near universal' 93 per cent national penetration rate masked regional, class and racial disparities: the rate for African Americans and Hispanics was 81 per cent; just 71 per cent of rural black

(a)

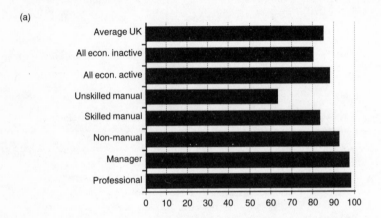

(b)

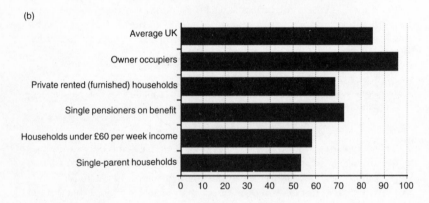

Figure 5.4 UK households with telephones: (a) by socioeconomic status, 1988; (b) by housing and social status, 1991

households in Louisiana had a phone (Bowie, 1990; 143).

In the UK, in 1988 the South East region (89 per cent) had a significantly higher level of household connection than Northern Ireland (74 per cent), the North (75 per cent) and Scotland (78 per cent) (Regional Trends, 1991). As Figure 5.4 shows, penetration rates were also correlated very strongly with socioeconomic status. But even these regional and class disparities mask much wider disparities that only become apparent at the micro-spatial scale within cities. Figure 5.5 maps the distribution of household connection to telephones found in a 1986 study of socioeconomic conditions in Newcastle upon Tyne wards. This shows that household telephone penetration rates varied between wards from a low of 55 per cent to almost complete penetration. A 1992 study of conditions on one Newcastle council estate – Cruddas Park – found a connection rate of only 26 per cent. The limited data that is available indicates that, contrary to the myth of universal access to telephones, there is considerable variation in levels of connection between regions and localities and even at the level.

As market-led telecommunications develops, the prospects are that the patterns of access and exclusion to phones and telecommunications will become more starkly differentiated. As competition between tele-com companies strengthens for the lucrative niche markets like mobile,

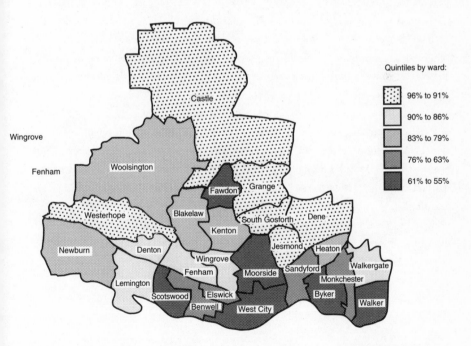

Figure 5.5 Proportion of households with use of a telephone in Newcastle upon Tyne, 1986

affluent social groups and long-distance communications, those at the margins seem likely to be further disadvantaged. For example, when a cable company lays its infrastructure down a British street and begins to offer competing telephone services, BT only target each of the households which spend more than £75 per three-month period to try and tempt them to remain as BT customers.

(Cornford et al., 1994)

The unevenness of access becomes more stark when advanced technologies such as telematics are considered – cable networks, videotex, personal computers and electronic mail. Such services are increasingly expensive both in capital and revenue terms. Future changes look set to mean that social life 'will depend much more on capital-intensive investment in the home to ensure a decent functioning and well-being' (Nowotny, 1982). The main factor excluding people, not surprisingly, is cost. Figures 5.6 shows how access to home personal computers is related strongly with occupational class in the UK and how this pattern tends to reinforce uneven access to the telephone. Figure 5.7 shows how access to personal computers at home and at work in the United States is very strongly correlated with household income. In addition, in the USA, white schools tend to have twice as many computers than schools for minorities (Davis, 1994). A recent comparison in *MacWorld* magazine found that a black school in Silicon Valley had one computer for every sixty students whereas a white school a few miles away had one for every nine (quoted in Davis, 1993).

Within this highly polarised picture, Doreen Massey speaks of a highly complex and uneven 'power geometry' of time–space compression within which different social and cultural groups have extremely uneven levels of control and access. She argues that 'different social groups and different individuals are placed in very distinct ways in relation to these flows and interconnections' (Massey, 1993; 61). She identifies three main groups within this broad and uneven picture: those in control of the 'time–space compression', those who communicate a great deal but are not in control, and those who are on the receiving end or are excluded from these processes.

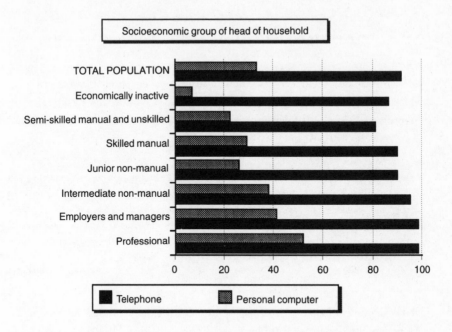

Figure 5.6 Citizen access to telephones and computers by occupation in the UK, 1991
Source: Devins and Hughes, 1995

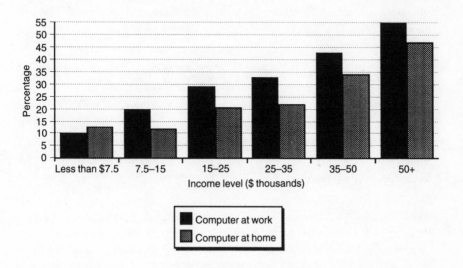

Figure 5.7 Citizen access to computers by income level in the USA, 1993
Source: Office of Technology Assessment, 1993

AFFLUENT SOCIAL GROUPS AND THE
'TRANSNATIONAL CORPORATE CLASS'

There is substantial evidence that a new 'transnational corporate class' is emerging which is the primary agent of operating the global economy, and which relies on intense access to global telematics networks on a continuous basis to 'command space' (Sklair, 1991; 62–71). Massey writes that:

at the end of all the spectra are those who are both doing the moving and the communicating and who are in some way in a position of control in relation to it. These are the jet-setters, the ones sending and receiving faxes and e-mail, holding the international conference calls, the ones distributing the films, controlling the news, organising the investments and the international currency transactions. These are the groups who are really, in a sense, in charge of time–space compression, who can effectively use it and turn it to their advantage.

(Massey, 1993; 61)

These powerful élite groups, who tend overwhelmingly to be white and male, are not totally abstracted from local and national contexts. But they tend to rely on the intense use of global transport and telematics networks that link the main global information capitals together into networks. As telecommunications liberalisation focuses investment, innovation and competition on to these places, and on to the affluent districts within them, so these social groups also benefit most from being the target of 'cherry picking' by competitive telecommunications companies in the most profitable domestic telecommunications markets.

Increasingly, these processes seem likely to support the development of 'personal information networks' through which tailored services are developed geared intimately to the intense and communications needs of affluent profession-als (Noam, 1992; 408). Mobile phones and computing, linked into advanced telematics services, can then be used to support ever greater degrees of mobility while control over complex business and personal lives may be maintained. Eli Noam speaks of the prospects of:

individually tailored networks arrangements. [This] does not mean a physically separate system, except for inside wiring and maybe the last mile [telephone] circuits, plus some radio-mobile links and terminal equipment. The rest could consist of virtual networks, provided by a whole range of service providers and carriers, and packaged together to provide easy access to an individual's primary communications needs: friends and family; work colleagues; frequent business contacts, both domestic and foreign; data sources; transaction programs; frequently-accessed video publishers; telemetry services, such as alarm companies; bulletin boards, and so on. Contact to and from these destinations would move about with the individuals, whether they are at home, at the office, or moving about.

(Noam, 1992)

Again, these developments in electronic space match the selective colonisation of exclusive urban places as the homes for these élite groups. The exclusive areas of cities that are the home for these groups are increasingly linked together by social and technical networks while being secured from the rest of the cities in which they are placed. Manuel Castells (1989; 348) describes how 'the new professional–managerial class colonizes exclusive spatial segments that connect with one another across the city, the country and the world; they isolate themselves from the fragments of local societies, which in turn become destructured in the process on selective reorganization of work and residence'.

The high levels of mobility of these groups are linked in to their relentless travel for commuting, work and leisure, and, increasingly, to the nomadic shifting to resort areas for periods of teleworking (see Chapter 4). The key point to stress here is that, rather than substituting for travel, evidence suggests that the intense use of telecommunications networks by these élite groups is related positively to high levels of travel (see Chapter 7). This is clear from the following quote from an information technology executive exploring ways of targeting products on the most profitable section of the social telecommunications market:

Information network providers and information service providers should examine the information needs of the most information-hungry individual of all, the traveller. The daily information and entertainment needs of a traveller are typically multitudes greater than those of an average residential customer. . . . Better yet for the telecom service provider, travellers are geographically concentrated in [certain] districts. It's far more likely that business and leisure travellers will pay for information, entertainment, personal services, and teleshopping on a daily basis and do so in a more manageable geographic area as compared with widely dispersed residents.

(Darby, 1994; 231)

These groups tend to dominate use of computer networks and mobile telecommunications (phones and, increasingly, computers) which add to the flexibility of combining transport with telecommunications. They also dominate markets for mobile phones and computers, which add a further intensity to their mobility (Spector, 1993; 406). A recent Finnish survey, for example, found that mobile phones 'facilitate more flexible work arrangements, a more intense family contact network, increased sociability . . . and a possibility to combine an active, mobile way of life with the types of contacts previously related only to staying at home' (Roos, 1994; 27).

COMMUNICATORS WITHOUT CONTROL

Massey identifies a second group in this uneven picture: those who, like the workers in the tele-mediated back offices in less developed countries, communicate a great deal but have no control over this communication. In the last chapter we noted the decentralising logic where routine back-office functions and teleworking are moving from high-cost, big-city locations. This trend is being encouraged as a way of bringing much-needed employment to disadvantaged urban neighbourhoods. Outside of the large-scale back offices, though, there are dangers of 'telecommuters' being isolated and easy to exploit in their own homes. There is also increasing evidence that the patterns of teleworking are developing to reflect the wider polarisation of labour markets underpinned by economic restructuring, with a highly contrasting work experience at different ends of the social hierarchy. Those at the upper end of the teleworking experience – those in control – tend to be men, while those at the lower end – those who are being controlled – tend to be women. These trends tend to mean that the potential telematics may offer for positive employment shifts towards flexibility and overcoming the isolation of being disabled or housebound tends not to be realised. Zimmerman, for example, argues that:

Telecommuting promises two very different types of work experiences for those at the upper and lower ends of the occupational scale: data entry clerks and secretaries will handle routine tasks under continuous computer scrutiny of their performance and hours, while professionals will have discretionary hours and unrestricted freedom to use computers for personal tasks such as home accounting and database access.

(Zimmerman, 1986; 30)

EXCLUDED GROUPS

At the other end of the scale, there are those groups who do not participate at all on these networks: those who, as Massey (1993; 62) puts it, are simply on the receiving end of time–space compression. Peter Golding traces the stark social inequalities in access to new telematics and media equipment and services:

Entrance to the new media playground is relatively cheap for the well-to-do, a small adjustment in existing spending patterns is simply accommodated. For the poor the price is a sharp calculation of opportunity cost, access to communication goods jostling uncomfortably with the mundane arithmetic of food, housing, clothing and fuel.

(Golding, 1990; 90)

As we saw above, the make-up of the groups excluded from the telephone or from access to personal computers is fairly easy to identify. They comprise poor socioeconomic groups, low educational attainers, people who are not computer literate, and other generally disadvantaged or marginalised groups such as the disabled, ethnic minorities, women and remote rural communities. These groups and communities will tend to face further marginalisation from the sources of power, employment, information and services within market-based telecommunications regimes. While many women, disabled and housebound people might be expected to gain from potential home-based and flexible computer work, it is difficult for them to obtain the necessary skills.

The exclusion of women from the male-dominated world of computers is particularly stark (Frissen, 1992). Over 90 per cent of computer scientists in American universities are men and 'women are not well represented in most computer networks', covering perhaps 10–15 per cent of the World Wide Web usage (Shade, 1993). Given this context, Valerie Frissen argues that:

Compared to men, women are still in a very marginal position in the design and production of IT, which in a way also excludes them from power. Women furthermore seem to be afraid of using these technologies to their own advantage. Finally, the effects of information technology seem to be negative, or at best not positive enough, where women are concerned. In short, the exclusion of women seems to be the main message.

(Frissen, 1992; 45)

Many disadvantaged groups are clearly at the margins of the information society, if access to these technologies is considered as a necessary support for participation in such a society (Nowotny, 1982). This is despite the fact that, in principle at least, many of these groups – the housebound, the disabled, old-age pensioners and single-parent families – stand most to gain from space-transcending communications technologies. There are a growing number of examples where the use of computer networks had genuinely empowered disadvantaged social groups by overcoming disabilities, isolation and supporting access to new, high-quality employment (Downing et al., 1991). As we shall see in Chapter 9, many urban policy initiatives are addressing the barriers and supporting new patterns of social access to telecommunications in cities.

THE 'COMMODIFICATION' OF INFORMATION AND THE 'PAY-PER REVOLUTION'

It is very important that any exploration of social access to telematics looks beyond the hardware of the networks and technologies themselves at the burgeoning range of services that are provided over them. Telephones and telematics terminals are gateways into much more than interpersonal communications – they also support entrance into a rapidly growing world of information services, welfare services and value added services (such as teleshopping, telebanking and other tele-transactions). As Bowie (1990; 133) suggests,

the terms and conditions for access to information technology increasingly define one's right of access to information *per se* . . . information that is particularly useful, relevant, timely information, is increasingly tied to complex electronic technology. . . . The gap between the information rich and the information poor is a gap between the privileged and the powerless.

As innovations in these areas blossom, fuelled by waves of private innovation, being denied access once again becomes more important and more disadvantaging. But these concerns arise against a backcloth of growing privatisation of information services and a shift from free, public sources of information towards 'information commodities' – to be traded, bought and sold on a bit-by-bit basis in the so-called 'tradable information sector' (Openshaw and Goddard, 1987). Telematics and computing allow information to be controlled, processed and managed with unprecedented sophistication and precision. This means that highly individualistic market solutions become possible where previously services often had to be offered at generalised charges or as free public services. As we saw in the last chapter, this has led to a burgeoning economy in tradable information where firms process raw data and resell it in marketplaces for profit – often via on-line databases. In this context, Theodore Roszak notes that 'more and more information is falling into the hands of a profit-making information industry, turning what might be a public benefit into a private business. The irony here is acute. At the very moment in history that new technology makes the distribution of information potentially cheaper and easier than ever before, interests are at work using that same technology to restrict flow' (Roszak, 1994; 179). As Openshaw and Goddard (1987; 1424) argue, there are real dangers of being '"locked out" of the information society. As information becomes a vital resource, it is being commodified and traded at a price, with access being determined by one's ability to pay'. Market-based mechanisms for allocating access to information services means inevitably stark 'information inequality'. Dordick *et al.* consider this be the 'policy crisis for the 1990s'. They suggest that 'in the nineties the issue will be one of equality of access to information . . . there exist two classes of people and businesses: the information users and the information used' (Dordick *et al.*, 1988; 237).

In many public libraries, for example, attempts to maintain free public access to electronic information sources are being thwarted by 'publishers who wish to maintain control of their materials; owners of the new electronic media . . . see their role much differently from that of traditional publishers. To protect their rights to the material, electronic publishers increasingly seek to control access to information, selling use of the material rather than the product itself – libraries are charged on a pay-per-view basis' (Shulman, 1992; 14). The proprietors of many electronic networks and information providers such as government departments, universities and libraries are charging some of their services on a bit-by-bit basis (Schiller and Fregaso, 1991; Roszak, 1994). With the growth of digitised and multimedia information services, there is even the longer term prospect that physical libraries will be electronically substituted. One British librarian predicted recently that 'the function of libraries as places which you physically visit to access information is going to be superseded by "virtual" or "electronic libraries" based on telematics networks' (quoted in Roszak, 1994; 182).

There are also clear trends towards the commodification of other media services – with 'pay-per view' television taking advantage of the control capabilities of telematics to turn previously public goods into private commodities. The prospect, with the centralisation of capital on to giant, global media conglomerates, is a systematic commodification of all information via telematics: data, images and television, for sale in private markets.

This process is tied up with what Vincent Mosco calls the 'pay-per revolution' – the uses of the control capabilities of telematics technologies to move from aggregated general pricing or public regulation of services to precise and individualised pricing. He argues that:

We see the evidence of pay-per society all around us. There is pay per call in telephone, pay per view in television, pay per bit or screenful of material in the information business. Advertisers refer to pay per reader, per viewer, or per body when they place an advertisement. In the workplace, word processors know about pay-per keystroke. And so on. . . . The essence of what is happening is this: new technology makes it possible to *measure* and *monitor* more and more of our electronic communications and information activities. Business and government see this as a major instrument to increase profit and control. The result is a pay-per society.

(Mosco, 1988; 5)

Rather than leading to universal access to all information in all places, as many utopianists would argue, Mosco argues that commodification and the 'pay-per revolution' again serves to accentuate information inequalities and creates great problems in urban society. It supports the shift towards privatisation and liberalisation because new technologies make profitability possible in previously public services. It supports the monitoring and exclusion of those who cannot afford access into the 'network marketplaces'. And it underpins new stark inequalities in access to goods and services in urban society:

The growing class divisions between those who can afford computers and pay-per-view television in stereo and those who cannot read or afford the price of a phone is reinforced by the growth of a two-tier workforce . . . we are dividing into a society characterized by a high-tech minority at the top and a mass of people at the bottom whose work has suffered the ravages of automation and deskilling.

(Mosco, 1988; 12)

ACCESSING SERVICES IN ELECTRONIC SPACES AND URBAN PLACES

We argue that the polarising effect of these mechanisms on social access to telecommunications services tends to perpetuate wider problems that disadvantaged groups experience from their marginal position in labour markets, housing markets and consumers' service markets within cities and the difficulties they already experience in exercising their rights as citizens. Parallel trends towards social and spatial polarisation can therefore be identified both in the urban places and electronic spaces. Unevenness in access to telecommunications networks and services compounds and reflects the wider processes of social polarisation driven by the processes of urban economic restructuring and labour market change analysed in Chapter 4. Lash and Urry capture the parallels between the social and class makeup of urban places and electronic spaces:

Access to information and communication networks . . . is a crucial determinant of class position. The 'wild zones' of very sparse lines, flows and networks tend to be where the underclasses, or at least the bottom third, of the 'two-thirds societies' are found. . . . Similarly the unusually densely networked centres of the global cities tend to be where the top fractions, in the corporation headquarters, business and finance and legal services, of today's new informational bourgeoisie are primarily located.

(Lash and Urry, 1994; 319)

Banking and shopping services, as well as information services, are also becoming more tele-mediated. This means that it is important that social access to both electronic and physical services are considered. There seem to be important substitution trends here, whereby electronic financial and retailing services can be brought in (which are geared toward affluent consumers) and substituted for physical services (which tend to be withdrawn from the more marginal and disadvantaged areas). This may even apply to other welfare and health services. With the growth of telemedicine, which allows surgeons to manipulate surgical robots remotely, Mitchell (1995) predicts that 'if you are a skilled surgeon . . . you might just want to stay well away from dangerous places like battlefields or the South Side of Chicago'.

More pronounced, so far, are substitution trends in financial services. We saw in the last chapter how the British First Direct phone-banking service (which is restricted to people earning over £15–20,000 per year) was linked to a wider removal of physical Midland Bank branches, largely from disadvantaged inner city areas. A recent survey in Birmingham, England, found that recent trends towards the closure of banks in poorer areas of the city had left five of the thirty-nine city wards without any banks and six with only one (Leyshon, 1994).

In the United States, more advanced trends towards the 'cherry picking' of lucrative markets and the 'social dumping' of marginal consumers have been noted by Susan Christopherson. As US banks have restructured to address profit crises, they have attempted to

withdraw or increase revenue from routine transactions-intensive markets and focus on markets with the potential for higher value-added transactions. As banks have established minimum account balances and imposed fees on small accounts, increasing numbers of people, especially in poor neighbourhoods and communities, have had to forego basic banking services such as current accounts.

(Christopherson, 1992; 28)

The emphasis now is on 'creaming' the market by closing branches in 'unfavourable' areas and targeting affluent consumers both with enhanced branches and improves services and, increasingly, with phone-banking services like First Direct.

Similar trends can also be discerned in food shopping (Christopherson, 1992). Wider spatial shifts in shopping towards out-of-town centres reliant on access to cars exacerbate this growing social unevenness in access to basic services. In the USA, careful locational decision-making ensures that new out-of-town super-stores are located as close as possible to the areas of maximum market potential:

Bloomingdale's and Nordstrom, for example, apply techniques based on the use of Geographical Information Systems (GIS) to 'look for sites with a 10–15 mile radius that contains at least 10,000 households with an average income of $60,000, or for smaller concentrations of high-income households' (Christopherson, 1992; 282). In the UK, meanwhile, 10,000 small local shops closed between 1954 and 1994 while 750 superstores were opened in carefully selected locations with the highest market potential (Vidal, 1994). As giant supermarkets enhanced their dominance of the UK food market, sales through local food shops dropped 15 per cent in only 12 months between June 1993 and June 1994 while doorstep deliveries fell 12 per cent. A food executive warned that 'there are parts of the inner cities where doorstep deliveries will disappear in four to five years' (quoted in Cowe, 1994).

For the growing number of people who cannot physically get to the shops – because of frailty, age, disability or lack of transport and/or telecommunications facilities – the only option is to rely on the goodwill of others (Ducatel, 1994). Ironically, these people have potentially the most to gain from teleshopping and the use of electronic forms of retailing to overcome physical barriers. However, the teleshopping services that emerge are, as we have seen, very likely to be targeted on mobile and affluent consumers. Murdock and Golding examine the ways that the electronic spaces of new teleshopping services are likely to interact with these changing patterns of traditional shopping in urban places:

Home shopping is a case in point. The poor are already disadvantaged in regard to shopping. Since most do not have access to a car, they are unable to take maximum advantage of the choice and price advantages offered by supermarkets in city centres or hypermarkets on the edge of towns. Instead they are confined to the relatively limited choice and high prices of local shops. . . . But run on a straight forward commercial basis, [teleshopping] services are invariably geared towards servicing the better off, thereby extending the advantages they already enjoy.

(Murdock and Golding, 1989; 186)

THE HOME AS A LOCUS OF URBAN SOCIAL LIFE

As we saw in Chapter 2, utopian treatments of the information society invariably predict the emergence of a new home-centred society where work, leisure, information and communications are all effortlessly carried out through the telecommunications equipment of the ('smart') house (see, for example, Mason and Jennings, 1982). This leaves the rest of the time for enjoying the (usually rural) surroundings and a 'family-centred' life; it also incidentally solves many of the problems associated with concentration in the industrial city. While these scenarios can be discounted as being utopian and socially blind, there are signs that trends towards greater home-centredness and the privatisation of social life in western society do exist (Smart, 1992; 53). As families get smaller, new technologies allow the home to emerge as a centre for communications, receiving information and entertainment, obtaining goods and services, and even linking in with workplaces and employment. Advances in telecommunications, and, more particularly, the way they are being socially shaped and marketed to be individualised services to households, can be seen directly to support this shift toward home-centredness. The ever-increasing telecommunications, media and computing equipment that is finding its way into western homes (see Figure 5.1) is providing the technical underpinnings to this growing home-centredness.

There are three areas to consider here: home teleworking, interactive home tele-services and 'smart' home technologies. First, some argue that progress towards home-working has been slow because of the psychological or physical problems involved (Forester, 1989; Gold, 1991). The Tofflerian notion of mass, full-time home teleworking has receded: 'empirical evidence as yet gives scant support to any major structural change in social and spatial patterns due to homeworking' (Gold, 1991; 339). Nevertheless, the flexible use of the home for a portion of the week for tele-based work is growing. In the USA it is estimated that 13–14 million people regularly do paid work at home using personal computers and 6.6 million of these can be classed as 'teleworkers' (Hillman, 1993).

Second, it is clear that most experiments in developing interactive, personalised or 'multimedia' home entertainment, commercial and information services have, so far, been 'strikingly unsuccessful' (Mulgan, 1991; 62). But the wider diffusion of more user-friendly personal computers and related technologies into the home are leading to a fast-growing user base amongst particular social groups for certain home 'tele' services. In France, the state-backed Minitel videotex service now equips 30 per cent of all households and offers 20,000 services. Many experiments in new forms of home entertainment and service use are currently emerging.

In the longer run, as we saw in Chapters 3 and 4, there are many predictions for much more sophisticated and important home-based electronic activity. Sheth and Sisodia (1993), for example, predict the development of 'information malls' – whole 'virtual' service spaces accessed from the home. These, they argue:

will truly become one-stop non-stop service providers. They will be able to handle voice mail, storing and forwarding voice messages worldwide. They will allow for customised newspapers culled from information sources around the world and updates on a real-time basis. They will fuel the growing trend towards work-at-home. They will permit home consultation with a variety of professional service providers, from physician to lawyers to educators. They will permit on-demand, node-to-node or one-to-many videoconferencing.

(Sheth and Sisodia, 1993; 382)

According to this argument, such information malls will emerge as sophisticated electronic spaces, accessible via home telematics. They will use video and 'virtual reality' technologies to offer the tangible sense of purchasing products, accessing services, experiencing education, undertaking transactions, communicating and obtaining information from within the household – as though in a real 'place'. They are, in fact, only one of many predictions that electronic urban spaces will develop to be analogous and tangibly similar to the physical spaces that they replace or complement. To Sheth and Sisodia, the versatility and flexibility of these electronic spaces will threaten the retailing cores and out-of-town shopping centres of cities. They suggest that 'an important part of an information mall's identity will, in fact, be that of a virtual shopping mall; to the extent that such virtuality can effectively substitute for the "real thing", information malls will supplant shopping malls' (Sheth and Sisodia, 1993; 381).

Finally, most experiments in 'smart home' technologies – which integrate consumer appliances via telematics and artificial intelligence – have been simple technology promotions (Moran, 1993). The supposedly imminent onset of the 'smart' or 'electronic home' – a household where all domestic technologies become integrated via telematics into a multi-function household communications system that is also linked with outside networks – has long been promised. Such an integrated system would allow the remote control of consumer appliances, new energy management systems, new environmental control systems and new security systems (Moran, 1993). Catherine Richards, for example, dreams of 'an environment where computers have become completely absorbed into domestic space. Where the inhabitant is tracked by invisible sensors. So it's like you come home from work and the system reads the chemicals you're giving off and knows you've had a bad day. It programs the TV to put on a soothing CD-ROM and tells the fridge to fix you a drink' (quoted in Beard, 1994). Much more prosaic 'intelligent home' demonstrations projects have been developed by public and private organisations in most western nations: the Smart House was built by the US National Association of Home Builders in 1985; Thorn-EMI built an 'automated house' in the UK to demonstrate its products in the early 1980s; and in Japan an intelligent house was built at Tsukuba Science City in 1992 (Cancelieri, 1992). But there remains a good deal of scepticism about the real demand and economic viability of such technologies and broadband services – let alone about their wider implications for urban social life. In many cases, such technologies seem to be an example of 'technologies in search of applications'. Cable, computer and telecommunications companies seem to be investing without clear ideas as to the real usefulness and viability of the end result. Noll, for example, argues that 'neither the market, the technology, the economics, nor the consumer are ready for broadband to the home' (Noll, 1989; 200). Nevertheless, there are some signs that some 'smart' household technologies are emerging, but as 'islands of automation' rather than integrated systems (Wise, 1992). The greatest demand seems to arise for home security systems, alarms – with domestic alarms linked to private security firms or police offices – and for entertainment systems – for example, with remote-control video recorders.

So, while definite trends towards tele-based home-centredness can be observed, we are sceptical of the claims that this is some sort of complete revolution in the nature of home life. We would agree with Tom Forester (1989; 224) when he argues that 'a combination of home computers, consumer electrical goods, videotex services, and home security systems, even in a "smart house", wired with heating and lighting sensors . . . hardly add up to a revolution in ways of living'.

THE HOME AS A DOMESTIC 'NETWORK TERMINAL'

In the modern city, the home emerged as 'the last reserve space'. It was, to quote Helga Nowotny, 'a social space of special significance which has come to signify for us the last sanctuary in a bewildering outside world' (Nowotny, 1982; 102). But this sense that the home is isolated from the rest of the social world is changing rapidly. Many homes, as we have seen, are being incorporated into more and more networks on increasingly global scales. These blur the dividing line between what is public and what is private. To Putnam, this means that the best way to consider the home is as a 'terminal'. He writes,

in speaking about the modern home, we are talking about more than technologized comforts. The modern home is inconceivable except as a terminal, according the benefits of, but also providing legitimate support to a vast infrastructure facilitating flows of energy, goods, people and messages.

The most obvious aspect has been a qualitative transformation of the technical specification of houses and their redefinition as terminals of networks.

(Putnam, 1993; 156)

This is not the result of some simple technological 'logic' however. Nor can we attribute this process entirely to broad political and economic forces. Rather, Roger Silverstone has shown that new communications and information technologies are entering homes through complex and diverse processes of social construction and 'domestication' (Silverstone *et al.*, 1992; Silverstone, 1994). The consumption of these new technologies, and the services that are accessible through them, are

pursued by consumers who seek to manage and control their own electronic spaces, to make mass-produced objects and meanings meaningful, useful and intelligible to them. This is a process of 'domestication' because what is involved is quite literally a taming of the wild and a cultivation of the tame. In this process new technologies and services, unfamiliar, exciting, but also threatening, are brought (or not) under control by domestic users.

(Silverstone, 1994; 221)

Often, though, these technologies create conflicts – particularly gender conflicts – over who has control and access within the household (Mulgan, 1991; 69).

But these processes operate within the context of broader political, social and economic changes. The shift towards information labour, home and telecentre-based teleworking, flexible labour markets and the cultural shifts towards globalisation provide the context for the changes underway. Evidence suggests, however, that beyond the hype, there remains 'deep resistance to electronic interactivity beyond a small number of enthusiasts' (Mulgan,1991; 69). Most of the development of technology in the home still centres on stand-alone media and entertainment systems such as CD players, personal computers and TVs. Most cable services are not interactive – they merely pump more channels to be passively consumed than terrestrial broadcasting. Telephones remain by far the most important interactive home communications system.

We must also be conscious of the varied prospects for genuine tele-based services to the home. In the United States, home tele-services seem to have a greater chance of reaching 'critical mass'. In Europe, resistance seems to be much deeper. A recent study by the Inteco consultancy of 12,000 people across Europe, funded by 30 retailers and telecom companies, found that there is deep consumer resistance to interactive shopping; entertainment and security services were marginally more attractive (Bannister, 1994).

The main problem, then, is that, in mass markets 'the audience does not yet provide the revenue which supports the services' (Curtis and Means, 1991; 24). Thus the focus, as we saw above, is on 'demassification' and the 'cherry picking' of markets by offering higher value-added services for more lucrative consumers and those small markets where genuine prospects exist. The Inteco study also concluded that the only attractive markets at present comprise 'active, relatively high income families where time is at a premium, where someone in the family will be learning to play a musical instrument or be keen on gardening or some form of home study. [These] will be the fertile ground for multimedia seeds to root' (Inteco, 1994; quoted in Bannister, 1994).

Despite uncertainties in demand, commercial efforts to enhance the role of homes as terminals for many technical networks, offering 'pay-per' services (at least for these affluent groups), are intensifying. These are driven by the view that there are huge commercial rewards that may be in store from commercial services offered over the much-vaunted prospect of multimedia, interactive television or the information superhighway. Because of this, the still rather marginal shifts towards home banking, home- and centre-based teleworking, and tele-based access to services may eventually grow, as new technological capabilities come into the social mainstream – or, perhaps more likely, the financial reach of socially affluent groups.

Already, 'narrowcasted' electronic newspapers have been developed in the United States, accessible by computer and geared to the particular programmed interests of consumers (White, 1994). 'Video on demand' technologies are about to enter the market, offering selected videos played down phone lines as compressed signals. In a few places, cable networks are beginning to offer value added services rather than just TV. A growing range of consumer services such as 'electronic bookstores' are developing, based on the Internet (which is emerging as a domestic service for technology enthusiasts). 'Smart', 'interactive' or 'high definition' TVs have been developed which, when combined with signal compressions technologies, offer the technological potential for user-friendly and interactive services.

A range of full-blown commercial trials are now in the offing for interactive, sophisticated home telematics systems. Time Warner and US West are already operating an interactive TV trial in Orlando, Florida, for 4,000 homes. A system called 'stargazer', developed by Bell Atlantic, will soon offer education, entertainment, information and shopping on a trial basis to 2,000 homes in Northern Virginia. But the commercial and commodified nature of these emerging systems makes the widely used analogy of the electronic highway profoundly misleading.

If the dream is of consumers who 'spend at the touch of a button', those without money are likely to have little to do with these developments. Brody (1993; 32) writes that 'what is emerging has less the character of a highway than a strip mall, focusing on services with proven demand that can rapidly pay for themselves . . . cable operators are seeking to establish a presence in the lush, unregulated territory of enhanced services'. They also seem to be supply-driven – often crude attempts by technology and telecommunications companies to increase the use of existing networks and technologies and so improve profitability (Noll, 1989).

'ELECTRONIC COTTAGE OR NEO-MEDIEVAL MINI-FORTRESS?'

While there are uncertainties over the viability of these systems, there does seem to be an overall trend towards the mediation of more work, travel and consumption via home-based telematics which intensifies the home-centredness of urban society. These trends suggest that homes may be progressively disembedding from their immediate social environment within urban places through their linkage into electronic spaces. Manuel Castells, for example, argues that

homes . . . are becoming equipped with a self-sufficient world of images, sounds, news, and information exchanges. . . . Homes *could* become disassociated from neighbourhoods and cities and still not be lonely, isolated places. They would be populated by voices, by images, by sounds, by ideas, by games, by colors, by news.

(Castells, 1985)

In this context, Barry Smart asks the important question: will home-centredness emerge as a benign and positive development as in the Tofflerian 'electronic cottage' scenario (see Chapter 3) or, conversely, will it 'encourage a retreat from public life and pubic space and serve thereby to increase the sense of insecurity which has become and increasing feature of the "post-modern" urban environment' (Smart, 1992; 53). The trends are complex and varied and belie easy generalisation. But there seems to be a shift, particularly in the middle classes, towards 'cocooning' – their withdrawal from public spaces in cities and the use of home-centred and self-service technologies and network access points in their place (see Gershuny and Miles, 1983). Fear of crime and social alienation with urban life are key supports to this trend. A recent Channel 4 television programme (1994; 5) commented that access to telematics networks for shopping, leisure and work 'looks safe indeed compared to urban decay. Paranoia, violence and pollution are eating away at the soul of America, driving it inward – to the protection of the home, private security, entry codes, and video-surveillance-controlled gated fortresses'. In this vein, Manuel Castells warns of a dystopian (American) urban future where 'secluded individualistic homes across an endless suburban sprawl turn inward to preserve their own logic and values, closing their doors to the immediate surrounding environment and opening their antennas to the sounds and images of the entire galaxy' (Castells, 1985; 19).

There are two key issues here. First, it is becoming increasingly clear that there are dangers that these home-based technological innovations will simply exacerbate existing trends towards individualisation and polarisation within western cities (Robins and Hepworth, 1988). Relying in the telematics field on 'a totally individualised society completely ruled by market mechanisms' (Kubieck, 1988) seems to threaten to undermine the public, civic sense of cities as physical and cultural spaces of social interaction. Certainly, the reliance on technical networks for social interaction and entertainment and the withdrawal and cocooning that goes along with this, encourages fear of crime and a shift toward the 'tribalisation' of socioeconomic groups in cities.

Second, as we have seen, these trends threaten to exclude and marginalise already disadvantaged groups who are unable to afford participation in these technological futures. Women, for example, may become ensnared in domestic space by a pervasive shift towards home-based teleshopping.

SOCIAL SURVEILLANCE AND THE CITY

A fourth key set of issues arise when telematics are used for the purposes of social surveillance within cities (Gandy, 1989). Telecommunications, combined with computers and media technologies, are at the most fundamental level control and surveillance technologies. One of their essential uses is to support new degrees of control and surveillance across time and space boundaries: this, indeed, is what they were first designed for (see Beniger, 1986). We have already seen how these control capabilities support new economic processes such as globalisation and new ways of allocating goods and services such as the 'pay-per revolution'. We have also seen how they are central to new ideas about smart homes and information malls which look set to support further centralisation of social, economic and cultural life on to the home. In all these areas, the control and surveillance capabilities of telematics across time and space boundaries are of central importance.

But the ways in which telematics allow vast quantities of real-time data about all aspects of social and cultural behaviour to be collected, processed, stored and continuously monitored raise important broader questions about the issues of social surveillance and monitoring in cities. Of course, such surveillance and monitoring is nothing new (Lyon, 1993). It is an essential component of all advanced and complex societies and all large organisations. Surveillance is necessary to maintain 'cohesion and coordination of the economic and social order' (Robins and Hepworth, 1988; 169). Control technologies have allowed large and complex industrial societies to develop in the first place. To the critical theorist, Vincent Mosco, new technologies merely deepen and extend old processes of surveillance and monitoring. He writes:

The computerised credit card, the home computer, and the sophisticated television system that permits home banking, shopping, opinion polls, and so on also allow corporations to collect massive amounts of information on users. One analyst estimates that within five years 40 million so-called smart cards for automatic banking, shopping and other services will be in circulation in the United States alone. How much money you have, what you like to buy, your views on capital punishment, your preferences for president or for laundry detergent – the new technology is used to draw detailed marketing profiles of individual households for what is called . . . precise targeting of potential buyers.

(Mosco, 1988; 6)

To Mosco, telematics-based surveillance and monitoring by firms and governments is the inevitable parallel to the 'pay-per' revolution whose product is the continuous collection of digital information in all walks of urban life: social and welfare services; employment; leisure and consumption; and criminal justice and law and order (Lyon, 1993).

Increasingly, then, the inhabitants of urban places are each subject to many electronic 'images' held in various places through the electronic spaces within and between cities (see Lyon, 1993; 70). Records in firms and government agencies are becoming digitally encoded; employment performance is increasingly monitored in real time; individual consumer and spatial behaviour is tracked through credit card and 'smart card' transactions and road transport informatics (see Chapter 7). New trends towards smart cards allow individuals to actually carry around their own digital images for the purposes of private or public transactions.

Lyon (1993) argues that telematics allow the physical limitations which traditionally reduced the scope of surveillance to be transcended. To William Mitchell, 'life in cyberspace generates electronic trails as inevitably as soft ground retains footprints' (Mitchell, 1995). This can bring dystopian concerns about emergence of some Orwellian Big Brother society. It is now also argued that the whole of modern society is growing to resemble some all-seeing electronic version of the surveillance-based prison known as the 'panopticon' which Jeremy Bentham developed in the late eighteenth century (see Lyon, 1994; Chapter 4; Foucault, 1977). But the great diversity of different surveillance systems under development means that it is too simplistic to assume some easy transition in this way. There are three key areas where the enhanced surveillance capabilities of telematics raise important questions for urban social life: in consumption, in social control and the workplace, and in the public spaces of cities and the fight against crime.

SURVEILLANCE AND CONSUMPTION

In the shift from mass Fordist styles of consumption to highly segmented, 'demassified' and increasingly personalised consumption markets, targeted marketing is replacing mass advertising as the key tool to stimulate consumption (Wilson, 1986). Mouftah (1992; 29) stresses that 'the last decade's demographic changes are making the personalization of information virtually a necessity, due largely to the fragmentation of consumer markets'. Mass-produced products and media services are being replaced by carefully tailored products and services that are designed to meet the needs of small market niches – as with Cable TV channels – or, ultimately, individuals (as with the personal newspaper example above) (Curtis and Means, 1991). These shifts are driven not only by the competitive force of firms; the increasing reflexivity and aesthetic dimensions to consumption is also a key factor in driving the growing segmentation of markets.

These shifts require much greater personal information about consumers – their incomes, reading, viewing and consumption tastes, habits, views and orientation. This information is also extremely valuable to firms trying to predict volatile market changes. It helps reduce bad debts through more sophisticated credit referencing. In its own right it is extremely valuable as a commodity to be traded within the burgeoning 'network marketplace' to the information agencies and commercial firms. Some of the fastest growing companies are in these parts of the 'information business' (Roszak, 1994). The largest private 'information bureaus' in the USA, for example – TRW, Equifax and Trans Union – make large profits by maintaining detailed birth, family, address, telephone number, social security and salary history, credit and transactions, mortgage, bankruptcy, tax and legal records of US citizens (Eder, 1994). Blacklists are built up from electronically integrated databases and photographs which are then resold to many different firms who use them to control access to credit, goods, services and even premises such as shops. Government agencies, too, are increasing their processing and marketing of their data as commodities to be sold to private firms (Roszak, 1994).

These reasons mean that the sophisticated and real-time capabilities of telematics-based systems are increasingly being used to capture data about consumers. In the area of home telematics, Robins and Hepworth write that, 'it is the nature of interactive telematics as process and control technologies that electronic transactions (television viewing, teleshopping, remote working) *must necessarily be recorded. The system is inherently one of surveillance and monitoring'* (Robins and Hepworth, 1988; 169. Emphasis added).

A good example of this is the emerging generation of video-on-demand (VOD) technologies, which allow consumers to order selected videos for personal transmission down phone or cable lines to their homes. These produce a continuous stream of information for a cable or telecommunications company about the media preferences of individual households. The telecommunications company, Bell Atlantic, are developing a computer system linked to VOD which will 'monitor the movies that a person orders and then suggest others with the same actors or theme'. The system would also 'enable advertisers to send commercials directly to customers known to have bought particular kinds of merchandise. Thus, people who bought camping equipment from a video catalogue might start seeing commercials for outdoor clothing' (Hunt, 1994). In the same vein, an interactive and pay-per-view cable system known as QUBE, which ran between 1980 and 1982 in Columbus, Ohio, was scanned every two minutes by its owners, Warner Brothers, to yield aggregated data on consumer profiles and watching patterns – which were then resold (Meehan, 1988). The information also proved useful in the development of new products and services. Such sources of revenue can be lucrative to cable companies keen to diversify their economic base.

Such monitoring allows access to services to be precisely controlled and consumption to be precisely monitored. It produces continuous knowledge of the flows of goods and services through massively complex systems of transaction (see

Chapter 7). It supports the development of intimate degrees of understanding of the structuring of markets, the social makeup of places, and individual consumption habits. The whole point of smart home technologies, too, is the monitoring and surveillance of all the environments of the home to collect information. The usefulness of these stand-alone technologies may be multiplied when they are linked together. Wilson even suggests that it might be possible to create a new 'era of cybernetic' consumerism by integrating domestic, home-based and electronic/cash-free retailing and credit systems, logistics systems such as JIT, and with the information gained from junk-mail response. This would lead to an extremely efficient 'cycle of production and consumption, since every consumptive activity will generate information pertinent to the modification of future production' (Wilson, 1986; 26).

The Bell Atlantic example above shows that this is not just some dystopian vision. Huge research and development expenditures are investigating precisely this shift towards personalised marketing as a boost towards more flexible, targeted and lucrative production, distribution and consumption systems. Already, at the Massachusetts Institute of Technology, a computer system known as 'Doppelgänger' has been designed to be 'embedded' into the home and so help the consumer decide the nature of the media they choose. A proponent of the system argues that 'the computer should "know" its user. . . . [The Doppelgänger system] is a system that accumulates information about people through passive sensors watching the "viewer"' (Hunt, 1994; 40). This allows it to personalise the outputs of the home's media devices to the tastes of the viewer. But it also, of course, entails the collection of unprecedented degrees of personalised information about the viewer which can easily be used for consumption targeting or resold as a commodity.

These developments lead Vincent Mosco to suggest that there is a lot more to home shopping and banking, and the shift to 'pay-per' facilities, than their attractiveness (or otherwise) as services. Much wider issues to do with privacy, civil liberties and social equity are at stake. Mosco argues that

buying into the pay-per society means more than instant shopping or movies. It means providing companies and governments with enormous amounts of information on how we conduct our daily lives. Since it is increasingly essential for us to use the new technologies to bank, shop or even work, we resign ourselves to the loss of privacy. If you need a job you keep quiet about the relentless monitoring.

(Mosco, 1988; 12–13)

Any widespread shift towards tele-shopping, information malls and even the use of virtual reality (see above and Chapter 4) will also rely on the detailed and continuous surveillance and scrutiny of customers' consumption habits. To 'try on' clothes or 'talk' to 'virtual assistants', and for truly personalised services, retailers will need to maintain intimate details of those doing the teleshopping. As Richard Tompkins argues, 'detailed analysis of customers' buying habits will allow retailers to build up much more accurate pictures of who their customers are and what they want' (Tompkins, 1994; 11).

Once again, these real and potential shifts challenge old and accepted boundaries between the public and private spheres of both electronic spaces and urban places. Previously private household spaces and activities can become the sites of intense and continuous electronic surveillance because of the fundamental nature of the interactive telematics networks that cross through them. These processes have major implications for privacy and civil liberties. Again there are contradictions and ironies here as 'the increasing privatisation of social existence is accompanied by decreasing individual privacy' (Robins and Hepworth, 1988; 170).

SURVEILLANCE AND SOCIAL CONTROL

A second area of growing telematics-based social surveillance is in social control and the workplace. The control and surveillance capabilities of telematics are being widely explored as tools for new methods of social control in cities that go beyond simple incarceration in prisons. In 1991, over 4.3 million Americans were under 'correctional supervision' at home (Gowdy, 1994). The burgeoning costs of the American prison programme, for example, are leading to the widespread use of electronic tagging for low-level offenders, who are free to maintain some semblance of daily life through 'walking prisons' (Winckler, 1993). 'Less dangerous offenders now are confined to the home, except to go to work and run errands, freeing jail space for more dangerous criminals' (Gowdy, 1994). Anklet transponders, linked to telephone modems, provide continuous monitoring of the location of offenders. Newer 'smart' systems promise a much more fine-grained and tailored control over the behaviour of offenders. For example, in a retailer, the 'arrival of an ankleted shoplifter would set off a silent alarm, and the system would identify the offender to the store management' (Winckler, 1993; 35). When linked to wider urban surveillance systems through city-wide radio networks – which will be available between 1995 and 2000 – the movements of all ankleted offenders could be correlated with the incidence of crime in time and space to actually help in conviction. 'Every place the offender went – and the time he or she was there – would be recorded and compiled and could then be cross-indexed against known crime scenes and times' (Winckler, 1993; 35). With inter-linked computers now threaded through the physical fabric of cities, through the computerisation of urban transport and road traffic systems, the potential for such real-time tracking of social movement is raising increasing concern. The Singapore mass transit system, for example, integrates a real-time tracking facility as a direct policy against 'subversion' (Jenkins, 1992).

Surveillance technologies may also be used to make political and security surveillance much more sophisticated. Trends towards the introduction of 'smart cards' – plastic cards with embedded microchips – allow individuals actually to carry around their own 'digital images' for the purposes of private or public transactions. Smart health and identity cards were introduced in France in 1994 and many nations are exploring their possibilities. The European Commission sees 345 million smart cards carrying passport, driving licence, police records and medical information as being the solution to fraud, drug trafficking, and terrorism within a fast-growing Single European market (Doyle, 1994). The 178 federal agencies in the United States are also moving towards the integration of their systems based on telematics and smart cards (Eder, 1994). In the UK, a nationally distributed computerised database of DNA signatures has been unveiled as the latest technology in the fight against serious crime.

The social polarisation of cities described above, matched by political shifts to the right, are fuelling an increasing use of telematics as a surveillance tool for improving social control over welfare systems. Once again, these trends seem most advanced in the United States. In California, for example, political initiatives exploiting fear of crime and the potent image of 'welfare fraud' has led to the electronically fingerprinting of welfare recipients (Davis, 1993). Los Angeles has been fingerprinting welfare recipients since 1991, and has a new plan to extend the system to welfare mothers and their children, adding 300,000 more sets of digital fingerprints to their files (Davis, 1993). Electronic fingerprints then provide a powerful, unique digital link between welfare and police computers systems. These trends lead Jim Davis (1993) to note caustically that 'computers are more likely to be used, by the police or the welfare agency, *against* a poor person, than they are to be used *by* a poor person'.

Other state efforts are being made to survey and monitor on-line social interactions, which are extremely difficult to access currently. Recent debates in the USA have centred on a new microprocessor known as 'Clipper' which is being mooted as part of the Clinton–Gore National Information Infrastructure plans

(Agre, 1994). The National Security Agency (NSA) and the FBI hope that a clipper will someday inhabit every phone and computer in America, allowing authorised access to the nation's telephone and telematics traffic for surveillance purposes. This would allow the 20–30 million people who communicate over the Internet to be monitored.

More general access to real-time information on social and economic behaviours is crucial for private firms and state agencies as they strive to cut costs, improve competitiveness, and increase efficiency. Nearly 30 million or 40 per cent of office jobs in the USA are now continually monitored via on-line telematics connections using private company telematics networks; between 1990 and 1992 more than $500 million was spent on surveillance software in the USA by more than 70,000 companies (Aiello, 1994). Such aspects as key stroke counting, pay-per-keystroke, electronic mail monitoring and the monitoring of the frequency and length of toilet visits are now common (Aiello, 1993; DeTienne, 1993). The information service and back offices we discussed in the last chapter make particularly widespread use of real-time monitoring systems. A Californian company markets its products to company managers as follows: 'look in on Sue's computer screen. . . . In fact, Sue doesn't even know you're there! Hot key again and off you go on your rounds of your company. Viewing one screen after another, helping some, watching others. All from the comfort of your chair!' (Quoted in Aiello, 1994; 500).

These innovations are an essential part of the wider drive towards increased efficiency and an intensification of information labour. They allow wages and benefits to be closely tied in with personal performance (DeTienne, 1993). The use of automated 'active badge' office systems is also growing fast. These track the movements of staff, maintain office security, support work-pattern analysis, and route communications (Fala, 1994). Future shifts towards expert systems in employment monitoring look set to intensify the extent to which computers are intrinsic elements of the management and monitoring process in many sectors (DeTienne, 1993).

SURVEILLANCE AND CRIME PREVENTION

A further set of issues relate to the use of telematics technologies to create continuous electronic images of urban places themselves – so supporting the drive towards the surveillance-based city. As cities become more home-centred,

Figure 5.8 'Generation Why', by Tony Reeve and Steve Way

Source: The Independent, 20 February 1995

neighbourhoods are becoming more defensive against incursions and perceived threats to property values. This process is driven largely by rising crime and more importantly, rising fear of crime. The security and surveillance industry is now the fastest growing in the UK and the USA; domestic houses are the focus of ever-more sophisticated alarm and infrared, movement and (real and mock) Closed Circuit Television (CCTV) sensors (Squires, 1994). This growth of investment means that 'quite literally, a person going about his or her daily routine may be under watch for virtually the entire time spent outside the house' (Squires, 1994; 396) – through CCTV systems covering shops, health and government institutions, Automatic Teller Machines (ATMs), petrol stations, road and bus stations, leisure and entertainment centres, workplaces, shopping precincts, sport stadia, schools, city centres, road networks and even whole cities. Here, 'technology is being used to single out those who "do not belong"' (McKie, 1994), or, more importantly, those who are perceived not to belong by those in control of these systems. This general tendency is linked with the social and spatial polarisation of the urban fabric. As cities become more socially fragmented, the distrust and fear of the 'other' in cities – whether this be other racial, social, cultural or spatial groups – becomes exaggerated. This then feeds back into the process of segregation in physical space (Loukaitou-Sideris, 1993). In these circumstances of fear and distrust, telematics-based surveillance systems are being mobilised as technical-fix solutions and as electronic guardians between the elements of the polarised matrix of the post-modern urban landscape.

Most common is the use of telematics to help protect and enforce the privileges of social élite areas and areas of economic investment – the corporate office enclaves and new consumption spaces of the post-modern city. In the global command centres, surveillance is especially intense. Allen and Pryke (1994; 469) argue that in the City of London, 'the spaces of security in the City are . . . a straightforward *extension* of dominated space. Security guards produce controlled spaces through their spatial practice, yet they do not occupy these spaces. They secure the formal spaces of finance through their own rhythms and rituals'. The debate on 'smart' and 'intelligent buildings' also stresses heavily the need for electronic surveillance and entry control systems. For example, the 'telehouse' development in London's Docklands – which offers back-up computing and accommodation facilities to global information firms in the City of London – boasts a 'sound-sensitive external fence which can detect a sparrow landing on it, infrared and videophone surveillance an cameras everywhere. Inside, customers [need] PIN numbers to head from chamber to chamber' (Quillinan, 1993; 14).

Increasingly, however, the improved capabilities for integrating large numbers of sensors together – usually CCTV cameras – is leading to the development of wide-area urban surveillance systems covering large districts or even entire cities. City and nationwide surveillance services, such as those offered by the Telecom Security firm in the UK, provide round-the-clock protection by linking household burglar alarms to distant monitoring centres via home telephone links. Thus, as we saw in the last chapter, the electronic flows generated by the growth of surveillance are themselves helping to support the new patterns of economic geography, where services are delivered to cities from distant (even inter-continental) locations.

This process reaches an extreme in the United States, where the extent and degree of ghettoisation and social segregation dwarfs that in most European cities. In Houston, for example, a six-mile network of underground sidewalk tunnels and tubes has been developed for the city's corporate office élites (Norfolk, 1994). These link the main business centres of the central business district, are closely monitored by a CCTV system and are hermetically cordoned off from the street altogether – which becomes ghettoised and undermined. As Simon Norfolk describes,

in Houston, white Americans still come to work. They can now walk safely through the downtown area without fear of crime or of rubbing shoulders with those they perceive to be the criminal classes. . . . The whole logic of the system is that owners can control who enters it. What happens *out there* has nothing to do with respectable, automobile-abiding citizens.

(Norfolk, 1994; 26. Original emphasis)

Other American cities present even more extreme pictures of electronic and physical 'fortressing'. Here, electronic monitoring and closed circuit television systems are bolstered by 'armed response teams' in the 'fortressing' of neighbourhoods (Davis, 1990; 223; Dillon, 1994). Crucially, the drive towards social exclusion and fortressing is a joint physical and technological process. The processes of privatising access to urban places, through post-modern architecture, private security guards, the building of controlled plazas and the walling of neighbourhoods, are supported by a sophisticated array of electronic monitoring and surveillance technologies – from computer communications systems in police departments to telematics-based alarm systems, infrared sensors, motion detectors and CCTV. Gated, walled, master-planned communities are increas-ingly common in the suburbs of all large American cities (and are increasingly on the agenda in the UK). Over one-third of all new communities in southern California are now gated; there are twenty-five alone in Dallas. Sales of homes in such places rose 17 per cent in 1992 (Dillon, 1994). But these processes reach their apogee in Los Angeles, as shown in the account by Mike Davis (1990). He offers the following sobering account of the fortressing and militarisation of American urban life:

Welcome to post-liberal Los Angeles, where the defense of luxury lifestyles is translated into a proliferation of new repression in space and movement, undergirded by the ubiquitous 'armed response'. This obsession with physical security systems, and, collaterally, with the architectural policing of social boundaries, has become a zeitgeist of urban restructuring. Yet contemporary urban theory, whether debating the role of electronic surveillance in precipitating 'postmodern space', or discussing the dispersion of urban functions across poly-centred metropolitan 'galaxies,' has been strangely silent about the miltarization of city life so grimly visible at the street level.

(Davis, 1990; 223)

In American cities, this technological mobilisation can reach military proportions, as a nexus emerges between crime fighting, the drug war and the pervasive shift towards the defence of the middle-class home and its immediate environment impinges into city politics. In Los Angeles, a dedicated geostationary satellite for the police department is being considered to back up a fleet of 50 surveillance helicopters with infrared sensors, and a military-style communications and computing system (Davis, 1990; 253). To Davis, this process means that the Los Angeles police department 'will try to acquire the technology for the electronic battlefield and even star wars. We are at the threshold of the universal tagging of property and people – both criminal and non-criminal (small children, for example) – monitored by both cellular and centralised surveillances' (Davis, 1990; 253).

As Box 5.3 shows, the increasingly widespread development of city centre CCTV networks in the UK presents a more prosaic but equally important example of the ways in which electronic surveillance interacts with changing notions of the physical space of cities. The tracking and monitoring capabilities of national identity 'smart' cards, home telematics and road transport informatics systems means that the potential to integrate networks into real time 'urban panopticons' seems close.

BOX 5.3 CITY CENTRE CLOSED CIRCUIT TV SYSTEMS
IN THE UK

City centre CCTV was placed on the agenda in the UK by the case of James Bulger – a three-year-old boy who was abducted and murdered in 1993 in Liverpool. His murderers, only a few years older, were tracked and, ultimately, arrested because they had unknowingly left CCTV images of themselves.

In the climate of fear built up by this and other cases, seventy-five towns and cities across the UK have installed comprehensive and high-definition CCTV surveillance systems. In 1994 alone, over fifty British local authorities built wide area CCTV systems; by November 1994, around 95 per cent of local councils were considering them (Davies, 1994). Increasingly, such networks are being seen as a prerequisite to cutting crime, improving both consumer and business confidence, and underpinning the competitiveness of city centres in the UK – all in a cost-effective manner. Backed up by central government finance, support and the abolition of the need for planning permission for cameras, the British market for such systems has doubled since 1989 from £170m to £300m. There are over 150,000 professionally installed CCTV cameras in British towns and cities; over 500 more are installed each week.

The aim of these systems is to allay public fears of crimes by giving visitors to city centres a sense of security; related objectives include the deterrence of crime and disorder and the improvement of the economic vitality of city centres. Research shows remarkably little concern about the privacy and civil liberties aspects of these systems. A recent survey in Glasgow – where the largest system, a 32-camera network was switched on in November 1994 – showed that 90 per cent of people supported the project, 66 per cent believed the system would make the city centre a better place and 40 per cent said it would make them visit the city centre more regularly.

The system in central Newcastle (see Figure 5.9) presents a good example. These sixteen cameras, installed in 1991, cover all areas of the city centre and are linked back by microwave to the city's main police station. The recordings are used for evidence in convictions. A police superintendent remarked recently that the system 'gives me effectively 16 full-time police officers on the beat 24 hours a day all taking notes' (BBC TV, 1993). By the end of 1994, the Northumbria Police claimed that the system 'will have effectively reduced crime in the City Centre by some 6,000 offences over the last 3 years. The provision of CCTV has enabled us to make quite literally hundreds of arrests covering a wide range of offences including public order, theft from the person, robbery, auto crime and even arrests involving the possession and supply of drugs' (Durham, 1994). The system is linked to a parallel radio-based alert system linking the security guards in the city's main retailers. In combination, the system provides a real-time tracking and alert capability for people moving through the city centre. In Newcastle, for example, a second CCTV system based on more modern technology is being developed for the entire western end of the city, as part of the City

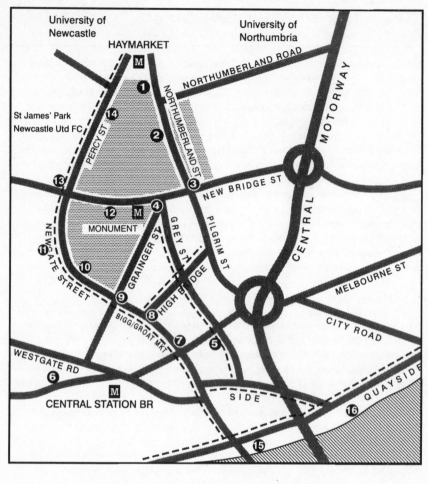

1 Ridley Place
2 Northumberland Street
3 New Bridge Street
4 Monument
5 Dean Street
6 Neville Street
7 Groat Market
8 Bigg Market
9 Grainger Street
10 Clayton Street

11 Newgate Street
12 Blackett Street
13 Gallowgate
14 Percy Street
15 Swing Bridge
16 Quayside

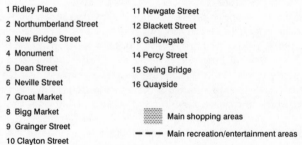

Main shopping areas

--- Main recreation/entertainment areas

Figure 5.9 Map of the Newcastle city centre closed circuit TV system
Source: Northumbria Police, 1991

Challenge urban policy initiative for the area.

Such systems bring concerns about civil liberties and an 'overspill effect' as crime simply moves elsewhere to the areas outside the widening coverage of surveillance systems. In one Scottish town, Airdrie, a CCTV system in the centre cut crime but crime levels rose in the peripheral areas (Dawson, 1994). New technological and political developments seem likely to intensify the threats that such surveillance systems pose to civil liberties and privacy. Microcameras and facial recognition technology is developing fast, both for in-store security systems and wider city centre networks. In a new experimental project, BT were recently reported to be working with the Massachusetts Institute of Technology and the major British retailer, Marks and Spencer, on digital image and television-based computer system to be installed in its stores (McKie, 1994). In this project, real-time cameras linked to image data-bases will alert security staff of the presence of convicted shoplifters in their stores through advanced facial recognition software. Accuracy is said to be 'greater than 90 per cent' (McKie, 1994). In the long run, BT anticipate major new telecommunications markets. For example, 'all commercial outlets in a town could be linked and an alarm be set off the moment a person who has been seen shoplifting in one store enters another' (McKie, 1994).

These systems, and their possible linkage into wider networks, raise important questions about privacy, civil liberties, the threat of error and the possibility for their misuse as tools of social control. They give their controllers an unregulated set of powers to determine who has access and who is excluded from the public spaces of cities, based on their own prejudices about the visual appearance of people. They may be used to exclude social groups not seen to be befitting to a commercial space. As Naughton (1994) argues, they bring the spectre of social manipulation and segregation to previously public spaces where people mixed freely – all driven by perceptions and stereotypes of the 'other' or 'others' who are to be excluded. These systems, he writes, are 'already used as a means of anticipating trouble. "See that crowd of boisterous teenagers over there on camera nine? Let's get someone there before they get out of hand". Or "What's that guy with dreadlocks going into Watches of Switzerland for?" The technology will become a way of singling out those who "do not belong" in a particular environment, and of taking preemptive action to exclude them' (Naughton, 1994). In the Newcastle system between 1993 and 1995, for example, 9 per cent of the incidents with which the system was used involved dealing with begging and vagrants and 19 per cent involved suspicious youths (Northumbria Police, 1995). Liberty, the UK civil liberties organisation, argue that 'no society which values freedom should permit the creation of this surveillance infra-structure. One of the responsibilities of living in a free society is to resist policies of "crime prevention" that may one day become tools of social control' (quoted in Davies, 1994).

CYBERSPACE AND THE CITY: VIRTUAL URBAN COMMUNITIES AND SOCIAL INTERACTION

A fifth and final key element in the relations between telematics and the social life of cities is the exploration of telematics systems as 'virtual' networks supporting new types of social interaction 'within' electronic space. Many commentators now argue that such virtual communities offer support for specialised social groups to function irrespective of geographical separation. These networks operate to support 'imagined communities' (Anderson, 1983), without the spatial and temporal barriers of urban physical space. They only actually 'exist' in an abstract sense within the managed flows of digitised data and the digital processing facilities of computers and telematics systems. In addition, the convergence of computing and media technologies with telecommunications is fuelling the growth of the 'virtual reality' industry, where electronic spaces actually create artificial environments within which virtual 'people' (from any location that is linked into the network) can interact as electronic constructions of them (or other) selves.

Because they can be constructed ad infinitum, utopianists stress that cyberspace can be seen as an infinite resource offering perpetual solutions to the finitudes of the real world. Don Mapes even suggests that 'we have to do away with our territoriality. The good news is: cyberspace is big. It's basically infinite. Earth is limited, it's finite. In cyberspace, if you don't like it, you can move out to the next frontier. There's always another continent in cyberspace' (Mapes; quoted in Channel 4, 1994; 19).

Particularly important as the main support for these virtual communities is the Internet – a global network of computer networks that has grown from the US military research and development telematics networks of the 1970s (Batty and Barr, 1994; Ogden, 1994). The Internet reaches over 150 countries, is used by around 30 million people and is growing at the rate of one million new users per month (Ogden, 1994; 714). It supports a massive range of specialised news groups, information and list services, bulletin board services and conferencing networks on a whole gamut of subjects, from Bauhaus architecture to zoological science. Specialised networks address the needs of many minority and disadvantaged communities or lobby groups: SeniorNet for the old (Furlong, 1989), PeaceNet for the peace movement (Downing, 1989), Greenet for environmental groups, etc. This movement has led some to speculate on the emergence of a 'global civil society' (Frederick, 1993) based on these networks. By 1997, the Internet is expected to have 120 million users around the world. Increasingly, Internet 'log on' is being offered as part of mainstream telecom services by large companies such as BT.

These shifts lead Eli Noam (1992; 409) to speak of a general trend towards the emergence of 'electronic neighbourhoods' on a global basis based on group interests. Rather than emerging as some convivial 'global village' however, he argues that they 'create the world as a series of electronic neighbourhoods. . . . In the past, [physical] neighbourhoods had economic and social functions. In New York, for example, there [was] Chinatown, Little Italy, the Garment District, Wall St [etc]. But now group networks can serve many of the functions of physical proximity. . . . They create new ways of clustering, spread around the world'.

But how do such virtual communities operating in electronic space interact with and affect urban places? There are two key issues here: the way the growth of virtual communities relates to the public sphere of community interaction in urban places, and the possible implications of the widespread emergence of truly virtual 'cities' as simulations and software constructions within electronic space.

VIRTUAL COMMUNITIES AND THE URBAN PUBLIC SPHERE

While telematics can support the disembedding of social relations away from specific geographical localities, many argue that they can also be used to strengthen the civic, public dimensions of cities by providing public fora for debate and interaction. Some argue that the multitude of specialised virtual communities on the Internet is evocative of a sense of convivial urbanism that has been lost in the physical and social transformations towards post-modern urbanism. Efforts to commercialise and regulate the Internet are growing fast, and a major struggle is developing between those wanting to protect the open, free and unregulated aspect of the networks and the major business and government interests trying to commodify and regulate it as a 'digital gold rush' (Ogden, 1994; 720). Nevertheless, the Internet is still often 'lauded by communications purists as a model for a truly anarchic society where information is freely exchanged, control and regulation are impossible to exercise and where there is no hierarchy' (Bell, 1994). Theodore Roszak (1994; 185) notes its 'spontaneously democratic and libertarian spirit' and argues that 'the coffee houses of eighteenth-century

London, the cafes of nineteenth-century Paris were rather like this: a gathering place for every taste and topic'.

To some, virtual communities offer potential for resuscitating the public sphere of cities and creating a new form of urban conviviality – a new urban vision even. Geoff Mulgan, for example, argues that 'given that the architecture and geography of large cities and suburbs has dissolved older ties of community, electronic networks may indeed become tools of conviviality within cities as well' (Mulgan 1991; 69). These debates lead Patsy Healey and colleagues to ponder whether the physical city is:

to be replaced with a *virtual* urbanity, a city of the mind, enabled by telematics? The hope of this new technological revolution is that it will provide channels through which knowledge and information can be democratised, dispersed around the diversity of relational webs in urban regions. Could this clever technology provide the basis for dispersing power out from current nodes, and empowering and articulating diverse democratic voices? Could it provide a route through which marginalisation could be reduced? If so, the constraint of space could finally be annihilated, allowing people to network globally with similar communities of interest, thereby mobilising the power to contest specific developments. The need for place-based political mobilisation would disappear.

(Healey *et al.*, 1995; 277)

But what is really encouraging the growth of virtual communities? Perhaps people explore these networks for the conviviality that they find hard to discover in contemporary cities because of the shift towards commercialisation, the packaging and theming of urban landscapes, and the wider fortressing of affluent neighbourhoods?

To some commentators, virtual communities represent the search by people who are alienated by the repressive and instrumental character of contemporary urban life. Howard Rheingold, a keen advocate of virtual communities, suspects that 'one of the explanations for the [virtual community] phenomenon is the hunger for community that grows in the breasts of people around the world as more and more informal public spaces disappear from our real lives' (Rheingold, 1994; 6). Rosenberg *et al.* (1992) argue that 'Multi-User Dungeons' (MUDs) – virtual role-playing games accessed from all over the world – allow people to 'create and dream and live in ways that they cannot in their real lives'. Ralph Schroeder (1994; 524) sees the exploration of 'virtual worlds' as heralding a 'new technology-centred era which will release human beings from the material constraints of their current lives' – including, ironically, the dehumanising effects of modern science and technology.

An example of these trends can be found with the famous *messagerie* services, which make up over half the traffic on the French Minitel system. These seem to thrive because a largely white and middle-class user base (75 per cent male) has developed a whole culture of interaction known as *téléconvivialité* based on anonymity and disguise (Poster, 1990; 121). Many of these interactions are – as with those on Freenets and the Internet – based in electronic analogies of physical urban spaces: 'electronic singles bars' and 'computer cafes'. Freedom and flexibility are the marks of these electronic spaces: 'there are no taboos, one can talk about anything, to whoever one wants, at any hour of the day or night. One can look for a soul mate, pleasantly converse without worry to a total stranger, reconnect with one's regular discussion partners' (Marchant, 1987; 119). Mark Poster offers fascinating speculations that people enter such spaces to escape from the highly regulated and repressive forces they experience in their physical day-to-day life in cities:

One can speculate that the Cartesian subject, trapped in a world of instrumental rationality of its own making, has discovered itself trapped in a dystopia. Playfulness, spontaneity, imagination and desire all are diminished from the public and private domains of career-building. Only the messagerie, with its fictional self-constitution and perfect anonymity, offers an apparent respite from what has become for many a treadmill of reason.

(Poster, 1990; 120)

While this search for 'playfulness' may be important, can telematics and virtual communities really substitute for the declining public spaces and sense of community in the real city? We have doubts about the extent to which convivial, face-to-face interaction and the public, democratic realm of urban places can be genuinely substituted by virtual communities. We are sceptical of the wilder claims for virtual communities. While it is easy to romanticise some long-lost 'public realm' or 'sense of community' in historic urban life, we argue that public interaction on streets and in public spaces offers much more than can ever be telemediated. Real face-to-face interaction, the chance encounter, the full exposure to the flux and clamour of urban life – in short, the richness of the human experience of place – will inevitably make a virtual community a very poor substitute for the kinds of urban communities celebrated by Jane Jacobs in her famous book *The Death and Life of Great American Cities* (1962). McBeath and Webb (1995; 7–8) argue that the experience of virtual communities is an illusion brought by the use of computer technology, offering 'a fantasy through which we can live in apparent proximity to others, talk to them and express feelings', but which 'ignores a dimension of community which we consider central to the concept, namely, its affective aspect, the dimension of the fellow-feeling bound to "being together". This is the emotional/feeling strand of solidarity'.

In fact, much of the social interaction on the Internet can be banal (A. Brown, 1994). The enthusiasm to fill the system with information often leads to a glut of low-quality, out-of-date information of questionable usefulness (Roszak, 1994; 165). Not surprisingly, given the social inequalities in access to computers we have noted above, network traffic is also often dominated by the technologically 'converted': a very small and extremely unrepresentative set of people, many of whom are paid to be on-line. The user population of virtual communities tends overwhelmingly to be a white, male technological élite. 'Cyberspace is a sexist, male dominated (non) place' (Burrows, 1995; 10). These networks tend, like other telematics systems, to be the preserve of highly mobile, affluent (and usually male) groups who are exploring these systems as a way of keeping in touch with highly specialised peer groups on a global basis. The poor, the excluded and the disenfranchised who have tended to suffer most from the polarisation and privatisation processes in contemporary cities tend to be overwhelmingly excluded from virtual urban communities because they do not have the the the skills and finance necessary to participate.

There are also severe dangers that reliance upon virtual communities to address urban atomisation and fragmentation may actually promote the further withdrawal of urban social interaction from urban places on to specialised electronic networks, so exacerbating the problems. Because virtual communities tend to be so specialised, there are risks that 'ethnic groups [will] collect in their own electronic communities, libertarians speak only to libertarians . . . inevitably, the effect will be to shatter local geographic communities and ultimately weaken the national community' (L. Brown, 1994). The clear risk is that 'telematically linked communities could fragment our larger society, enabling each of us to pursue isolation from everything different, or unfamiliar, or threatening, and removing the occasions for contact across lines of class, race and culture' (Calhoun, 1986). In fact, it seems most likely that virtual communities will simply develop to reflect wider inequalities within cities, as a highly segmented set of private spaces that are as tightly controlled and surveyed as the most dystopian vision of 'fortress' America. To quote Mike Davis (1993), the danger is that, far from resuscitating the urban public realm, *urban cyberspace* – as the simulation of the city's information order – will be experienced as even more segregated, and devoid of true public space, than the traditional built city'.

VIRTUAL REALITY (VR) AND VIRTUAL COMMUNITIES

Speculation, and not a little hype, now centres on the extension of these message-based electronic communities into networks supporting 'virtual reality' (VR) services. Here we come back to many of the ideas of home telematics, 'information malls' and the 'smart home' covered above. Virtual reality technologies combine digitised media equipment with artificial intelligence, computing and high-capacity telecommunications to provide truly interactive 'virtual environments' within which people can immerse them (or other) selves as electronic constructions interacting with the virtual environment (Biocca, 1992). This happens through audio and video sensors combined with gloves, suits and helmets which monitor human actions and integrate people sensorily into software environments. As these technologies become more and more sophisticated, William Mitchell (1995) predicts that 'you will be able to immerse yourself in simulated environments instead of just looking at them through a small rectangular window. This is a crucial difference: you become an inhabitant, a participant, not merely a spectator'.

Virtual reality games, sport shows and entertainments drive innovation here, and some cable companies are looking to deliver such services to cable subscribers. Peter Cochrane believes that these developments will mean that urban life after the year 2000

will be about 'being there'. Virtual reality will probably become a key technology. . . . Instead of buying a television set, a camcorder, a computer and a telephone, you will buy one terminal that will integrate all of those capabilities together. Then we will be able to enter the real world from a distance – to go to the Olympic Games, Wembley Stadium or Wimbledon or whatever from our living room or office. We will be able to have new experiences and go to places we've never been before. I would guess that we will see 40 years experience crammed into just a five-year period.

(Quoted in Harrison, 1995; 7)

The widespread adoption of such sophisticated electronic spaces would signify a further intensification of the separation of places from space, and the blurring of the real and simulated, public and private, within the post-modern city. Gale (1990) suggests that such technologies will inevitably change their users' perceptions of the 'real' world. Biocca (1992) argues that 'people spend approximately 7 years of their life watching television. If people eventually use VR technology for the same amount of time that they spend watching television and using computers, some users could spend twenty years or more "inside" virtual reality. Greater levels of temporal immersion may match greater levels of sensory immersion' (Biocca,1992; 14). Taylor and Rushton (1993; 40) even predict that such systems might lead to a kind of 'social autism' with 'people living in a virtual world in preference to the natural world'. It is hard to speculate on the ways in which such changes might affect cities, but there seems little doubt that the elaboration of a whole complex of virtual realities within which sizeable portions of urban populations spend sizeable portions of their time will alter the ways in which they approach and use the physical spaces of cities and the nature of social interactions more widely.

CONCLUSIONS

In a recent interview Edward Pilkington of *The Guardian* newspaper talked to the architect and urbanist, Richard Rodgers (Pilkington, 1994; 39). What, he asked, would be the fate for Britain's cities if a new set of urban ideals, and the mechanisms to achieve them, were not built up to address the growing sense of urban crisis in Britain? Rodgers's response was stark and simple: 'Blade Runner. The poor will be ghettoised in their estates, walled in by police and by the barriers of unemployment. The rich will be in their ghettos too, electronically and physically fortified. Everyone will be separated in his or her own security castle. There will be no society.'

In this chapter we have explored a wide range of areas where the social aspects of urban life are becoming more and more mediated by telecommunications and telematics. We have found the utopianist and deterministic rhetoric promising telematics to be the vehicle for a new, egalitarian transformation of capitalism to be profoundly misleading. The situation is much more complicated and diverse. Complex processes of social construction of technology are underway within which new social applications are being developed. But this is going on against the broader political economic backcloth supporting social polarisation and the commercialisation of both electronic spaces and urban places.

We do not suggest the inevitable and universal emergence of some urban dystopia. We stress, first, the diversity of the experience of cities in the development of social telematics, and second, the fact that remarkably little is known about the detail of the development underway. Not all cities are like Los Angeles or Las Vegas; just because the rhetoric emanating from these cities tends to dominate debates here does not mean that the situation is not more positive elsewhere.

Nevertheless, much of what we have found in this unprecedentedly broad review does seem to support Richard Rodgers's rather pessimistic outlook. As part of the ongoing economic, social and cultural change surrounding the shift to post-modern urbanism, telematics do seem to be helping to support the emergence of new, more highly polarised social and cultural landscapes in cities. The truly public dimensions of cities where citizens interact and encounter each other in physical space seems threatened. Urban trends seem to be supporting instead a shift towards tightly regulated private and semi-private spaces – both physical and electronic – oriented towards the exclusion of groups and individuals deemed not to belong. Most of these are directed towards market-based consumption and treat people very differently according to their ability to consume rather than according to universal notions of rights or citizenship. Related shifts in telecommunications policy away from public, social orientations and towards telematics markets and consumerism are also involved.

There are growing signs that the urban landscapes which result from these shifts are based on heightened polarisation and inequality, ever greater degrees of segmentation, and a shift towards a fortressed, home-based and paranoid urban culture. This in turn can fuel fears of the 'other' and lead to further surveillance, fortressing and exclusion. The highly uneven 'power geometry' which helps to shape these landscapes does seem to be increasingly geared towards allowing certain social groups unprecedented power to exercise control over space. Such groups are the lucrative market segments benefiting from the 'cherry picking' of telematics and service companies (see Figure 5.10). At the same time, marginal groups face being spatially trapped and further disenfranchised by a combination of economic, political and technological shifts. While the global capitalist classes seem to be uprooting on to global networks to feed and service the globalisation of economic relations, so it seems that the 'information poor' without even a basic telephone become more and more ghettoised beyond the physical and electronic margins of mainstream society. Those excluded from the telematics-based social mainstream are 'pushed into a lonely and isolated world of a poverty which is highly place-bound . . . , the forced isolation and anonymity of the depressed inner cities' (Squires, 1994; 398). The spaces of exclusion often then face the prospect of further marginalisation to be 'socially dumped' as the types of physical and electronic restructuring in services that we saw in the last chapter take place. Figure 5.10 outlines how current shifts towards market-based telecommunications are helping to underpin this combined process of cherry picking and social dumping in cities.

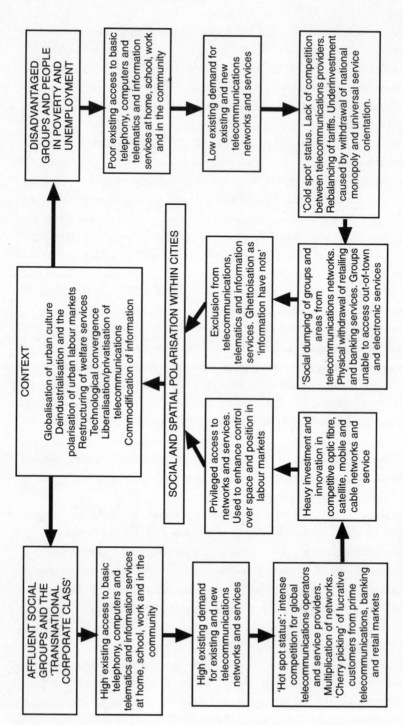

Figure 5.10 Cherry picking and social dumping: a schematic diagram showing how trends in telematics are underpinning the shift to more socially polarised cities

The social and electronic borders between these different zones seem to be emerging as the sites for the most intense investment in electronic surveillance systems and physical walls and gates. These trends reach their apogee in Los Angeles. Mike Davis (1993), one of the most powerful dystopian commentators of these urban trends, argues that a shift towards 'information apartheid' is occurring in Los Angeles. 'Southcentral L.A', he writes, 'is a data and media black hole, without local cable programming or links to major data systems. Just as it became a housing/jobs ghetto in the early twentieth century *industrial city*, it is now evolving into an *electronic ghetto* within the emerging *information city*'.

In virtually all we have seen, then, trends towards social fragmentation and the erosion of the civic, public aspects of cities seem to be prevailing over forces of social integration and what we would consider to be socially beneficial telematics innovation.

Yet all is not gloom and doom. The social constructivist approach teaches us that alternative dynamics of social telematics development are possible. Many examples of social innovation in both telematics and urban places are emerging which hope to help counter the polarisation being driven by wider commercial forces. These, and innovative urban and community development strategies, offer some hope that what Judith Squires (1994; 397) calls 'new enclosures' can be developed which can help resuscitate the public dimensions of cities, underpin progressive social change and help in the emergence of less depressing urban futures. The key task seems to be supporting these parallel processes of social innovation in urban places and electronic spaces so that new 'homes' for convivial and democratic urban social life can be constructed amidst what Squires calls the 'flux and flow of the globalisation of social relations'.

The task of shaping these new enclosures – such as social and economic development strategies and the construction of public electronic spaces – is a daunting one. Resources are scarce; wider forces promoting polarisation and fragmentation seem almost overwhelming; and pitifully little is known about the policy requirements in this new field. But here there are at least hopes that telematics can be used to help forge new models of conviviality and social cohesion within and between cities. We will explore some publicly supported examples of electronic public spaces in Chapter 9.

6

URBAN ENVIRONMENTS

INTRODUCTION

This chapter examines the relationships between the urban environmental crisis and the role of the telecommunications sector. The links between the physical environmental problems caused by modern urban life and the burgeoning interest in the role of telecommunications in cities remain curiously disconnected debates (Marvin, 1994). While researchers and policy-makers are developing a greater understanding of the role of telecommunications in many aspects of urban life, they have little insight into the environmental role of telecommunications in cities (Cramer and Zegveld, 1991).

Because of this neglect there are serious difficulties in addressing relatively simple questions about the linkages between the two sectors. This raises three key questions. First, why are the relations between the urban environment and the telecommunications sector so poorly developed? Do demands for the development of the 'sustainable city' compete or coincide with the concept of the 'information city'? Second, what are the environmental implications of telecommunications in cities? Are telecommunications clean, environmentally benign technologies that simply substitute information for physical resource flows along energy and transport networks? Or, alternatively, do telecommunications encourage more movement by both generating new demand for trips and enhancing the attractiveness of travel by improving the efficiency of transport networks? Finally, what is the potential for shaping the environmental role of telecommunications in cities? Are telecommunications technologies of little relevance to urban environmental management? Do telecommunications follow some predefined logic which largely determine their environmental impacts? Are urban policy-makers attempting to manage telecommunications according to wider environmental criteria?

TELECOMMUNICATIONS – THE NEGLECT OF URBAN ENVIRONMENTAL ISSUES

Unfortunately, there are few satisfactory answers to these questions. Almost all the research and policy interest in telecommunications has focused entirely on their importance in urban economies, social polarisation, urban culture and the changing physical form of the city (Brotchie *et al.*, 1991). Likewise, the environmental sector has focused on the physical problems generated by transportation, energy, water and waste networks, industrial production and by buildings in the commercial and domestic sectors (Douglas 1983; CEC, 1990; Breheny, 1992; Cooper and Ekins, 1993). There are very few examples of research that have explicitly looked at the potential connections between the two areas.

Instead, the vacuum has been dominated by a perspective which assumes that telecommunications are an environmentally benign technology that can easily be manipulated to improve environmental conditions in modern cities (Harris, 1987; Lee, 1991). As we saw in Chapter 3, this view is often most powerfully expressed by utopian and futurist commentators (Toffler 1981; Martin, 1981), the providers of telecommunications services (Tuppen, 1993) and the managers of transportation networks (Saloman, 1986). In many circles, it has become an accepted orthodoxy that telecommunications networks can simply substitute for travel, save energy, reduce emissions and solve the problem of urban congestion. Anyone can see that a telephone does not directly generate the same type of environmental problems as the old, polluting and noisy energy and transport infrastructures. Even the urban environmental policy community has tended to assume that telecommunications have little to do with the causes of urban environmental problems. Instead it is usually accepted that telecommunications have significant potential to provide a solution to urban environmental problems by substituting information for travel (Irvine, 1993).

The power of this perspective means that little effort has been made in developing more critical assessments of the relationship between telecommunications and the urban environment. Many of the most powerful actors who would need to participate in a more balanced debate have strong interests in presenting telecommunications as an environmentally benign technology. Telecommunications suppliers and the providers of new electronic services like teleworking and road transport informatics are anxious to establish the environmental credentials of telecommunications networks and the services they provide. However, there is concern amongst service providers that more critical approaches could provide evidence questioning the environmentally benign image of telecommunication networks. If, for instance, telecommunications actually help to generate new demands for travel should environmental policy attempt to increase the cost of telephone calls in an attempt to reduce new trip generation?

These potential contradictions also seem to be evident in the environmental sector. Urban environmental policy-makers are searching for new ways of understanding and dealing with the complexities of the physical city. They seem reluctant further to complicate this task by acknowledging that telecommunications may actually be a cause rather than a simple technical-fix solution to urban environmental problems. For instance, if telecommunications generate environmental problems how could policies designed to manage the physical environment of transport, and buildings begin to cope with the intangibility and invisibility of the telecommunications sector? In these contexts it is hardly surprising that neither sector has much interest in a more critical analysis of the linkages between telecommunications and the urban environment.

Existing paradigms are unable to cope with the challenges raised by telecommunication – environmental interactions. Each sector is having major difficulties critically analysing the potentially complex and contradictory relations between them. Instead the gap is filled by utopian, futuristic and technologically deterministic perspectives of telecommunications as a panacea for urban environmental problems. There is nothing less than a paradigm crisis in the way in which conventional frameworks are unable to explain or cope with telecommunication – environmental interactions. We argue that there is a desperate need to develop new conceptual and analytical frameworks that allow policy-makers and researchers to understand how parallel developments in telecommunications and the physical environment challenge urban environmental policy.

Indeed, we argue that it is impossible to develop effective urban environmental policies without an understanding of the complex role of telecommunications. At the same time, the telecommunications sector needs to develop a more complete understanding of how its activities generate new types of problems for urban environmental policy. We seek to develop a framework that acknowledges that the telecommunications sector and the urban environment are likely to have extremely complex and contradictory sets of interactions. The form, direction and implications of these linkages are likely to depend on the specific social and locational contexts in which telecommunications are used. While in some circumstances telecommunications may act as a substitute for physical movement at the same time they can generate new demands for travel. New perspectives and frameworks need to be developed that allow urban researchers and policy-makers to cope with these new complexities in the links between physical and electronic environments in contemporary cities.

TOWARDS THE DEMATERIALISATION OF CITIES?

A powerful body of ideas has developed around the notion that electronic flows and spaces can simply displace or substitute for physical travel and physical urban functions. New forms of electronic flows and spaces actually undermine the need for physical spaces and flows eventually leading to the dematerialisation of cities. Electronic forms of communication and a range of telebased services are seen simply to displace the need for physical movement between home and work, while urban functions will no longer have a physical presence as services are delivered in electronic form.

But the dematerialisation scenario is not a coherent theoretical or conceptual argument. Instead, there are a number of different strands which together may enable us to understand how electronic spaces and flows interact with the physical urban environment. But the main focus is on one form of interaction – telecommunications simply displace, substitute and lead to the eventual dissolution of the physical city. These apparent potentials have often underpinned powerful utopian and technological deterministic views of telecommunications role as an environmentally benign technology that can easily be manipulated to improve environmental conditions in cities. We need to address a series of questions about this view. Where do these arguments comes from? How do telecommunications lead to dematerialisation of the physical environment? Does this perspective mean that telecommunications are a solution to urban environmental problems?

CONVENTIONAL APPROACHES: ENVIRONMENTALLY BENIGN TELECOMMUNICATIONS

Conventional approaches to telecommunications – environmental relations are dominated by utopian, futuristic and technologically deterministic perspectives. A consistent theme in these assessments is the environmentally beneficial role of telecommunications technology. The first is that telecommunications technologies are inherently benign – they use very few resources, consume low amounts of energy, and are non-polluting. The second concept is that telecommunications can unproblematically substitute for physical transport flows and movement reducing the need for travel, lowering levels of pollution and urban congestion.

In 1976 McHale argued that the 'new electromagnetic spectrum industries are, by comparison with other heavy industries, *non-resource depletive, extremely economical in their energy use and have correspondingly low impacts on the environment*' (1976; 21; added emphasis). Parker goes even further than this and argues that telecommunications can overcome physical constraints on further economic growth:

In the approaching information age, the characteristic machine is one that processes information, augmenting not human physical energy but human information processing. The difference is a significant one. Since energy is used to manipulate symbols rather than physical objects, the consumption of energy and materials can be made arbitrarily small by using smaller and smaller physical representations of symbols in information machines. This means that *in an information age unlimited economic growth is theoretically possible even though a steady zero-growth state is reached with respect to energy and materials.*

(Parker, 1976; 5. Added emphasis)

These perspectives tend to assume that the production of telematics equipment is 'largely by environmentally benign processes' and that its 'operation is likewise benign, and it facilitates similar effects as it is used in other industries' (Harris, 1987; 397). Martin argued that teletravel will be the 'new gasoline', cities will spread out to support higher qualities of life, and pollution and stress will be reduced. In a passage that illustrates many of the predictions of futurists, he writes that 'the growth of telecommunications society would relieve some of the population pressure on the cities. It could greatly reduce the pollution, raise the quality of life, and lessen the drudgery of commuting' (Martin, 1981; 192). These findings have important implications for the telecommunication sector. Lee argues that the sector '*may be concerned comparatively less about environmental problems*' because telecommunications systems are 'in harmony with nature ... are environmentally sound, non-polluting and non-destructive of the ecology' (1991; 30, added emphasis). At the same time the sector could offer significant environmental benefits: 'teleworking should not be perceived to be a mere passing trend, but as a policy which can actively tackle the problems of a dwindling supply of natural resources: it reduces commuting and also reduces pollution' (Eubanks, 1994; 42).

SHAPING ENVIRONMENTAL PERCEPTIONS OF TELECOMMUNICATIONS

Telecommunications suppliers are keenly aware of the 'new opportunities' that can develop if they take a 'proactive' stance on environmental issues (Taylor and Welford, 1994). BT, the largest telecommunications company in the UK, has developed its own environmental policy (see Box 6.1) which provides useful insights into the ways in which telecommunications companies want to position themselves in the environmental debate. In a 1992 survey on Business and the Environment 46 per cent of people interviewed felt that BT 'caused no environmental damage'. The company has 'never experienced any sustained pressure by environmental pressure groups and has not had to invest heavily to reduce polluting emissions or otherwise react to environmental legislation' (Tuppen, 1993; 24). Despite these perceptions, BT, together with other large telecommunications companies, have undertaken environmental audits of their companies' activities (Chittick, 1992; Tuppen, 1993; Taylor and Welford, 1994).

BOX 6.1 BT ENVIRONMENT POLICY STATEMENT, MARCH 1991

In the pursuit of its mission to provide world-class telecommunications and information products and services, BT exploits technologies which are basically friendly to the environment. In the sense that the use of the telecommunications network is often a substitute for travel or paper-based messages, BT is contributing positively to environmental well-being and conservation of resources.

We recognise, however, that in our day-to-day operations we inevitably impact on the environment in a number of ways and we wish to minimise the potentially harmful effects of such activity wherever and whenever possible.

As part of our continuing drive for quality in all things we do we have

therefore developed a comprehensive policy statement which will enable us to set the targets by which our efforts towards sustainable
environmental improvement can be measured and monitored on a regular
basis. In this way we aim to protect the health of our own people and our
customers while contributing to the future well-being of the environment.

We have undertaken to help every BT person to understand and to
implement the relevant aspects of this policy in their day-to-day work
through the regular communication of objectives, actions plans and
achievements. At Board level, the Deputy Chairman has specific responsibility for policy development, coordination and evaluation of performance. BT is committed to minimising the impact of its operations on the
environment by means of a programme of continuous improvement. In
particular BT will:

*   meet, and where appropriate, exceed the requirements of all relevant
    legislation – where no regulations exist BT shall set its own exacting
    standards;
*   promote recycling and the use of recycled materials, while reducing
    consumption of materials wherever possible;
*   design energy efficiency into new services, buildings, products, and
    manage energy wisely in all operations;
*   wherever practicable reduce the level of harmful emissions;
*   minimise waste in all operations and product development;
*   work with BT suppliers to minimise the impact of their operations on
    the environment through a quality purchasing policy;
*   protect visual amenity by careful siting of buildings, structures and
    the deployment of operational plant in the local environment and
    respect wildlife habitats;
*   include environmental issues in BT training programmes and encour-
    age the implementation by all BT people of sound environmental
    practices;
*   monitor progress and publish an environmental performance report
    on an annual basis.

(British Telecom, 1991)

Telecommunications suppliers are usually extremely large companies and have
a significant environmental impact in their own right. For instance, BT annually:

consumes 80,000 tonnes of paper (including Yellow Pages) annually, or 17% of total UK newsprint,
printing and writing paper . . . uses 160,000 nickel-cadmium rechargeable batteries a year in
operational equipment such as payphones, testers and hand drills . . . operates 10,000 company cars
and uses about 100 million litres of unleaded petrol and 25 million litres of diesel for its operational
fleet of vehicles, representing 1% of the fuel used in the UK by light vans and goods vehicles.

(IEE 1992; 212)

Therefore, any assessment of the environmental role of telecommunications first needs to recognise that the companies themselves are likely to have very significant impacts on resource use in the economy. Consequently, environmental auditing techniques and principles have been applied to the operation of telecommunications networks. In 1992 BT published its first environmental performance report (British Telecom, 1992a). This report monitored a series of policy initiatives designed to minimise the company's impact on the environment. These measures include the recycling of resources, minimisation of waste, energy efficient design in new buildings and services, inclusion of environmental issues in training programmes and the reduction of harmful emissions. Most of this work has tended to focus on the direct environmental impacts of particular materials rather than the wider use of telecommunications technologies.

But this has not always been the case. BT conducted research on the energy costs of different forms of communications technologies – air travel, car and the telephone. These confirm the work of earlier studies that the telephone (not surprisingly) consumes significantly less energy than other forms of communications. These findings have also helped telecommunications suppliers argue that their networks may be able to make wider contributions to sustainability. BT's Environmental Statement argues that as 'telecommunications network is often a substitute for travel or paper-based messages, BT is contributing positively to environmental well-being and conservation of resources' (British Telecom, 1991).

Telecommunications operators seem to have powerful incentives to promote telecommunications as an environmentally friendly technology that could provide ways out of our current environmental problems – particularly in regard to the potential for substituting for transport. BT's Environmental Manager argues that:

Telecommunications technology is likely to play an increasingly important role in offering a more environmentally sound alternative to travel. . . . Apart from a saving in energy the switch to telecommunication services would have other environmental benefits such as reduce noise levels, fewer new roads and lower levels of urban pollution.

(Tuppen, 1992; 81)

It is, however, surprising that even when the telecommunications sector itself has sponsored urban environmental initiatives, such as BT's sponsorship of Environment Cities in the UK, there has been little consideration of the potential role of telecommunications (Royal Society for Nature Conservation, 1990).

Despite this apparent contradiction some environmentalists have shown interest in the environmental benefits of telecommunications. For instance, one of the UK's leading environmentalists, Jonathon Porrit 'hails this fusion of communications and computing technologies as one of the "tools for sustainability"' (Irvine, 1993; 5). Although some environmentalists seem prepared uncritically to accept the environmentally benign role of telecommunications, with the exception of teleworking initiatives in California (State of California, 1990), there is little evidence that telecommuting has yet become part of urban environmental management policy. The message appears to be that telecommunications could have a positive role but there is considerable uncertainty about how this role is translated into specific initiatives.

THE DEMATERIALISATION OF THE INFORMATION SOCIETY

There is a wider set of debates about changes in the level of resource, materials and energy consumption associated with shifts from an industrial to an informational economy which may provide some support for the contention that telecommunications reduce environmental damage. Although there have been few attempts to examine this shift from an urban environmental perspective, it is clear that the schema embodies important, and often unjustified, assumptions about the relationship between physical and electronic environments. In the latest stage of development it envisages an enhanced role for electronic over material flows and resources.

There have been attempts to measure these in shifts in studies that have examined the increasing dematerialisation of resource and energy flows in developed countries. For instance, there is evidence of a delinkage between economic growth and energy consumption in the developed world. In the past the coupling has been rigid with 1 per cent growth in economic growth requiring the same growth in energy consumption. Since 1973 economic growth has not been linked to a similar level of growth in energy consumption in OECD countries (Hansen, 1990). In 1988, for example, UK primary energy consumption was 3 per cent less than in 1973 whereas GDP was 31 per cent higher (Toke, 1990; 672). Despite the difficulties of obtaining adequate data and ensuring comparability between countries, Bernardini and Galli argue that studies have 'provided evidence of the postulate of declining material (and energy) intensity over time and with increasing GDP' (1993: 433).

Dematerialisation is the product of three sets of factors each having important implications for the changing relationships between the physical and electronic environment. First, the changing nature of final demand in a post-industrial period of development in which the gradual shift of output towards the production of goods with higher unit value and lowering material content may lead to a declining intensity of materials and energy use. In particular the shift towards information-based services based on knowledge-intensive goods means that income is increasingly spent of goods that have low materials content to value. Instead, value comes from sophisticated services sector based on immaterial factors such as design, image, quality, flexibility, safety, PR and environmental compatibility. In the UK, the service sector is now responsible for 60 per cent of GDP yet only consumes around 25 per cent of total electricity demand. Manufacturing is responsible for less than 25 per cent GDP but over 33 per cent of electricity demand (Toke, 1990; 672). Second, current technological innovations may lead to improvements in use of materials and energy – such as miniaturisation, design and quality control and improved logistics – thereby cutting the generation of waste encouraging the recycling of materials and reducing energy consumption. Finally, new technologies may aid the substitution of energy-intensive materials by alternative materials such as glass and plastics.

The telecommunications sector itself provides a useful example of how dematerialisation may work in practice. There appear to have been considerable environmental improvements associated with the production of telecommunications equipment:

In 1955, telecommunications cables were made almost entirely of copper, steel and lead. By 1984, close to 40 percent of the materials used were plastics. If substitution of lead by polyethylene for cable sheathing had not taken place, consumption of lead by AT&T alone might have reached a billion pounds per year, an amount to create considerable anxiety from the point of view of environment, given the toxic properties of lead.

(Ausubel, 1989; 85)

While this form of 'dematerialisation' (Herman, et al., 1989) may decrease concern about the environmental impact of metal production – it now creates new concern about the production and disposal of plastics (Ausubel, 1989; 88). Telecommunications cables are now undergoing a second round of dematerialisation as metals are replaced by fibre optics and in the process creating new environmental concerns about the disposal of redundant fibre optic cables.

Shifts in the importance of different types of infrastructure network are clearly an important part of the dematerialisation of energy and materials. The early infrastructure networks of railways, highways and water supply accounted for a large part of materials intensive economic activity in the nineteenth and twentieth centuries. But successive waves of infrastructure have tended to be less material intensive. In the longer term, development will be based on new infrastructure based and built around natural gas, information systems, telecommunications and satellites. The physical requirements related to this growth are 'certain to be less material-intensive' than the industrial phase of development (Bernardini and Galli, 1993; 447). These trends may imply a declining rate of growth in use of material and energy worldwide and perhaps even absolute decline. But this is highly dependent upon the development trajectory adopted in newly industrialising countries.

Although there is considerable uncertainty about the validity of the dematerialisation concept it may signal declining materials and resource inputs and outputs through the city. There have, however, been few attempts to apply the dematerialisation concept to an urban context. Instead dematerialisation focuses on broader societal shifts in the use of materials and energy resources without specifically working through the implications for cities. For instance, does dematerialisation suggest that new telecommunications infrastructure will substitute for physical flows along heavy, polluting transportation infrastructure? How will dematerialisation affect the physical form and structure of cities?

TELECOMMUNICATIONS AND URBAN DEMATERIALISATION?

Although there are no easy answers to these questions, there are potentially three closely related aspects of urban dematerialisation which are associated with the application of telecommunications technology. The first is the so-called trade-off between telecommunications and transportation. The second is the potential for specific measures and initiatives that encourage trade-offs through telebased initiatives such as teleworking. Finally, there is the potential for the substitution effect to break up cities and create a more decentralised form of urban life – the electronic cottage principle.

Telecom — transport trade-offs

Early research on the potential for trade-offs between telecommunications and transportation networks was commissioned by the US government in the mid-1970s in response to the energy crisis (see Nilles *et al.*, 1976; Harkness, 1977; Meyer 1977). Much of this early work simply compared the energy costs associated with communications through the telephone and physical forms of communication such personal travel by car, train and aeroplane. Figure 6.1 presents a logarithmic assessment of the energy costs of different forms of communication. The energy savings associated with communications by telecommunications created much excitement about the potential for trade-offs between telecommunications and transportation. Rather than people or information goods and services needing to physically move between different locations electronic images, messages, data and faxes could directly substitute for these trips. These findings helped to fuel utopian and futuristic images of new clean and environmentally benign societies in which telecommunications could simply substitute for travel, generating significant energy savings, reductions in travel time, lowering levels of emissions and urban congestion.

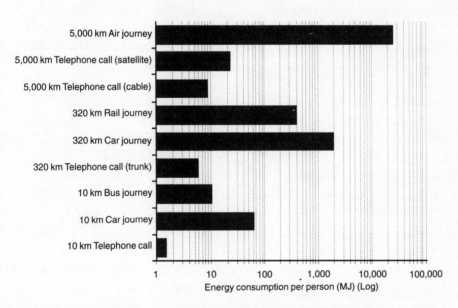

Figure 6.1 Energy consumption of transport and telecommunications

Source After Tuppen, 1992

Telecommuting as environmental policy

Much of the early interest in the role of telecom–transport trade-offs focused on the potential for the displacement of work-related commutes. In 1976, Jack Nilles invented the term 'telecommuting' to describe home- or neighbourhood-based working using computers and telecommunications technology. There has also been interest in the trip displacement potential of other telebased activities such as tele-education, teleshopping and telebanking. These new forms of telebased communication were seen as a solution to the problem of congested urban environments and long commutes to centralised offices. Studies evaluating the environmental potential of telecommunications usually adopted a perspective that is closely aligned to these technocentric approaches – a form of appraisal known as technological assessment (see Harkness, 1977).

Technological assessment was used to take a longer term review of the positive and negative effects of the environmental impacts of telecommunications. This approach claimed to identify the affected parties and unintended impacts in a neutral and objective fashion. Much of the early work was based on the assumption that telecommunications would simply substitute electronic flows for the transportation of people and information freight (such as videos, letters and books) along more polluting road, rail and air networks. There were a number of attempts to demonstrate the potential substitution effects of teleworking on travel patterns and estimate energy savings. In the US context these demonstrate that telecommuting has the potential to save between 1–3 per cent national energy consumption which is not necessarily as great as the early proponents of teleworking might have expected (Kraemer and King, 1982; Kraemer, 1982).

As the threat to the US energy supplies receded in the early 1980s so did official interest in funding research activities on telecommuting. It was not until environmental problems reached crisis proportions in the late 1980s that research once again started to examine the energy savings potential of telecommunications (Nilles, 1988). As Box 6.2 illustrates, these potentials have stimulated considerable interest in the environmental potentials of telecommuting, teleworking and teleconferencing (see State of California, 1990). Again there is now considerable interest in the role of information networking as the 'alternative fuel'. A recent study (Boghani et al., 1991, reported in Heilmeier, 1992) found that if telecommuting, teleshopping, teleconferencing and electronic document interchange were substituted for just 10 per cent to 20 per cent of transportation in the USA then substitution could eliminate:

- the daily drive to work for 6 million commuters;
- about 3 billion shopping trips per year;
- nearly 13 million business trips per year;
- more than 600 million truck and airline delivery miles per year.

There is still substantial interest in the environmental role of telecommuting to save energy, reduce emissions and lower levels of urban congestion.

BOX 6.2 STRUCTURAL DECENTRALISATION AND TELECOMMUTING

Prospects for telecommuting in Los Angeles have recently been ensured by Regulation 15, which was enacted and is being implemented by the Southern California Air Quality Management District (SCAQMD). The critical element in Regulation 15 is the so-called AVR or average vehicle ridership, which is determined by reporting to and from work at that site during peak hours. Thus, the higher the AVR, the more efficient commuters are. According to this regulation, every business with at least 100 employees in the central city must attain an AVR of 1.75, those in more outlying areas must attain an AVR of 1.5, and those in even more remote regions must attain an AVR of 1.3. Each business may use a variety of methods to meet its AVR requirement, including providing various incentives for carpooling, bicycling, and, most importantly for this report, telecommuting. A business may even have to destroy some parking space to comply with the regulation.

Should a business not file an acceptable plan to meet its AVR, it is subject to a fine of up to $25,000 per day and possible prison sentences for company management. These stiff penalties are designed to compel the business community to implement this regulation.

At least four telecommuting pilot projects existed in California even prior to the passage of Regulation 15. Pacific Bell Telephone itself had 500 formal and 500 informal telecommuters in 1989. The Southern California Association of Governments (SCAG) undertook a pilot project with eighteen telecommuters and found it to be an 'effective work option'. In addition, the State of California undertook a pilot project in which about 200 government employees spent between one and five days a week working off site. ... Finally, the Los Angeles Department of Telecommunications has awarded a contract to JALA Associates, a leading telecommuting consulting organization, to design a pilot telecommuting project for the city government of Los Angeles. Pat Mokhtarian, a leading transportation expert, will be assisting JALA in evaluating the success of the project. The goal of this city telecommuting project is to explore the potential of telecommuting in alleviating traffic congestion, energy waste, and air pollution without jeopardizing productivity, as well as to serve as an example for the private sector.

The advocates of telecommuting see it as a low-cost alternative to other traffic congestion solutions such as building new highways or the recently undertaken subway system. They believe that these traditional remedies are costly and disruptive and that they contribute to the centralization of the downtown area. Many, including Pat Mokhtarian, believe that if a fraction of the money allocated for the city's subway system were instead spent on telecommuting, the result would be a much greater reduction in traffic congestion. Advocates of telecommuting also argue that it avoids the difficulties of large construction projects while providing enormous potential benefits. These benefits include less travel time for commuters, reduced costs for child care, and more flexible hours.

(Lion and Van De Mark, 1990; 101–125)

The dissolution of cities?

This perspective has helped to underpin the view that telecommunications are an environmentally benign set of technologies inevitably and logically leading to the dematerialisation of cities. The dematerialisation concept is based on the assumption that telecommunications would dissolve the very 'glue' that holds cities together. As we saw in Chapter 3, cities were originally established to make communications easier by minimising space constraints to overcome the time constraints of travel. The boundaries of the early city were based on walking distance with urban functions located in close proximity. Improved modes of transportation facilitated the dispersal of cities as motorised transport enabled functions to be located further away from the urban core but within reasonable travel times. The substitutionist perspective argues that telecommunications are able to displace and substitute for physical transport.

Without the constraints of travel time and distance telecommunications could help facilitate the dispersal or break-up of the city. Toffler argues that the electronic cottage could be the basis of new form of decentred home life in which the potential of telebased work and services means that home and work can be separated by long distances (1981). Toffler (see Chapter 3) assumes that these trends will have important environmental implications because transport becomes less significant. Taken together, these two aspects open up the potential of new forms of decentralised urban environment in which functions can be many miles apart but communication takes the form of electronic flows. In this scenario electronics create the potential dispersal of urban space and substitution of physical movement along telecommunication networks.

These debates have now developed in a more sophisticated and critical level that goes beyond the substitutionist thesis and argues that cities are increasingly being reconstructed in electronic space. For instance Virilio (1993) argues that telecommunications now mean that the distinction between departure and arrival is increasingly lost (see Chapter 2). He captures this concept by arguing that 'where motorized transportation and information had prompted a *general mobilization* of populations. . . modes of instantaneous transmission prompt the inverse, that of a *growing inertia*' (1993; 11; original emphasis). In a physical sense this creates an increasing sense of inertia as people are able to 'travel' so fast electronically that they do not actually need to move at all.

Telebased services allow the world to be shrunk so that the distinction between a global and personal environment is increasingly lost as individuals interact through the capacities of telematic receivers and sensors. Virilio is insistent that this ultimately signals the end of the car because speed of telecommunications will inevitably displace slower forms of infrastructure. He does not speculate about the implications of these transformations for the physical form of cities but does agree with Toffler that the home becomes the basis for electronically mediated forms of interaction (see Chapter 5). But this is a highly dystopian vision of the future with human interaction through electronic media subject to a form of 'domiciliary inertia . . . and behavioural isolation' (Virilio, 1993; 11).

THE END OF THE URBAN ENVIRONMENTAL CRISIS?

These perspectives seem to signal some form of dematerialisation of the city in terms of the dissolution and substitution of physical flows and physical spaces. Although it is not clear what the result will be at this stage most attention has focused on the home as a battle ground for the future. Toffler presents a utopian vision of the electronic cottage recreating a convivial form of family life which contrasts strongly with Virilio's vision of an inert and sedentary society. But both strongly suggest that in their longer term assessment of the relationship between electronic and physical environments, telecommunications will lead to the dematerialisation of the city. These would seem to suggest that environmental problems are simply dissolved through a very simple one-way relationship between the physical and electronic which leads to the substitution or dispersal, fragmentation, dissolution of the city. But is all the traffic one way?

COMPLEMENTARY PHYSICAL AND ELECTRONIC INTERACTIONS

Approaches to telecommunication–environmental interactions that suggest that telecommunications will lead to the dematerialisation of the city sit rather uncomfortably with the heightened level of concern about global environmental change. If the substitution thesis is correct, surely we would expect to find comparatively less concern about urban environmental issues as telecommunications lead inextricably towards the dematerialisation of the city?

Clearly there are some wider dematerialising tendencies at work in the economy as the energy and materials inputs into some products and services are reduced. But there are also conflicting tendencies, especially the growth of movement and mobility within and between cities, which are a crucial focus of the current round of concern with the urban environment. This apparent paradox may be a temporary one. It may only be a matter of time before the new 'electronic fuel' starts to replace transportation, leading to significant energy savings, savings in travel time and reductions in levels of urban congestion.

An alternative view would pose more serious questions about the validity of the substitution thesis and the eventual dematerialisation of energy and resource flows through the city. While there may well be potential for trade-offs between the physical and electronic movement of people and information goods and services, many aspects of modern life will still require physical flows of goods and services. Information cannot simply substitute for flows of water, waste, electricity and gas, oil and the movement and distribution of many different types of materials, food and finished products. All these resources will require energy to power physical movement, distribution, collection and processing of materials. Also, as Chapter 4 argues, many more 'symbolic' functions will continue to require concentration based on the need for face-to-face interactions that cannot be easily telemediated.

How are these two different and apparently contradictory perspectives to be reconciled? What, if any, are the connections between the two views? Will telecommunications inevitably lead to the dissolution of cities? Are there alternative links in which telecommunications help to generate new demands for movement and mobility? Why should the development of telecommunications applications simply reverse unsustainable patterns of resource consumption and waste production in cities?

While telecommunications are being applied to many aspects of urban life, we argue that there is no simple logic which determines that their impact will inevitably be environmentally benign or even positively beneficial. The conventional perspectives are based on highly technocratic and élitist views of technological change which assume that telecommunications could simply be modified or shaped by an external assessment of its potential environmental effects. These approaches fail to look at the reasons why particular technologies have been developed in response to specific economic and social pressures and which interests are served by these applications.

More critical approaches have started to raise uncomfortable questions about the relationship between telecommunications and the urban environment that acknowledge that telecommunications will *reproduce* or perhaps *intensify* already unsustainable patterns of urban life. For instance O'Riordan argues that the technological determinists are 'usually found among an urban-dwelling élite who thrive on the sophisticated communications of the electronic global village and the jet age invisible college' (1981; 1). Salomon, looking more specifically at the proponents of the transport–telecoms substitution thesis, goes on to argue that they represent 'a coincidence of strong vested interests on behalf of producers of telecommunications technologies and the quest of transport planners to solve problems via technology-oriented solutions rather than measures that require changes in consumer or individual behaviour' (1986; 235). Teleworking offers a quick technical fix to the problems of urban traffic congestion. But it also offers the potential for new products and services for telecommunications providers and may also gives business an opportunity to shift many of the costs of employing workers on to the home – including the additional energy costs of heating and lighting the home during the day.

Clearly, there is a desperate need to start critically unpacking the link between telecoms and the environment. The role of telecommunications is not necessarily deterministically benign or inevitably contributing to the unsustainability of cities. Instead a critical approach needs to examine the context within which applications are being developed and at least acknowledge the potential of contradictory environmental implications. Our blend of the political economy and SCOT perspectives, developed in Chapter 3, opens up the potential for developing a more critical debate about the environmental role of telecommunications in cities. These go beyond the simple deterministic rhetoric of the utopians and start to acknowledge that there may be more complex and contradictory linkages between telecommunications and the urban environment. Despite the relative paucity of work there is enough evidence to demonstrate that utopian and highly deterministic perspectives fail adequately to capture the linkages between the two sets of issues.

COMPLEMENTARY INTERACTIONS BETWEEN PHYSICAL AND ELECTRONIC CITIES

The most important assumption of the substitution thesis is that the linkages between telecommunications and transport are one-way – telecommunications simply displace or substitute for trips based on the physical movement of people or information goods and services. There is, however, considerable evidence to suggest that the relationship is much more complex than this assumption suggests. Instead of telecommunications substituting for transportation the two forms of infrastructure have developed in unison with complementary relationships between each other (Salomon, 1986; Grubler, 1989; Mokhtarian, 1990). The interdependence between telecommunications and transportation is extremely complex.

Telecommunications products and services are often initially sold as substitutes but are actually used as part of a more complementary system. Therefore it is very difficult to estimate the relationship between telecommunications and travel demand. For instance, the growth of mobile telephones and more flexible work practices is likely to increase the total volume of traffic on the roads but reduce peak time congestion (Marsh, 1994). The time saved in commuting to work could be replaced by additional recreation and leisure travel. More worryingly for the substitutionist thesis is that easier and cheaper communications increase the number of actors in social or commercial network generating more demands to travel for personal interaction, while telecommunications enhances the efficiency, reliability and, ultimately, the attractiveness of travel.

There are different types of interface between electronic and physical flows not solely based on telecommunications simply displacing the need for transport and physical place. Our alternative framework acknowledges the potential for synergistic relationships between electronic space and urban place and physical flows. In this scenario electronics can generate changes in the physical environment rather than simply displacing or substituting for it. These linkages can take a number of different forms:

• New forms of infrastructure network can have unintended consequences. Each round of innovation based on new forms of infrastructure network has eventually led to new forms of unanticipated environmental consequences. For instance, the car was initially welcomed as an environmental innovation because it could help tackle the problems of disposing of huge quantities of waste from horse-borne modes of transport.

• Telecommunications can act as significant inducers of physical movement because they increase individuals' conceptual spaces and personal contacts. This can then generate new demands for higher levels of physical interaction with particular places or individuals through personal travel.

• Telecommunications can enhance the effectiveness of physical transport systems by improving their reliability, capacity, safety and reducing the costs of travel. This then helps accommodate new demands for physical movement.

• Telecommunications underpin new land-use changes that can lead to increasing separation of home and workplace, so creating more dispersed patterns of settlement and longer travel trips.

• The physical environment can itself generate demands for new types of telecommunications services that help overcome the problems of urban congestion. For instance, mobile telephone and data communications can help drivers stuck in traffic to keep in touch with their office and clients.

Production, use and disposal of telecommunications

Coolidge *et al.* (1982) examined the environmental problems associated with the use of the complete package of telematics technologies, including computers and IT equipment as well as telecommunications. The most important environmental costs associated with telematics technologies which they identified are: microwave radiation; indoor pollution, ozone, and other materials produced by telematic equipment; and human factors such as design failures and the psychological and physiological stress in the modern office.

Another related area is the energy consumption of computing equipment. It has been estimated that 'worldwide, computers consume as estimated 240 billion kilowatt hours of electricity each year – about as much as the entire annual use of Brazil. That, of course, represents a not insignificant share of total greenhouse emissions' (Young, 1993; 43). Much of the energy is consumed when the computers are switched on but not being used. In response, the United States

Environmental Protection Agency launched the Energy Star Computers programme in 1992 to improve the energy efficiency of computing equipment. If
these products can capture two thirds of the US market by the year 2000 they
could reduce CO_2 emissions by 20 million tons, an amount equal to emissions
from 5 million cars (Young, 1993; 45). There have also been attempts to create
new guidance to reduce the environmental, health and safety impacts of the use
of telematic equipment in offices. This evidence starts to dent the view that the
information society is totally benign environmentally.

BT's own environmental audit identified disposal as a key issue for their
company in relation to certain toxic chemicals and the disposal of redundant fibre
optic cables. Although this review does not provide a comprehensive evaluation
of the environmental impacts of telecommunications equipment, it clearly shows
that there are contradictory environmental impacts associated with the telecommunications sector. Each round of technological development may provide
new solutions to environmental problems associated with older methods of
production, but then the innovation raises new types of environmental problem
(see Ausubel, 1989). The utopian stress on substitution effects tends to ignore the
more complex issues associated with the production, use and disposal of
telecommunications. This indicates that much more care needs to be taken before
simply assuming that telecommunications are an inherently *benign* technology.

The 'inducement' of transportation

Although the trade-off thesis focuses on the potential for telecommunications to
substitute for transportation, it has long been recognised that telecommunications
can generate or induce new demands for physical movement. Electronic flows are
able to act as powerful generators or inducers of movement in both physical flows
and spaces. The ease of electronic communications based on decreasing cost and
increasing availability of telecommunications networks can therefore generate new
demands for physical transport (Salomon, 1986; Mokhtarian, 1990).

Telecommunications networks can help increase a person's conceptual spaces
(Wise, 1971; 27). Mokhtarian argues that by 'making information about outside
activities and interaction opportunities more readily accessible, telecommunications creates the desire to participate in those activities and opportunities, thereby
stimulating travel to engage in them' (1988; 283). As email, fax and telephone
effectively increase the number and geographical scattering of participants in a

business or recreation network, this can then create demands for higher level forms of interaction between the participants in the network based on interpersonal interaction. This creates a demand for physical proximity leading to new forms of physical travel that might have not taken place without the telecommunications linkage.

Wise (1971) has offered assessments of the magnitude of induced and substituted movement. He argued that commuting would fall by 20 per cent but social travel, as leisure time would be greater, would increase by 30 per cent and would not be substituted by increase in home-based entertainment. The concept of a more or less constant personal travel time budget means that 'the natural result of reducing some travel by whatever means, is that additional travel will be created to compensate – to fulfil the travel time budget' (Mokhtarian, 1988; 283). Here, telecommunications do not necessarily stimulate travel but by reducing travel create time for increased travel for other purposes. Thus, Wise predicted an overall increase in travel with the influx of telecommunications:

Although the use of telecommunications will substitute for movement in many instances, the substitution does not occur without an equally important expansion in the number of trips made. It can, in fact, stimulate physical movement in many ways, both directly or indirectly. . . . Simply, . . .*electronic communications may alter the purposes of travel but not necessarily the frequency.*

(Wise, 1971; 31. Emphasis added)

These conclusions are supported by long-run analyses of the relationship between telecommunications and travel. Little evidence of substitution effects have been found. Instead, they seem to grow in unison. Figure 6.2 charts the growth of passenger travel in kilometres transported and the total number of messages sent by the communications sector (letter, telegram, telex, telefax and telephone) in France. Transport and communications have developed in unison, increasing by a thousand times since the beginning of the nineteenth century. But transportation planners have a tendency to dwell on the potential substituting effects – perhaps because they offer a technical fix and are easier to model than the potential to generate travel. It is clear that electronic communications breed new demands for physical flows. Although it is difficult to assess the strength of these effects it is apparent that telecommunications do not simply displace physical travel. While substitution effects may occur these are likely to take place in the context of an overall increase in all forms of communication, including personal travel, in which telecommunications may simply take a larger share (see Salomon, 1986; Mokhtarian, 1990).

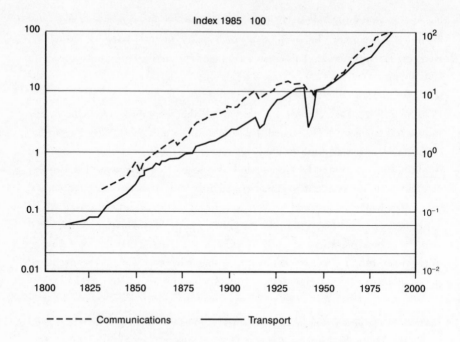

Figure 6.2 Growth of passenger transport and communications in France

Enhancing the efficiency of transport networks

Telecommunications contribute towards the enhancement of transportation networks by increasing the efficiency, safety and attractiveness of different transport modes. Throughout history 'transportation has always been in the vanguard of new communications applications, because control messages must exceed the speed of the transport itself for effective adjustment to delays, crises and accidents' (Boettinger, 1989; 288). Transportation in all its forms is controlled by telephone and data lines to dispatch and control traffic. Improvements in these control systems have important implications for increasing the effective capacity and reducing the cost of transportation networks.

The new control, supervision, and data acquisition role of telecommunications can increase the attractiveness of travel (Cramer and Zegveld, 1991). For instance, new computerised systems for booking and payment of travel make it very easy to obtain information and pay for air travel. In turn, more effective methods of managing travel networks can help increase the efficiency of transport networks at all levels – road, rail, air travel – so lowering costs and increasing the attractiveness of travel as an option. It has become increasingly clear that the new technologies of road transport informatics provide ways of overcoming the problems of congested road networks and increasing the effective capacity of these networks at a fraction of the cost of constructing entirely new transport infrastructure (Hepworth and Ducatel, 1992; Giannopoulos and Gillespie 1993).

After reviewing the environmental potential of new forms of road transportation informatic systems such as advanced traffic management, demand management and road pricing, Gwilliam and Geerlings conclude that:

Developments in information technology have the potential to improve efficiency of operation of most modes and to replace the movement of people by less environmentally damaging information. Exploitation of that potential, however, is beset by great behavioural problems. Technologies that simultaneously reduce unit environmental impact and reduce operating cost or improve operating capability may be exploited by users to increase total traffic volume or performance (e.g., greater speed, heavier and more sophisticated vehicle) to such an extent that the net environmental effect is negative.

(Gwilliam and Geerlings, 1994; 314–315)

In this sense the retrofitting of electronic networks over relatively old and polluting physical transport networks can enhance the efficiency, capacity and attractiveness of these networks. The problem is thus not just, or even primarily, that of developing new technologies 'but more a matter of securing the effective implementation within an appropriate policy framework of these technologies that are, or will soon, become available' (Gwilliam and Geerlings, 1994; 314–315). We therefore need to be extremely cautious about assuming that telecommunications have benign implications for the environment.

Telecommunications and urban sprawl

Telecommunications technologies have actively facilitated both the centralisation and decentralisation of cities (Pool *et al.*, 1977). In the early part of the century the development of skyscrapers was underpinned by the telephone without which communication in multi-storey towers would have been a logistic nightmare. Later the telephone with increased access to public transport facilitated movement to the suburbs. These trends have continued with the new telecommunications infrastructure allowing functions to disperse out of cities to new locations in rural areas. Increasing separation between functions such as work and home, leisure, recreation and shopping has helped to generate increasing and longer trips. As Herman *et al.* argue 'the spatial dispersion of population is a *potential materializer*' (1989; 337. Emphasis added).

Over the long term we might see complex three-way relationships between telecommunications, transport and land use. An increase in telecommuting might induce changes in employment and/or residential location patterns. If employees become telecommuters and only have to travel to work two days a week, they may move to more affordable desirable housing many miles from work. This may effectively increase the distance of their total weekly commute.

Research on the environmental implications of different types of land-use patterns has found that highly dispersed and highly centralised settlement patterns use more fuel for transportation (Breheny, 1992). Both these centralised and decentralised land-use patterns are underpinned by the communication abilities of electronic telecommunication networks. For instance, Newman (1991; 342) found an 'exponential increase in gasoline use as density falls'. In this instance, electronic flows displace physical distance so allowing functions to be located more flexibly. But this in turn generates longer physical movements and the need for more energy and waste infrastructure. The concept of the environmentally benign electronic cottage is therefore much more complex than originally conceived. Telecommunications can play a central role in underpinning the development of resource-intensive land-use patterns.

Urban congestion – generating new demands for electronic communication

While telecommunications can generate new demands for movement within and between cities, the state of physical flows and spaces in cities are also able to create demands for new forms of telecommunications networks and services. It might be expected that increasing levels of congestion would stimulate a shift to alternative forms of transport and/or communications but this is not necessarily the case. The development of mobile radio-based telephones and data transmission has been given an important boost by the problem of urban congestion. Without mobile data communications drivers stuck in traffic would normally waste time and resources. Mobile communications systems mean that they are able to reorganise their work according to traffic conditions and can even complete work while sitting in traffic jams (see Figure 7.6, p. 306).

Consequently, physical congestion can actually stimulate new kinds of telecommunications services which can become ways of overcoming the limits to effective working. Another example would be auto route guidance technologies. These enable car drivers with such systems to find alternatives to congested routes. The telecommunications systems have been developed in response to physical congestion, not by substituting for trips but by overcoming the physical constraints for those car drivers fitted with the system. These both enhance the attractiveness of transport networks and induce new forms of travel. Rather than telecommunications simply displacing physical movement, the relationships are exceedingly complex.

THE ENVIRONMENTAL ROLE OF TELECOMMUTING?

Clearly, then, we have to be much more cautious in assessments of the environmental benefits of teleworking. As we can now see, the whole issue of trade-offs is very complex and findings need to be treated with caution (Mokhtarian et al., 1994; 2). There have, however, recently been a number of assessment of the environmental implications of telecommuting that have attempted to deal with the complexity of transport–telecommunications inter-actions. Much of the recent work on the potential environmental benefits has originated in the USA. There are three factors which explain this level of interest.

- A 1992 survey found that 3.3 per cent of the US workforce – 4.2 million employees – were telecommuters in 1992, a 27 per cent increase over 1991. Projections indicate that there may be 10 million telecommuters by 2000 and more than 30 million by 2010.
- Cities in the USA have the most severe transportation problems which have generated innovative ways of managing traffic and attempting to reduce environmental emissions.
- The United States Department of Energy argued that telecommuting could reduce highway travel and 'could significantly help ameliorate the growing problems of traffic congestion, air pollution and petroleum dependence' (Department of Energy, 1994; ix).

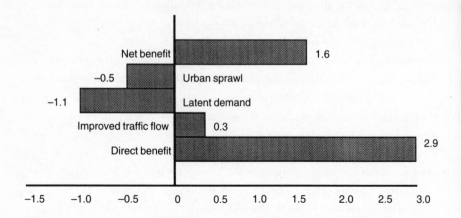

Figure 6.3 The direct and indirect effects of telecommuting on fuel consumption in 2010
(in billions of gallons of petrol)
Source: Department of Energy, 1994

Despite this, potential assessments of the environmental potential of tele-commuting are extremely complex. Figure 6.3 reports on the latest US Department of Energy assessment of the petroleum savings from telecommuting in 2010. The direct benefits are substantial reduction in fuel use because of trip substitution and the improved traffic flow that results from lower congestion and higher vehicle speeds. But the indirect effects halve the fuel savings. Tele-commuting enables workers to live further from their workplace in more attractive or cheaper locations, thereby increasing travel distances when they do commute. More worryingly, the people who avoided using their car because of congestion will now start driving because telecommuting has lessened congestion. Consequently urban sprawl and latent demand will tend to offset reductions in fuel use and emissions.

Further confirmation of the complex level of interactions has come from a recent study evaluating the results of a number of telecommuting initiatives (Mokhtarian *et al.*, 1994). The study confirmed that teleworking can reduce the number of commuting trips to work places. But they then went on to find that the number of person-miles saved is smaller than the reduction in the number of commuting trips because of the need to make non-work trips such as shopping, which were formerly part of the commute to work. There is an even smaller reduction in the number of vehicle-miles travelled because many of the new telecommuters formerly travelled to work by public transport or shared cars.

These vehicle trips still take place even though the telecommuting is based at home. There is an even smaller proportional reduction in energy consumed by the household because of the need to increase energy consumption in the home during the day and the decrease in the fuel efficiency of the remaining short trips. This also means that the reduction in total emissions is smaller than the reduction in fuel use because of the slower speed of remaining trips and the higher proportion of cold starts.

These findings have been confirmed by an assessment of the environmental benefits of teleworking as a CO_2 reduction strategy in the UK (see Figure 6.4). A recent study found that the total energy reductions are small – a 0.6 per cent reduction in consumption for every 1 per cent of jobs transferred. Nevertheless, for the individual teleworker, the energy savings are significant and over a year average 19.6 GJ – the equivalent of 135 gallons of petrol – for each teleworker (CEED, 1992; 11). In the UK the average length of commutes is much lower than in the USA so that the energy savings were relatively low, particularly when compared to the increased demands for heating and powering the home (British Telecom, 1992b).

Decentralisation of work also created new difficulties with the recycling of products and services that is much more feasible in centralised offices. In addition,

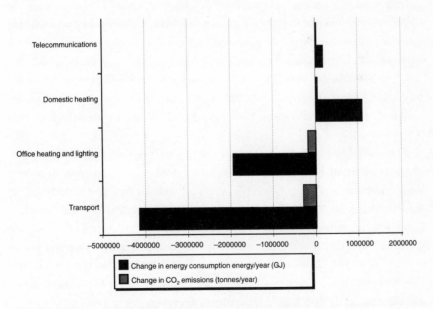

Figure 6.4 CO_2 effects of 1 per cent of the UK population teleworking
Source: Derived from British Telecom, 1992b

working from home may save commuter trips but it can also allow workers to decentralise the home, resulting in longer trips for leisure, recreation and shopping. The time saved through not commuting may also create the potential for other trips that might previously have not taken place or been combined with the commute to work.

Consequently, the assumed environmental benefits of the telecom–transport trade-off look highly suspect. Teleworking can reduce overall travel, reduce congestion, reduce traffic emissions and increase fuel efficiency, but these benefits are a lot smaller than has been previously thought. We could also expect a decline in benefits as telework grows because early adopters are more likely to have long commutes and are more conscious of the potential energy savings. While they may have a role in particular locations the energy-saving benefits are not as significant as first thought. Teleworking increases energy consumption in the home, makes the recycling of office products more difficult and may create additional time for alternative forms of travel.

But the most damaging problem is that substitutionists ignore the traffic-generating potential of telecommunication. The relationship is much more complex and contradictory than they suggest. There are also other issues that need to be considered – shifting employment costs on to homeworkers, increasing flexibility for companies, psychological problems of isolation, surveillance and control. Environmental benefits should not be used as a front carefully to hide the social, economic and political issues raised by telecommuting. This framework suggests that electronic–physical interactions are much more problematic than the simple substitutionist view that telecommunications simply displace the need for travel and cities. Assessments of the environmental role of substitution are extremely complex and contingent on the particular locations where experiments have taken place.

The central question is whether absolute or relative substitution takes place. There is potential for telecommunications to displace trips and this does have some environmental benefits, particularly when teleworking or teleconferencing is substituting for a long commute. But it appears that rather than absolute displacement, telecommunications is simply taking a larger relative share of growth in all forms of communications. If the amount of communications was to remain stable and electronic forms of communication took an increasing share then absolute displacement would take place. But if all forms of communications increase absolutely then little displacement effect takes place.

It appears that telecommunications and travel have increased in parallel and are not currently substituting for each other. Long-run analyses of the relationship between telecommunications and travel have found little evidence of a substitution effect and instead they argue that 'due to the diffusion of new telecommunications technologies there will be no reduction in passenger travel, instead considerable growth is likely to occur' (Grubler, 1989; 258). Substitution effects may offer some potential in the Californian example. But this only works because of the long commutes induced by highly decentralised land-use patterns which were facilitated by telecommunications networks in the first place. The substitution effect is only likely to become more significant if the costs of transport increase to the point where telecommuting becomes a more attractive alternative to travel.

But this will not solve road traffic congestion which is predicted to double in many developed countries but may be able to reduce peak time congestion when combined with road transport informatics. 'All things considered, however, the long-term role of telecommuting as an energy – and air pollution-reducing policy will probably be less than its role as a congestion reducing policy' (Mokhtarian et al., 1994; 27). There have been proposals for telecommuting schemes that attempt to reduce peak demand on public transport networks and to accommodate more cheaply growth in traffic than investment in upgrading and increasing rail infrastructure. For instance, Roarke Associates (N.d.) estimated that the peak-time commuters in the South East of England could be provided with places in telecommuting offices at less than half cost of providing for demand by investment in rail infrastructure. Consequently the relationship between telecommunications and the urban environment is not as simple as the substitutionist perspective would imply. Instead, electronic and physical transformation proceed in parallel producing complex and often contradictory effects on urban flows and spaces.

ELECTRONIC MONITORING OF THE URBAN ENVIRONMENT

Although there is considerable uncertainty about the environmental implications of the synergistic relations between electronic and physical environments there is one set of relationships that may have environmental important benefits. Figure 6.5 sets out a typology of the potential relationships between electronic monitoring and the physical environment. These relations focus on the role of electronic networks in the development of new capacities for collecting, analysing and distributing information on the environment conditions in the physical world remotely and in real time.

We use the generic term 'monitoring' to refer to a range of roles that includes the measurement, monitoring and pricing of the physical environmental. Telecommunication technologies now present major new opportunities for monitoring the physical environment, further increasing the complexity of electronic and physical interactions.

There are close links between the physical environment and the development of electronic monitoring technologies. Information on the physical environment has huge economic importance, particularly for the agricultural, building and construction, energy and transportation sectors and for environmental policy (Maunder, 1989). Before the development of telecommunications and sensor technologies, environmental monitoring relied on manual systems of data collection and the transportation of data and the results. There were often considerable delays between an environmental event – such as an air pollution episode – and knowledge about the extent, duration and seriousness of the event. New telematic-based systems of monitoring, analysis and modelling now offer the potential to examine physical environmental conditions in real time. These new capabilities provide the tools for monitoring environmental problems that largely lie outside the normal bounds of human perception – gaseous emissions, toxins and climatic changes.

Consequently, telecommunications now provide the technologies through which we can map and measure environmental change. As the focus of concern

Electronic role	Physical role
Computer information networks	Exchange of information on environmental conditions
Remote, real-time monitoring systems	Air, water, noise pollution monitoring
Satellite monitoring systems	Global climate change
Smart metering technologies	Water, gas, electricity tariff and load control and road pricing
Intelligent buildings	Environmental management of internal environment in home and offices
Supervisory, control and data acquisition systems	Optimising environmental performance of water, energy and transportation networks
Modelling of physical environment	Regional and urban pollution modelling

Figure 6.5 Electronic monitoring of the physical environment

has shifted from the local and physical urban environment to issues connected with global environmental change, telecommunications are the technology through which our understanding of physical phenomena are mediated. Without electronic communications it would be impossible to develop an understanding of the urban environment and its implications for global environmental change.

Closely linked to these changes are the development of technologies for simulating and forecasting environmental change. The 1973 oil crisis gave a considerable impetus to the development of technologies that could simulate new military environments, particularly flight and battle simulation. As the cost of these technologies has fallen we now have a range of tools that allow policy-makers to generate and test different environmental scenarios. There is a major new area of activity attempting to forecast and model the land-use impacts of climate change and sea level rise – particularly on cities which are often located on coastal plains.

It is possible to consider a hierarchy of linkages that could build into complex electronic systems for the monitoring, pricing and control of physical flows and spaces in cities. There have been a number of initiatives to develop environmental information systems for carrying data about environmental conditions and legislation for access by community groups and policy-makers (see Box 6.3) (Cassirer, 1990; Young, 1993). The information on these networks is closely linked to the development of new forms of environmental monitoring technologies (Environmental Resources Limited, 1991). New technical capabilities have been developed to monitor a range of air pollutants, noise levels and water pollutants at different spatial scales, remotely and in real time.

BOX 6.3 ENVIRONMENTAL INFORMATION

Electronic mail has become a vital tool for those who work on environmental and social issues. Thousands of activists and organisations around the world are now using computer networks to co-ordinate campaigns and exchange news.... Networks put enormous resources and reliable, inexpensive global communications at the fingertips of ordinary citizens. They allow people to sift through large collections of environmental data for the information they need.

The Right-to-Know Computer Network (RTK Net), operated jointly by two Washington DC, nonprofit groups, offers more than 800 users on-line access to the US government's TRI database on industrial releases of toxic chemicals. Information can be combined with other government environmental databases – such as Superfund hazardous-waste clean-up sites and pending environmental litigation – and a variety of census data, including the ethnic and economic characteristics of the local population. RTK Net users have produced some particularly compelling studies. For example, Florence Robinson of the North Baton Rouge Environmental Association, a community group, has been using the TRI data and census information to demonstrate that Louisiana's minority communities live with disproportionately high levels of toxic pollution. She has reported her findings in a series of papers and testified on the subject at a US Civil Rights Commission hearing on environmental racism.

RTK Net was created because the federal government was not making TRI data available on-line at low or no cost. The same data are available on a federal system – Toxnet, administered by the National Library of Medicine in Bethesda, Maryland – but for $18–20 per hour, which few community-based activists can afford. The federal system also lacks RTK Net's on-line community of activists, who share tips on how to conduct searchers and spiritedly debate a variety of topics, including environmental racism and the health effects of toxics.

(Young, 1994; 99–116)

Major innovations in smart metering technology in the water, energy and transport sectors have created the potential for more effective monitoring and control of resource flows along the networks (Sioshansi and Davis, 1989). For instance, flows of physical resources can be charged for in real time according to changing levels of demand (see Chapter 7). During periods of peak demand electricity tariffs can be increased encouraging customers to shift demand to non-peak periods. Smart meters can also help customers to set targets for reductions in energy and water consumption which can then be monitored through the meter. It is also possible for the utility to control the use of appliances in the home by sending messages through the meter switching off appliances during periods of peak demand. Together this set of applications may be able to smooth out peak demands for resources, avoiding expensive investment in infrastructure supply such as roads, power stations and waste treatment plants.

At the level of individual buildings telecommunications systems have been developed for controlling services such as heating, ventilation, lighting and security control systems. These can optimise the environmental performance of the building by turning out lights when spaces are unoccupied, taking advantage of cheap tariffs and providing information that can help reduce the input of resources and waste generation. Information networks are now being embedded in the building structure to monitor and carry data on wind loading and structural damage to the building envelope in addition to monitoring the internal environment (Huston and Fuhr, 1993). These new monitoring and control capabilities designed for physical spaces can also be applied to whole networks. Utilities and transport authorities are increasingly making use of telecommunication networks for optimising the performance of fixed physical networks of transport, energy and water, waste flows (Dupuy, 1992; Laterasse, 1992). Used as part of wider demand management strategies these systems may have the potential for improving the environmental performance of infrastructure networks.

There are clearly a range of complex interactions between the electronic and physical environment. These applications further compound the difficulties of simply separating out physical and electronic space. Take the example of a sophisticated urban air-quality monitoring system (Environmental Resources Limited, 1991) (see Box 6.4). Electronic sensors can remotely and in real time monitor levels of air quality in different parts of the city. If levels of emissions reach certain limits warnings can be sent over telecommunications networks – public information systems, email, radio and TV – to ask people to leave their cars at home and travel by public transport. However, if emissions continue to rise past statutory limits urban traffic control systems can prevent further traffic from entering the city while driver information systems instruct drivers to switch off their engines and large sources of air pollutants such as power stations can be instructed to reduce operations.

BOX 6.4 THE AIR QUALITY MONITORING PROGRAMME IN BERLIN

THE MONITORING NETWORK

It is Government policy to monitor any pollutant which could pose a threat to human health. Currently Nox, SO_2, O_3, CO and particulate matter are measured continuously. The monitoring network consists of 38 measurement stations, the majority of which are located on a 4 by 4 km grid system. Others are situated in highly polluted areas such as major roads, industrial sites and densely populated areas. Four of the sites are reference stations located to the north, south, east and west of the region, and their function is to determine the quality of the air outside the city. The purpose of positioning the monitoring sites in a grid format is to aid the licensing of industrial sites. German Law requires that baseline air quality monitoring should be obtained prior to a license being issued. In order to encourage development within the city, it is the policy of the local government to monitor in such a manner that baseline data are readily available on license application.

DATA RETRIEVAL, STORAGE AND VALIDATION

Data are transferred to a central computer by means of dedicated telephone links. [This] enables frequent transfer of data, every three minutes, and allows real-time observations of pollution levels at the central bureau.... Hourly and daily averages are calculated for input to the alarm and public information systems.

PUBLIC INFORMATION

There is a high public demand for information concerning pollutant concentrations in Berlin.... Pollution information is available via the radio, television, teletext and an electronic public advertisement board on one of the main shopping streets in the city. Information is relayed to the meteorological institute every hour to be broadcast with weather reports. There is a three stage alert system in operation in Berlin.

Pre-stage alert – the public is informed of the pollution situation and voluntarily requested to reduce emissions from vehicles and domestic heating.

Stage I alert – vehicle usage is restricted except those which are fitted with catalytic convertors; power plants are restricted to using fuel containing less than 1 per cent sulphur.

Stage II alert – vehicle usage is prohibited and industrial activities are restricted.

(Environmental Resources Limited, 1991; A7–A11)

We can see that there is a whole range of complex physical and electronic interactions and trade-offs. Presumably commuters could take the option of teleworking from home rather than commuting to work by car. These interactions can, however, become even more complex when computing and Geographical Information Systems using environmental and land-use data are used to model the urban environment. GIS systems have been developed for new applications in the management of the urban environment such as modelling waste transportation, mapping acid rain and countryside information systems (see Cornelius, 1994). These systems are used electronically to generate scenarios which can simulate the physical impacts of particular land-use planning and traffic management options.

Such environmental monitoring technologies have now started to raise fundamental issues for urban environmental policy. While our capacity for measuring, monitoring and even modelling the physical urban environment has increased dramatically, our capacity for making changes in the physical environment lags a long way behind (Harris, 1987; 401). There is a multitude of problems here, including the difficulties of interpreting data, the problem of gaining access to commercially confidential data and understanding how individuals respond to electronic information about the physical environment. But perhaps the main problem is the failure to conceive the city in terms of both electronic and physical space. Without this perspective, urban environmental policy will continue to focus on a one-dimensional view of the problem and develop little understanding of the complex role of telecommunications technology.

We need to develop a better understanding of the complex relations between electronic and physical flows and spaces. In particular, we need to focus on the recursive character of the interactions in which physical flows are simultaneously substituted and generated. In many senses our ability to monitor, measure, map and model urban environmental change goes way beyond the ability of environmental policy to control the forces generating the environmental problems. Here we have the apparent paradox of telecommunications technologies as key generators of environmental problems but they also provide a new window on to urban and global environmental change. The key question is how far environmental policy can use the new networks for controlling environmental change.

CONCLUSIONS: BLURRING BOUNDARIES BETWEEN PHYSICAL AND ELECTRONIC CITIES

This chapter has shown how the relationships between telecommunications and the environmental debate is much more complex and contradictory than is often assumed. Rather than simply substituting electronic for physical flows telecommunications technologies have a number of effects which can also create new physical spaces, generate new physical flows and increase the effective capacity of infrastructure networks. Consequently, it is not possible to make any simple assessment of the environmental role of telecommunications. Where does this leave us? Is it possible to define a role for telecommunications in environmental policy?

While there do appear to be dematerialisation effects in the production of manufactured goods, there are contradictory trends around demands for increased mobility. Telecommunications technologies are firmly implicated in both sets of changes. They help increase the efficiency of production processes and reduce the need for material inputs, but also allow physical spaces and flows to be reconstituted, so generating new forms of environmental problems. Policy-makers need to acknowledge these contradictory effects. While telecommunications have the potential to substitute and monitor physical flows, they have a powerful role in generating new physical problems through dispersal of land uses and the generation and enhancement of travel.

Although the technodeterminists tend to dress telecom–transport trade-offs in a heavy veneer of environmentalism, they ignore the potentially contradictory environmental effects. A similar logic appears to be at work in other sectors such as intelligent buildings and road transport informatics, where research pro-grammes in Japan, the USA and Europe are developing technical standards for, and demonstrating applications of, these new technologies. Environmental factors are often cited as a key justification for the development of these technologies, but less evidence is presented to demonstrate how these environmental benefits will be achieved. Instead, consortia of government standard agencies, telecoms companies, car makers and producers of consumer products are much more interested in creating new markets and services, by adding value to existing products and increasing the use of telecommunications networks, particularly premium radio-based services, rather than optimising the environmental perform-ance of the road network or buildings. While they may well have environmental potentials, the companies developing these products and services have very little interest in exploring their negative environmental impacts – especially if this was to suggest that public policy should attempt to restrict or redirect tele-communication services or applications.

We need to develop new ways of looking at urban environmental policies which begin to recognise and unpack the complexity of the relationship between electronic and physical flows. Here we have attempted nothing more than a preliminary analysis of these relationships. But these interactions are extremely complex. Urban environmental policy needs to develop new conceptual frameworks that start to acknowledge the contradictory and complex role of telecommunications. Without such changes we will fail to develop a more complete understanding of urban environmental problems.

The boundaries between the physical and electronic urban environments are increasingly blurred. Telecommunications are firmly implicated in the generation of physical environmental problems. At the same time telecommunications technologies provide the tools for monitoring and mapping invisible environmen-tal change. We need to develop conceptual frameworks that examine how vast telecommunication control networks can be used to reduce the environmental impacts of cities.

7

URBAN INFRASTRUCTURE AND TRANSPORTATION

INTRODUCTION

Urban infrastructure networks – transportation, gas, electricity, water and telecommunications – provide the basic infrastructural foundations to the operation of modern economic and social systems. They provide access to energy, water and communications services through extensive physical networks of pipes, ducts, pylons, cables, radio, roads, rail and airline links. Such infrastructures are the conduits or 'technological systems' (Hughes, 1983; Preston, 1990) which link firms, organisations, and households into wider economic and social structures.

Utilities and transportation networks are the very 'glue' that holds together modern society. Cities and urban systems, in particular, are today intensely dependent on dense and interwoven lattices of technical networks. Indeed, without them, virtually all aspects of the functioning of the modern 'networked' city would be impossible (Tarr, 1984; Tarr and Dupuy, 1988). For most utility and transport providers, cities play a central role by providing the base for over 80 per cent of all demand for their services.

Revolutionary changes are currently underway in the relationship between urban technical networks and telecommunications networks. The public provision of infrastructure services has increasingly been challenged by the shift towards private and often competitive provision of networked services. We argue that the new capabilities of converging sets of computer and telecommunications technologies for supporting information processing and communications are playing an absolutely central role in the development of these new 'logics' of urban infrastructure development and management.

A technological revolution in urban infrastructure is now paralleling the wider political, economic and environmental trends we have examined in previous chapters. Telematics networks can open up what were previously considered to be monopolistic networks to new forms of competition. Electronic road pricing systems provide the necessary infrastructure for privatising the road network, while smart metering technologies enable electricity customers to choose their energy supplier. Infrastructure providers are now investing heavily in telematic systems to improve the efficiency and profitability of their networks with real-time pricing, remote monitoring and the provision of value added services.

This new revolution in urban infrastructure is one of the most fundamental changes in the development of urban networks since the start of century. Telematics provide essential tools when dealing with the massive and complex nature of infrastructure networks and are being applied pervasively across the urban infrastructure sector. As Gabriel Dupuy argues, 'virtual' systems of computer networks are now beginning to match very closely the 'real' hard networks of information, energy, water and transport flows (Dupuy, 1992; 67). The result is a movement towards the real-time management and development of urban infrastructure networks: decisions can be made on up-to-the-second information based on the real demands, flows and supply operating on a network. All aspects of the management, development and control of urban infrastructure networks are becoming increasingly reliant on parallel systems of computer networks (Dupuy, 1992; Madden, 1992).

Jean Laterrasse argues that these radical changes are generating new conceptions of the contemporary city. In particular, he focuses on the notion of the 'intelligent city':

The concept of the intelligent city, a foreseeable and even logical extension of the concept of the intelligent house or building, is in style. As an image it certainly brings to mind the related movement ('telematic' systems) of the computerization of urban network management, and the administration of cities, and the development of telecommunications. But beyond the image of the thinking computer and telecommunications as nervous system, the concept of the intelligent city, in order to be relevant, implies the emergence of new functions of auto-organisation, of control, of material and immaterial flux, and of adaptation to a changing environment and this in turns lets us glimpse a greater mastery in the solution of the problems of our metropolitan centres, and a better overall functioning of the city.

(Laterrasse, 1992; 1)

But what are the implications of this weaving together of infrastructure and telecommunications for contemporary cites? Do improvements in the operational efficiency of infrastructure networks simply produce wider societal benefits? Or do they give the providers of infrastructure new methods of controlling access to services, facilitate the privatisation of networks and encourage new forms of environmental damage? These questions raise serious issues about the nature of urban life and the nature of cities, yet they remain curiously unexplored.

While telecommunications and telematics applications will certainly support radical changes in the management and provision of infrastructure systems, both the urban studies and policy-making communities have often tended to neglect the study of urban technical networks in the governance of the contemporary city. Research on water, waste, energy and transport networks has generally been left to engineers, technologists and economists with relatively little consideration of the wider economic, social and environmental roles of networked services.

It is still often assumed that monopolies would provide networks in the places that needed them, at the correct time, serving domestic and commercial customers equally. But, as we shall see, assumptions about spatial and socio-economic equity in both the distribution and levels of access to urban infrastructure networks are increasingly dubious. Telematics systems are playing key roles in creating new spatially fragmented 'patchworks' of utility and transport services with more differentiated levels of socioeconomic and spatial access to the networks by different classes of customer.

This chapter attempts critically to examine these changes by focusing on three core questions. First, what are the linkages between new innovations in telecommunications and established infrastructure networks? How do electronic flows impact on the development of material flows along conventional infra-structure networks? Second, what are the technical, socioeconomic, commercial and/or regulatory forces driving the application of telecommunications to infrastructure networks? Finally, what are the implications of these changes for urban policy-makers? How can urban policy-makers 'shape' these applications to capture wider benefits for their cities? Do telecommunications actually encourage increased levels of social polarisation, environmental degradation and economic marginalisation?

URBAN INFRASTRUCTURE NETWORKS AND TELECOMMUNICATIONS

Until relatively recently the city has largely existed on the same 'bundle of technologies that was available in 1910, even though one of these – the automobile – has greatly diffused since then' (Hall, 1987; 5). The development of early water and energy networks helped to overcome the environmental and public health constraints on the development of the concentrated city (Tarr, 1984; Tarr and Dupuy, 1988). They facilitated the importing of clean water and the exporting of waste through pipelines, while energy networks provided powerful, dense and flexible methods of distributing coal energy through wires and pipes in the form of electricity and gas. In turn, energy systems helped assist the development of mass transit and telephone networks, thus facilitating greater physical separation between home and work, and different parts of industrial production processes (see Chapters 4 and 8).

Networked services underpinned many of the key aspects of the social, economic and environmental development of cities – they facilitated both the concentration and dispersal of urban form. In order to develop an understanding of the role of telecommunications in the restructuring of networked services in the contemporary city we require a framework that highlights both the parallels and complex interaction between electronic and physical flows over infrastructure networks. Much of the utopian and futurist writing has assumed that as the latest form of urban infrastructure telecommunications will substitute for older energy and transportation networks (see Chapter 6). As we shall see, these perspectives fail to grasp the significance of both the similarities between, and mutual interdependence of, different forms of networked infrastructure.

CONCEPTUAL PARALLELS

The most obvious sets of linkages concerns the close conceptual parallels between infrastructure networks. Taking a broad interpretation, all infrastructure networks require movement to occur, whether this is in the form of flows of energy, water, people, freight or electronic impulses. Water and energy services consist of one-way flows between production and consumption nodes. Transportation and telecommunications are much more complex, involving a multiplicity of interactive flows between many nodes which are both consumers and producers of communication. Despite these differences, movement may take one, or a combination, of three forms:

- Movement of people – to meet face-to-face;
- Movement of resources – such as freight, water, energy;
- Movement of electronic impulses – either in the form of electronic current along wires and cables or in the form of radio waves through the air.

Using this framework we can identify further conceptual parallels between all forms of networked technical services. The first is based on the close physical linkages between telecommunications and existing infrastructure networks. Many of the earliest networked systems were innovative users of basic telecommunications technologies (Dupuy, 1992; Hepworth and Ducatel, 1992). In the industrialising city water, energy and transport systems used the telegraph and telephone to facilitate the control, monitoring and acquisition of data over increasingly dispersed and complex technical networks (Beniger, 1986). These early telecommunication networks were often superimposed on to existing transport, energy and water networks. In a more modern context cable networks usually follow urban road networks.

The second link, which has recently assumed much importance, concerns the commercial partnerships between existing infrastructure providers and telecommunications networks. Increasingly, existing networks operators have either diversified into telecommunications or sold the rights to lay new telecommunications infrastructure over their existing transport, water or energy network to a third party. The new operator gains considerable economies of scale by avoiding the huge costs of establishing an entirely new telecommunications network, while the existing owner of the network generates a new income stream and can gain access to the new network.

Third, there are the close analytical parallels between telecommunications and infrastructure networks. All infrastructure services involve the development of networks which can be physically based on wires and pipes or virtual networks based on wireless telemetry applications. Each part of the network typology is usually based on some type of local, intercity and perhaps even international network (Schuler, 1992).

The final aspect is the close regulatory parallels between telecommunications and other infrastructures. In historical terms the infrastructure network sector has usually been considered as a set of 'natural monopolies' requiring some form of regulation to protect customers from over-exploitation. At least in their earliest stages of development, many infrastructure networks had monopoly status. This is now, however, a highly contested issue and, as we shall see, regulatory and technological advances have seen the increasing liberalisation of the transport and utilities sector. But this trend is developing unevenly between countries and across different infrastructure networks.

FORMS OF INTERACTION

Although there are close conceptual links between telecommunications and other forms of infrastructure, the relationship between electronic flows along telecommunications networks and the physical flows of people, goods and services along water, energy and transport networks are not necessarily symmetrical. The linkages between electronic and physical flows are extremely complex. Although much of the utopian and futurist literature focuses upon the potential for substituting electronic for physical flows, we have already shown in Chapter 6 that this perspective fails to grasp the dynamic and contradictory nature of the relations.

Here we argue that the interactions between telecommunications and infrastructure networks can be most accurately characterised as one of interdependence, complementarity and synergy. Rather than substituting for material flows the role of electronic flows can be characterised as one of synergistic generation (Mokhtarian, 1990). Although this term was originally developed in the literature on transport–telecommunication interactions we want to demonstrate how it provides a framework for analysing the relations between electronic and material flows along infrastructure networks.

The first form of interaction is enhancement. Salomon defines enhancement as 'situations in which additional telecommunications generate additional travel between two nodes, which would not have occurred had there not been a communications channel' (Salomon, 1986; 226). Increasing use of cheaper telecommunications services helps to increase the number of people in a network and knowledge about the availability of goods and services. This information can then result in an increased desire to travel in order to have a higher level of personal interaction. The development of mobile car phones, for instance, may actually stimulate more travel as new trips can be generated during a car journey.

As Mokhtarian argues, there have been close links between telecommunications and travel since the earliest development of the telephone: 'It is rather profound that the first words spoken by Alexander Graham Bell into the telephone he had just invented were, "Watson, come here I want you"' (Pool, 1977; 12). 'Even though the trip was just a few feet down the hall in this case, the very first use of that most universal telecommunications product, the telephone, was to generate a trip for the purpose of face-to-face communications' (1990; 238). After reviewing the pervasive nature of the enhancement effect of telecommunications, Mokhtarian concludes that: 'had the telephone not been invented, the entire economy would be drastically different than it is today, and much of the transportation we are asked to envision would not simply happen' (1990; 239).

The second issue is the use of telecommunications to increase the operational efficiency of infrastructure networks. There is now increasing interest in the use of telecommunications to help to increase the effectiveness of existing transportation and energy networks for the distribution of goods, people and resources (Dupuy, 1992; Laterrasse, 1992). Telecommunications can help to increase the capacity, efficiency and safety of existing networks. In the transportation sector there are numerous examples – traffic system control, travel booking and payment systems, airport landing systems and in-vehicle navigation systems. As Latterasse argues, these systems 'aim at increasing, or in any case more efficiently using, the treatment capacity of flow over the existing infrastructure and endowing them with greater operational facility, especially so as to absorb peak period effects' (1992). Telecommunications-based control systems increase both the reliability and efficiency of these networks, which in turn may then make them more attractive to users and perhaps increase usage of the network.

The third type of interaction concerns longer term effects on land-use patterns. Telecommunications have complex and contradictory effects on urban land-use patterns, neighbourhoods and the design of the home. These vary across different sectors, in some cases supporting the reconcentration of activities but in other areas facilitating further decentralisation. These effects can generate new demands for investment in all forms of urban infrastructure in new places – decentralised land-use patterns result in new water, energy and transport networks, which may result in more trips on less congested roads because of more dispersed land uses. Home-based working may actually increase domestic demand for energy services during the working day. At the same time reconcentration increases demand for the rehabilitation of older services and the search for new types of solution to urban traffic congestion.

Overall the relationship between telecommunications and urban infrastructure can therefore be characterised as one of synergistic generation (Mokhtarian, 1990). For instance, focusing personal transport, transport of information and telecommunications – Mokhtarian concludes that the 'more that one or another form of communication takes place, the more that all forms of communication are stimulated' (1990). This is not to deny that there is no potential for substitution. But while telecommunications could take a larger share of all communication and the relative share of transportation declines, the absolute level of all forms of communication will continue to rise. Any displacement takes place not through a decline in transportation but through the growth of all forms of communication in which telecommunications simply take a larger share. These findings have been supported by studies of the historical development of physical flows along networked infrastructures. New infrastructure networks tend to replace older networks more through the growth of the whole system rather than the destruction or displacement of the older systems (see Ausubel, 1989; Grubler, 1989).

Between 1910 and the early 1970s the basic set of urban networked technologies have been relatively stable and enduring. Rather than electronic flows through telecommunications networks substituting for physical flows of resources, goods and people through transport and utility networks, the two have developed in parallel. During this period the main logic of network provision and management was oriented around the development of large standardised networks which integrated the early local networks into vast national power and communication systems. Precisely because these systems were slow to change they have imposed powerful 'restraints on the freedom with which economic forces or public policy can reshape the city. Infrastructure has thus served both as force for development in one period and as a barrier to change in another' (Tarr, 1984; 6). These constraints provide a crucial context for understanding the role of telecommunications network in reshaping the relations between cities and their networked infrastructure services in the 1990s.

THE NEW URBAN INFRASTRUCTURE CRISIS

Contemporary relations between infrastructure and telecommunications need to be placed in the context of an emerging crisis in the provision and management of urban infrastructure systems that developed from the early 1970s, placing serious constraints on the development of the city.

- Water, waste and energy networks designed to meet the needs of the industrial city were reaching the end of their useful lives requiring massive investment in rehabilitation and renewal (OECD, 1991).
- Existing infrastructure networks were placed under increasing pressure as they were asked to meet new demands. Increasing levels of traffic congestion and poor environmental conditions placed new costs on industry – the CBI

estimated that congestion in the UK cost industry £15 billion per annum (Confederation of British Industry, 1989).

• There were difficulties financing the renewal of old networks and the construction of new types of communications infrastructure from central government budgets.

Together these trends led to increasing concern with the physical signs of an 'urban infrastructure crisis' which could potentially threaten the continued development of the city.

This latest round of restructuring is closely linked to increasing demands for new forms of transportation and telecommunication infrastructure that can compress the 'time and space' constraints involved in both production process and participation in global markets. While most attention has focused on the communications infrastructure associated with these processes they are also linked to demands for more flexible and high-quality forms of energy and water infrastructures. More flexible work patterns and production processes create new demand patterns for energy supplies which are no longer based on bulk supplies during the conventional working day. The reliability, costs, flexibility and quality of energy and even water supplies have become important issues in the operation of these new forms of production processes and financial services.

The increasingly obsolescent and congested infrastructure networks that were inappropriate to the demands of new forms of production process and organisation have contributed towards a wider financial crisis in the state provision of infrastructure services. Constraints on public expenditure meant that the state could no longer simply underwrite the rehabilitation of older networks and the provision of new infrastructure services of the type and in the places were new demands were developing.

In response, many governments have divested themselves of responsibility for the provision of the key infrastructure services through the privatisation and liberal-isation of the infrastructure sector. Centralised and monopolistic infrastructure providers have often been seen to be part of the problem – being 'slow moving' and 'inefficient' because they were not open to competitive pressures. The new orthodoxy saw the privatisation of key infrastructure sectors and the development of new competitive markets in energy, transport and telecommunications services as the solution to the problems of public and private monopolies.

At an international level, there is the increasing opening up of infrastructure markets to competitive pressures from global infrastructure providers. The European Union is implementing policies to create integrated supra-national infrastructure networks to develop competitive markets in telecommunications, transportation and energy services. The recent shift to a standardised voltage across Europe is designed to facilitate new competitive markets in electricity production and supply.

There are, however, some contradictory trends here which seem to be pushing governments into a closer role in the provision of infrastructure services. The private sector has greater difficulty underwriting the capital costs of highly expensive transportation infrastructure such as new airports, international railway links and urban transit systems. Although governments are developing new ways of privatising key aspects of transportation infrastructure the state often still has central role in the provision of large-scale road and transportation infrastructure. However, as we shall see, new forms of telematics systems could create the potential for privatising most aspects of the transportation infrastructure. Consequently in Europe, Japan and the USA national governments, telecommunications providers and car manufactures are now playing a key role in developing standardised systems of transportation informatics for new products and services.

Although utilities and transportation infrastructure clearly play fundamental roles in the operation of modern cities, their roles in economic, socioeconomic and spatial restructuring of cities has often been neglected. The new trajectories of urban infrastructure provision are only slowly emerging as previously monopolistic infrastructure markets such as energy, telecommunications and transport services are opened up to increasing competitive pressures and companies develop around new ways of managing their networks and the territories they serve. Until relatively recently there has been little critical research or policy interest in the relationship between these infrastructure networks and telecommunications.

More recent critical analysis recognises the complexity and contradictory nature of these linkages. These perspectives acknowledge that the pervasive role of infrastructure networks means that they are important both to the economic development of capitalist production and to the reproduction of social relations. Transport and utilities span the 'production–consumption nexus' within cities (Swyndegouw, 1989). Infrastructure networks, as large technical and institutional systems, are therefore closely related to the wider societal context within which they evolve (Gökalp, 1992). It follows that infrastructures are important factors in the economic, social and spatial development of cities, regions and space economies. The cost, quality, availability and reliability of these network services, how these vary over space, the technologies they employ, and how their development is regulated are all factors that have important influences upon the development and restructuring of cities and regions. Therefore, infrastructure networks are key elements in the wider process of urban management and governance. Telecommunication applications have a key role to play in the restructuring of urban infrastructure networks.

TELEMATICS AND THE URBAN CONTROL REVOLUTION

We are currently in a period of rapid change both in the technologies and regulation of urban infrastructure networks. Not since the early twentieth century have such rapid transformations taken place in the provision and management of infrastructure networks. In response to rapid economic, social, environmental and political change, established infrastructure networks are being radically reoriented by privatisation, liberalisation and the widespread application of telematic networks. Dupuy, in Box 7.1, provides a useful illustration of how telematics are fundamentally changing many of the key aspects of infrastructure provision with a move to the real-time planning of services. These rapid changes are forging a new landscape of infrastructure services which is having profound implications for all aspects of contemporary urban management and governance.

BOX 7.1 TELECOMMUNICATIONS AND THE MANAGEMENT OF URBAN NETWORKS

New information technologies are making spectacular progress in most fields of activity in modern society. The management of urban transport and communications networks, energy and other systems is no exception.

RATIONALISING URBAN NETWORK OPERATIONS

Providing users with services through a network means, in fact, putting production/processing units into contact with consumer units or users (or user installations) in time and in space and as efficiently as possible. It involves information, provision of the required services, invoicing etc.

In an ideal system, network operators would have 'real-time' control over all their production and consumer units, but network operations as they exist today run into numerous obstacles. Production and use occur at different points, use varies over time, some production/processing units or sections of the network may be unavailable, communication with users may pose problems, and a variety of potential hitches may adversely affect supply in regard to demand....

In long-term planning, increasing use is made of modelling techniques. Designed to optimise system design, investment and so on, models are becoming more practical thanks to increasingly extensive computerised databanks (operating data on previous years, costs etc) [and] much higher data-processing capacity, making possible the comparison of numerous variables. This is a most important point. Any change in a network can produce a combination which is impossible to process manually or at sufficient speed even with early-generation computers....

Medium-term planning has derived even more benefit from advances in computing ... examples include production planning methods, setting the output levels needed to supply a network in a given area at a given time. Electricity supply and major urban heating networks have been or are being equipped with this type of system, in which computerisation replaces manual methods of collection and processing. For the time being, apparently, this does not apply to the same extent to water/ sewage systems, except for the running of production, storage or drainage equipment (drinking water treatment, rainwater reservoir or storm drains).

The third and probably most promising case is that of very short-term planning, and sometimes incorrectly described as 'real time'.... Rainwater drainage networks are a typical example. Some years ago, they were managed solely on a long-term planning basis, the size of waste pipes being based on maximum rainwater level forecasts. Since then, the objective has become optimal network capacity, using real-time management techniques during wet weather. Local control systems open flood-gates to drain water held in temporary storage areas. But for a network covering a wide area, centralised control makes it possible to even out rainwater distribution with the help of radar images supplied several minutes earlier.

(Dupuy, 1992; 51-76)

A new era of control technologies are being harnessed by infrastructure managers to play a central role in the response to the crisis in urban infrastructure that developed in modern cities from the early 1970s. New telecommunications infrastructure are being retrofitted into the city and increasingly being layered over more established water, energy and transportation networks. This complex convergence provides the operators of networks with more effective ways of controlling access to their networks, focusing on the needs of premium customers, providing new methods of surveillance and control, increasing the effective capacity of congested networks and providing new products and services.

THE CONVERGENCE OF URBAN
INFRASTRUCTURE

These trends signal an important convergence between telematics and urban infrastructure. Figure 7.1 provides an illustration of these processes in the UK. The first element in the reorientation of urban networks is the convergence between previously separated infrastructure operators and networks. In the emerging infrastructure and utility marketplaces, cross-investment between utility and infrastructure operators is burgeoning. The new pressure to secure profitability is leading investors to investigate new complementary arrangements between infrastructures that previously were completely divorced in their development within separate public monopolies. This cross-investment is part of a wider trend towards diversification away from core businesses, a process driven by the aim to improve financial performance in utility and infrastructure companies (Brewer, 1989). In part, this is a response to the vulnerabilities imposed on networked services by their local dependence.

Infrastructure operators are now developing strong interests in telecommunications and telematics. Many energy, water and transport operators are going one step further and are amongst the first to invest in new public telecommunications systems to compete for custom with the established operators. Existing utility and transport companies are ideally placed to cross-invest into these new utility markets. They possess the necessary large amounts of capital. They already own or control strips of the land, ducts and leeways between the most lucrative business centres that can be used quickly and cheaply to construct new telecommunications networks. In many cases their own sophisticated internal telecommunications networks already have spare capacity that can now be simply resold to outsiders. Finally, they also possess established computer systems to handle billing and customer service as well as expertise in the construction of networks within urban areas.

As we saw in Chapter 1, as liberalisation spreads to be a global phenomenon, this process of cross-investment is taking on a global scale. For example, the current building of urban cable networks in Britain is fuelled overwhelmingly by investment from North American cable and telecommunications investors and French municipal service companies. Increasingly, privatised infrastructure companies face competition at home as well as engaging in competition abroad. This cross-investment and convergence is being paralleled by rapid convergence between computer, telecommunications and broadcasting companies, as the industrial repercussions

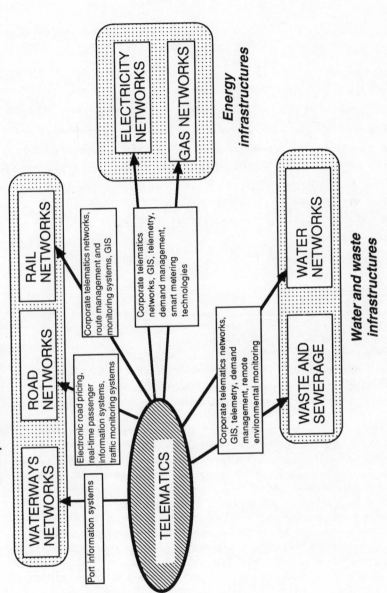

Figure 7.1 The convergence of urban infrastructure

of technological convergence between these technologies gather pace. The convergence of urban infrastructure services is also paralleled by the pervasive application of telematics systems within the utilities and transport sectors.

UTILITIES AND TELEMATICS

In most parts of the developed world (Nortof, 1991; Hill, 1992) utilities are responding to the new era of deregulation, privatisation, competition, uncertain demand and rising environmental concern by 'turning to information technology as an effective, strategic tool to reinvent their systems. Used appropriately, technology can help a utility reduce costs, streamline operations, improve customer satisfaction, provide better management controls, and manage and deploy assets prudently' (Hill, 1992; 23–24). As Box 7.2 shows, utilities are utilising telematics applications for almost every aspect of their planning, operation and management.

BOX 7.2 UTILITIES AND INFORMATION TECHNOLOGY

In the case of utilities, the range of applications of IT is very wide, covering both the production of a service (for example, electrical power) and its distribution. In this sector IT in the form of data processing of distribution and billing information has been in widespread use for some time, and the main changes likely to emerge here are in closer integration. Associated with this is the trend towards higher levels of automation in data collection, for example by replacing manual reading of meters for electricity, gas or water by 'intelligent' systems which can record consumption and transmit the information back to a regional or central office computer system. In this respect utilities seem to be following the example established by the telephone system (an information service), where such billing methods have always been the pattern, and where current trends are likely to be precursors of the pattern that will emerge across the range of utilities. Once again this reflects the pattern we have seen in the preceding sectors, where the coordinating and communicating activities peripheral to the core production are those most affected by IT. On the 'core' production side, IT applications are increasingly in the area of automated and integrated control of plant, and that of better management of resources through more accurate monitoring and the generation of more control information; for example in the area of water conservation through better flow metering, more accurate valve control, faster response times, and so on. This will probably lead to a trend in demand for skills similar to that in manufacturing, ... that is, for fewer people with higher level and broader based skills working in increasingly automated plants. In these instances the relations between technological development in extractive industries and construction are likely to be mutually reinforcing. IT also plays an important role in improving the utilisation and conservation of resources. Through more accurate monitoring and control it is possible to balance loads optimally (for example, those of power station operations), even under conditions of rapid or unexpected fluctuations in demand. At the same time the provision of information to the consumer – for example via a local 'intelligent' meter – helps encourage conservation and more efficient use of facilities. A major new application area for IT is in the monitoring and maintenance of what are often geographically widespread and complex distribution networks. Here remote sensing, telemetry, robotics and expert systems are all finding increasing application in fault location and remote repair activities. In terms of industrial structure these sectors are already highly concentrated in many countries – often by virtue of monopoly or near monopoly supplier status. IT is likely to facilitate the decentralised operation of such organisations since it offers the opportunities of networking, and, as such, opens up the possibility of reducing the entry costs and problems for possible competitors. It is also worth noting that IT provides support for the trend towards the use of in-house utility production within larger organisations, since it enables accurate control and management information and experience to be packaged within a reliable control system.

(Miles et al., 1988; 119–121)

Newly privatised utilities are expecting telematics systems to give them a competitive edge and so help maintain profitability in increasingly competitive and

---

- information collection and management associated with administration, customer management, etc.
- improved monitoring of distribution networks and faster fault finding
- improved maintenance procedures via remote control, for example sewer repairs by robot
- improved resource management through more accurate information and control and self-optimising systems
- improved customer service and information flow via networks of 'intelligent' meters linked to regional centres
- modelling and simulation for load management and optimisation under fluctuating demand conditions
- weather forecasting to enable supply/demand management
- mobile communications

Figure 7.2 Telematic applications in utilities

Source: Miles et al., 1988

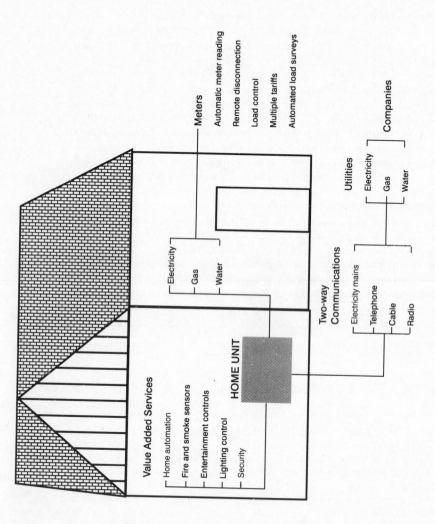

Figure 7.3 Smart metering technologies

uncertain markets (Madden, 1992). Telematics are being used to cut costs, improve the speed and responsiveness of large organisations and, above all, improve the degree to which enormous networks infrastructures can be controlled and organised in competitive ways. As Figures 7.2 and 7.3 illustrate, telematics can be used to help keep energy and water production more in line with demand; they allow a much more sophisticated control of plant to occur; they support the automation of meter reading through telemetry; and they support faster and more responsive customer services − particularly to large, lucrative customers (Miles *et al.*, 1988).

For instance, the budgets devoted by UK utilities to telematics went up by 147 per cent in one year between 1989 and 1990 − five times the average for all economic sectors (Madden, 1992). This makes utilities the fastest growing sector for telematics investment in the UK. Dave Madden highlights 'the vast scale of IT investments in the newly privatised water and electricity industries − and the pivotal role that investment is playing in creating a competing, and competitive, utilities sector' (Madden, 1992).

Typically, the fragmented IT 'islands' inherited from public utility monopolies are being transformed into sophisticated and integrated computer network infrastructures. These are emerging as the basis for all information, communications and transactions flows within the organisation and for supporting its relations with customers and suppliers. For example, National Power, the UK's largest electricity generator, is developing one of the largest computing and telecommunication projects ever undertaken by a European company. This will be used to improve customer service, flexibility, control and as the basis for developing the 'creative service products and complex tariffs' (Madden, 1992) now necessary to maintain competitiveness. The view behind the investment was that it 'would transform the generators' business by improving efficiency, changing the way it worked with customers and providing revenues in its own right' (Wilson, 1992).

TELEMATICS AND TRANSPORTATION

The second area is the development of transportation informatics. Transportation informatics are emerging as one of the most significant areas of application and growth in linking telecommunications to the management of urban infrastructure. As Giannopoulos and Gillespie argue, the 'informatics revolution' underpins innovation in both transport and communications:

A form of 'technological convergence' is thus affecting the relationship between the two, opening up new complementarities and potential synergies, most evident in the way telematics networks are becoming integral to an increasing array of transport operations. At the same time, however, technological advances in the telecommunications field are providing new opportunities for substituting electronic communications for physical transport.

(Giannopoulos and Gillespie, 1993; 30)

As Whitelegg graphically illustrates in Box 7.3, the urban transportation crisis – longer commutes, congestion, environmental externalities, and increased costs – has helped to stimulate increased interest in the application of informatics to almost every aspect of the transportation infrastructure.

BOX 7.3 CONFUSING SIGNALS ON THE ROAD TO NOWHERE

A reluctance to think about reducing demand for transport by sensitive and intelligent land use planning, public transport investment and genuinely sustainable transport policies has forced Europe, Japan and the US to embrace technology as the thing that will rescue us all, allow car ownership and use to rise and allow us to ignore ecology, habitat, health, global warming and the destruction of our cities and countryside by tarmac and concrete.

In Europe the European Commission has invested over $1,250 million in DRIVE (Dedicated Road Infrastructure for Vehicle Safety in Europe) and PROMETHEUS (Programme for European Traffic with Highest Efficiency and Safety) with the common aim of using electronics of 'telematics' as a means of getting more capacity out of our highway systems. In the US 'intelligent vehicle/highway systems (IVHS) does the same kind of thing and is budgeted to spend $218 million in this fiscal year. The explicit goal of IVHS is an effective doubling of infrastructure capacity though some commentators anticipate a three- to seven-fold expansion as a result of inboard computers, electronic guidance permitting vehicles to drive very close to each other and (supposedly) providing built in safety so that higher speeds with shorter distances between vehicles can be maintained over large parts of any journey.

The rapid development and application of new technology in these areas ignores the reality of transport's impact on the environment and the extent to which it transforms society to necessitate longer distance trips between home, work, shopping and recreational activities and the ever growing allocation of land for roads, car parking and service areas. Telematics can increase capacity on the highway system but what happens when all those cars pour into cities and search for somewhere to park? If we increase highway capacity between our cities then we must allocate even more land in cities to vehicles and in so doing transform cities into sterile wastelands dominated by noise, air pollution, danger and lack of people. Alternatively new technology offers the possibility of driving less wealthy people off the streets with pay-as-you-go smart card technology to allocate road space through ability to pay. Cambridge (UK) is already experimenting with such a scheme.

Technology largely ignores feedback loops that dominate traffic and transport developments. If we increase capacity on highways then we send signals to drivers that the next trip will be easier and so there will be more trips, people will live or work far from where they shop and where they send their children to school and all these decisions will involve more transport, more consumption of finite resources and more pollution. In the period 1985–1990 the distance travelled per person per week increased in Britain by 22 per cent.

We are all taking advantage of cheap transport and more roads to travel further. A benefit in the form of motorised transport that can save time and increase opportunities is simply being consumed by increasing the distances that we travel to accomplish basic everyday tasks. Technology makes this possible and stimulates more and more travel rather than encourage a fundamental reappraisal of accessibility versus mobility and the economic gains and quality of life gains that can be had from enriching the destination opportunities within five miles of where we live....

The technological sophistication of environmental concept cars and telematics is on course to deliver wall-to-wall cars in developed societies, fiscally ruinous car acquisition in Third-World countries and global environmental crises resulting from the massive scale of raw material exploitation and energy use to feed the appetites of two billion cars whether environmentally friendly or not.

(Whitelegg, 1993; x–xi)

Every major developed country is now in the process of developing and testing new road telematic applications. It is hoped that these systems will enable new, more flexible and responsive forms of traffic management and safety systems to be created to the benefit of all road users. Figure 7.4 illustrates the broad scope of road transport informatics technologies.

There are three key aspects to the family of transportation informatics technologies. These include traffic management systems for controlling traffic in cities; advanced logistical systems used by commercial freight companies; and smart vehicle technologies which use wireless technologies to communicate with traffic management and logistics systems. Together these applications could form an Integrated Road Traffic Environment (IRTE). The basic concept is that a common two-way communications infrastructure between vehicles and a network of control centres should develop a series of applications, each having different

Demand management
- electronic road pricing using automatic vehicle identification
- automatically enforced restrictions

Motorist communication and information
- electronic route planning aids
- traffic condition broadcasting
- on-board navigation systems
- electronic route guidance systems
- automatic vehicle location systems

Improved traffic control
- adaptive traffic control linked to route guidance
- integrated motorway and corridor control systems
- incident detection systems
- use of Artificial Intelligence (AI) in expert systems for traffic control

Automatic vehicle control
- automatic headway and steering control
- fully automated highways

Automated toll collection
- use of communications technology

Figure 7.4 The scope of road transport informatics technology

types of consequence for travel behaviour, including the time of travel, selection of route and travel mode. Examples of these applications include electronic road pricing, electronic route guidance, passenger information systems and smart highway corridors. Figure 7.5 illustrates the complex institutional and communications flows required for an electronic route guidance system. Together these applications potentially provide a 'transactional infrastructure for charging drivers for the use of road space; a system by which the costs and prices of using the transport system can be made clear; and a system to regulate and control how roads are used' (Hepworth and Ducatel, 1992; 92). A wide range of experimental road pricing projects are already in operation. Singapore and Trondheim in Norway use simple electronic turnpikes; more sophisticated comprehensive road pricing experiments are being planned in Europe through the DRIVE programme and in the United States through the Intelligent Vehicle Highway System. There are major political problems here, however, as public reaction to widespread commodification of road space would be fierce.

The development of new production processes, based on 'flexible specialisation' and 'just-in-time' (JIT) production methods created demands for new types of transport and telecommunications infrastructures (see Chapter 4; Hepworth and Ducatel, 1992; Giannopoulos and Gillespie 1993). The freight

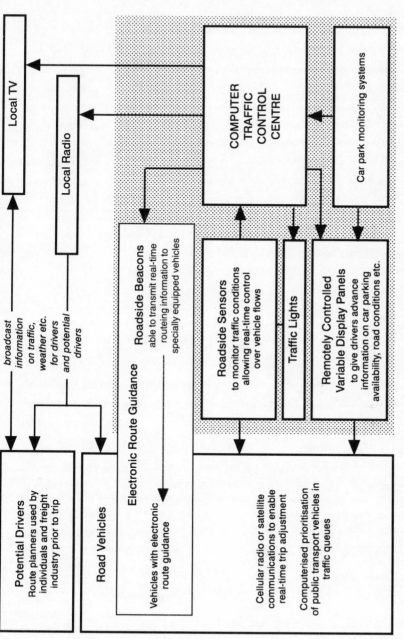

Local TV

Local Radio

broadcast information on traffic, weather etc. for drivers and potential drivers

COMPUTER TRAFFIC CONTROL CENTRE

Car park monitoring systems

Electronic Route Guidance

Roadside Beacons
able to transmit real-time routeing information to specially equipped vehicles

Roadside Sensors
to monitor traffic conditions allowing real-time control over vehicle flows

Traffic Lights

Remotely Controlled Variable Display Panels
to give drivers advance information on car parking availability, road conditions etc.

Potential Drivers
Route planners used by individuals and freight industry prior to trip

Road Vehicles

Vehicles with electronic route guidance

Cellular radio or satellite communications to enable real-time trip adjustment

Computerised prioritisation of public transport vehicles in traffic queues

Figure 7.5 Electronic route guidance

Source: Hepworth and Ducatel, 1992

'logistics revolution' has seen the merging of physical and electronic flows in which the transportation network itself becomes part of the production process controlled and monitored by new telecommunication infrastructures. These changes have had profound implications for the type of infrastructure that firms require – new software, radio-based communications with freight vehicles, electronic data interchange (EDI) between different parts of the production and distribution process. This telecommunications infrastructure in turn generated new demands for road space as firms used the road network as an integral part of just-in-time production and distribution systems.

The phenomenal growth of sectors such as financial services and media were both based on, and created their own demands for, global information networks. New global communication networks allowed these sectors to shrink both the temporal and spatial constraints on real-time participation in twenty-four-hour trading in global markets. The demand for advanced global telecommunications has been mirrored by the requirement for increased levels of high-speed personal travel between financial centres by air and train (see Chapter 4).

TOWARDS THE INTELLIGENT CITY?

The new control capabilities of telematics seem to offer important benefits for urban policy. Taking the example of the transportation sector, Hepworth and Ducatel argue:

The technical capabilities of RTI [Road Transport Informatics] appear to offer considerable promise as new instruments of urban development policy. Electronic road pricing, for example, could be used to influence consumer behaviour in favour of a better environment and improved public transport. Passenger information systems could be designed to regulate and promote urban bus services more effectively. The new logistical systems supporting just-in-time production do provide opportunities for managed industrial deconcentration and polycentric models of land-use planning... teleworking and teleshopping are potentially new instruments for overcoming the 'tyranny of geography' which puts jobs and the 'high street' out of the reach for less mobile sectors of the labour force and population.

(Hepworth and Ducatel, 1992; 178–179)

Similar types of urban policy potential exist in the application of telematics in the utilities sector. For instance, smart meters can help encourage the conservation of water and energy resources; monitoring and alert systems can highlight environmental dangers and emergency networks can be provided for vulnerable groups. But as the extract in Box 7.4 by Laterasse illustrates, there is relatively little evidence on which to base an assessment of the potential outcomes of the new urban control revolution.

BOX 7.4 EVALUATING THE INTELLIGENT CITY

It is not an easy matter to evaluate the impact of these systems on individual service functioning; and even less so for the city as a whole. Few rigorous studies on this subject exist; too often, the only information available has been provided by the companies marketing the system in question, and these assessments more often pertain to theoretical performances than to observed effects. A few thoughts on the subject can illustrate the problem before us.

From an economical point of view, what is at stake in systems of flow control appears at first quite obvious: at a cost in a range of from several thousandths to several hundredths of a percent of the cost of the basic equipment, such systems can generate effective capacity savings of from 20 per cent to 25 per cent....

However useful these systems may be, they do not for the moment make it possible to foresee a complete solution to the problems of urban traffic congestion. As for the automobile traffic, the endemic malady of French cities, the positive effects of control systems, in the absence of any accompanying measures, are going to be even more ephemeral ... – one is forced to ask whether we should congratulate ourselves for being capable of getting more and more cars to drive over the same space, or whether we shouldn't rather worry that any improvement in traffic, however relative, will end up as the expression of just additional flux and, finally, by a postponement, if not a complete aggravation, of the problem!

Concerning systems which inform users, with the exception of those whose effect can be directly assessed, for example in terms of how often a service is used, evaluation is as problematic as the two preceding categories of systems. The effects of what takes place between information and user behaviour in particular, is almost completely unknown. Few specific studies on this subject exist....

The idea of injecting data in order to regulate demand (in particular in order to level the peaks) is not, for all that, a new procedure; but setting up the same regulations in 'real time' raises problems which are just as serious as the static utilization of data. Projects for putting data systems, or car-computers, into automobiles, for example, once they are beyond the stage of simple guidance systems, and will include traffic data, are typically marked by the double system of data and control, and some of their advocates would be ready in the near future to substitute them for the systems for regulating traffic lights. This would imply that the collective good, in terms of traffic, is the sum of individual goods – and this remains to be proved. Moreover, as soon as there is a time lapse which is difficult to compress any further between the moment of data input and the moment when, after processing, it has reached the user ready to make (or modify) a decision, this implies that one would be capable of 'predictive' management, that is, not only to be able to predict user reactions, as has already been noted, but also, for example, to anticipate upcoming demand, or to evaluate the residual capacity of one branch of the network (Laterrasse *et al.*, 1990). Now, while this is relatively possible for water or energy systems, it is much more difficult, if not impossible, with the existing methods of modelling, for roads: this is because these systems, even more than with other urban networks, function in a completely open milieu, and controlling the entrances and exits is inconceivable.

This observation leads us to another: the model of the city-computer, in which all flux would be controlled solely by an intelligence centre, is not only unacceptable in concept, it is also completely unrealistic with respect to the inherent complexity of urban systems. This of course does not mean that the process of computerization of urban systems will not provide a partial solution to the problem, some of which we have tried to present above. But the concept of the 'intelligent city' implies at the very least that we would be capable of imagining an alternative to Orwell's Big Brother of 1984. But this is precisely the problem: we are obviously beginning to understand that centralized systems are not those which are the most suited for the urban context, and we are beginning to develop modular systems, with shared-out intelligence, which make it possible to combine the central level (control of large flux) with the local level (control of unknown factors); we are beginning, in addition, to perceive that the hyper-complex systems which are our metropolitan centres reveal, in particular, notably due to the more and more complex networks which have been developed in them, forms of auto-organization which need to be more carefully studied and taken into account.... It remains that this process is still in an embryonic state, and is hence difficult to formalize; it is more and more compartmentalized (each type of network is approached, essentially, in a separate manner), whereas the concept of the intelligent city presupposes transversal and globalized processes. Moreover, it is with respect to which functions, which end results that we would optimize the city-system which is rather by its nature multi-function, multi-actor, multi-criteria, to the point that the very notion of optimization raises a problem? While current research studies are not totally without an answer to these types of questions, they have, as yet, no entirely operational answers to give, either. And for all that, can research provide these answers?

(Laterrasse, 1992; 4–7)

The central question is whose 'intelligence' do the new telematics systems embody? Superficially, at least, they appear to offer urban policy-makers new ways of optimising the performance of cities. However, this ignores wider issues about who actually controls and benefits from these telematics applications. Our theoretical approaches lead us to be sceptical about the potential of telematics to meet these high expectations of a new 'technological mastery' over urban problems. An analysis of telematics applications in the infrastructure sector needs to consider five key issues.

TELEMATICS, PRIVATISATION AND LIBERALISATION

First, the new capabilities of telematics are helping to support the liberalisation and privatisation of utility markets and transportation networks. Virtual electronic networks are laid over established infrastructure networks, helping to undermine their 'natural monopoly' characteristics and so allowing private firms to operate them profitably (Robins and Hepworth, 1988). In the utilities sector, smart metering technologies allow affluent consumers to choose their electricity, gas or even water supplier by programming their own meter. Two-way communication then alerts their chosen utility that they now have a new customer whose resource consumption can be remotely monitored and debited – avoiding the huge transaction costs associated with manual metering reading. Telematics can open up 'monopoly' networks to competitive supply even in the domestic sector without the need for replicating new energy systems.

The implications are even more profound in the transportation sector. Here advanced transportation telematics provide the potential for privatising public goods like the road network. Hepworth and Ducatel argue that electronic road pricing (ERP) will:

create the physical infrastructure needed to privatise road space and will also create an institutional structure for administering a privatised road system. However, we suggest that the 'privatisation' of Britain's roads that ERP permits is based not only on charging for road use – consumption – but also making road space a tradable commodity which the private sector could provide. Also the 'privatisation' of road space is likely to become more attractive to private investors in the future because of the added value of road transport information generated as a by-product of electronic road pricing.

(Hepworth and Ducatel, 1992; 92)

What were previously considered to be public goods – because it was difficult to monitor and measure the exact consumption of specialised services – can now be privatised because of the new control techniques of telematics. These technologies are closely linked to the development of the 'pay-per-revolution' (see Chapter 5; Mosco, 1988). Telematics applications open up the potential of charging for infrastructure services in real time as transport and utilities are able continuously to monitor usage of road space, energy and water services. New metering technologies, in the form of smart utility meters or electronic road pricing, allow infrastructure providers to set multiple tariffs according to the cost of providing a service and levels of demand for road space. So, combined with the political movement towards liberalisation, the transfer from public monopoly to private marketplace can be seen to have both a technological and a political dimension.

NEW INFRASTRUCTURE PRODUCTS AND SERVICES

Telematics applications have the potential to create new markets for products and services in the utility and transportation sectors. Transportation informatics systems are 'all important new areas of market expansion for the IT industries and the vehicle and transport industries. The search for profits in these sectors accounts for the powerful technology/supply push behind the next transport revolution' (Hepworth and Ducatel, 1992; 3). Road transport informatics systems are being used to create a range of entirely new products and services – electronic route guidance systems, on-board navigation systems, automatic vehicle location systems.

With the reduction in arms spending, large companies are now transferring military technologies to the transport sector and creating new premium-based transportation services targeted at niche markets. Similarly, the utilities sector is using telematics applications to offer new types of products and services. The communications link to smart metering technologies can be used to offer alternative telephony services and form the basis of home automation systems with utilities offering add-on modules plugged on to smart meters for energy management and security applications (see Figure 7.3). Utilities could use telematics control systems to offer packages of energy services tailored to customers' specific needs but making the most profitable use of utilities' own production capacity.

INFRASTRUCTURE AND SOCIAL CONTROL

These new control capacities raise the prospect of increased surveillance, social control and the commodification of information. For instance, smart metering technologies enable utilities to build up detailed information on households' consumption patterns that go beyond total use of services (see Chapter 5). Smart electricity meters can identify what electrical appliances a household owns, how often they are used and for how long. It is not even clear who 'owns' this type of information but it opens up the potential for new forms of surveillance in the home. Transport telematics provide firms with increased surveillance and control opportunities over the labour process while route guidance and road pricing could be used to monitor the location and purpose of an individual's journey.

Communication with smart meters also allows infrastructure providers to decide the basis on which low-income households have access to these services – perhaps in the form of prepayment and are able remotely to 'switch out' unprofitable and marginal groups from the service. The information gleaned from all of this monitoring and processing becomes a valuable commodity in the burgeoning information services markets (Hepworth and Ducatel, 1992). Utilities are able to use information on energy use to identify and target households for new appliances and energy services.

INCREASING EFFECTIVE NETWORK CAPACITIES

Telematics technologies can be used to increase the effective capacity of infrastructure networks. While much has been made of the potential of telematics applications to help reduce or keep a downward pressure on resource flows through infrastructure networks there is relatively little evidence to support this in practice. Instead, there are increasing signs that telematics applications have the opposite effect. The application of road transport informatics in such areas as electronic guidance, real-time road monitoring and electronic road pricing effectively increases the capacity and reliability of the road network.

Of course those groups who have access to the networks during peak periods need to be able to afford the premium prices charged for electronic route guidance and road pricing systems. The new logistic revolution is closely linked to the shift from rail to road and increases in the number of trips by road within

THE THREE POINTS OF THE KENWOOD TRIANGLE REPRESENT ADVANCED TECHNOLOGY, QUALITY AND STYLE

THIS GUY'S STUCK IN TRAFFIC WITH A CELLULAR PHONE,

HE'S SPOKEN TO HIS OFFICE, HIS WIFE, HIS BOOKIE AND HIS GOLF PARTNER.

THIS GUY'S STUCK IN TRAFFIC WITH A KENWOOD MOBILE RADIO.

HE'S SPOKEN TO HIS OFFICE.

There are times when there is no substitute for a cellular phone. But equally, there are times when you end up footing the bill for a series of calls that aren't doing a thing for your business.

Kenwood's range of portable and mobile radio units can be tailored to local, regional or nationwide coverage. They can connect mobile to office or mobile to mobile, delivering total clarity over both short and long range. Best of all, because the system is tailored to your own user group, you have control over the costs: once you've made the initial investment in equipment, connection and line rental, there are no call charges to pay.

So before you equip your company with cellular phones, call 0923 212044 for details of the Kenwood Communications range. Because with Kenwood, you just pay for the talk. And not for the chat.

KENWOOD

HOME AUDIO, CAR AUDIO, COMMUNICATIONS EQUIPMENT, TEST AND MEASURING INSTRUMENTS, TELECOMMUNICATIONS

Figure 7.6 The use of mobile telecommunications to reduce the effects of road congestion – the example of a Kenwood communications advertisement for trunked mobile radio

and between cities. The application of telematics to infrastructure systems seems likely to increase their effective capacity and help facilitate increasing flows of resources in the context of expansion of the whole network. Relatively little thought has been given to the development of institutional frameworks that help use telematics to reduce flows and levels of pollution (see Chapter 6).

A more prosaic example of how telecommunications can increase the effective capacity of road networks comes from the relationship between traffic congestion and mobile telecommunications. As the Kenwood advertisement in Figure 7.6 shows, the flexibility of mobile phones and computers seems likely to enhance traffic congestion by converting the 'dead' time lost in traffic jams into 'live' working time in which executives can keep in touch with their businesses and clients. Thus, the experience of congestion becomes more bearable and much less costly in terms of wasted time and damaged links between organisations and individuals. This is a good example whereby telecommunications are involved with parallel shifts towards greater physical movement: in other words, the flexible use of electronic spaces on the move ties in with the flexible use of urban places (see Chapter 5).

RESTRICTING ACCESS TO URBAN INFRASTRUCTURE

Finally, telematics have important implications for levels of social access to infrastructure services and thus for physical movement within and between cities. The development of transport informatics is based on the need to minimise time–space constraints for powerful groups. Consequently, there are close linkages between control of space and the basis on which particular groups are allowed to have access to new technologies to overcome urban congestion. In this context, Swyngedouw argues that 'road pricing or other linear methods of controlling or excluding particular social groups from getting control over space equally limits the power of some while propelling others to the exclusive heights of controlling space and thereby everything contained in it' (1993; 323). Transportation and telematics applications reinforce and reconstitute control over space both within and between cities. Overcoming existing constraints is based on increasing both the level and speed of mobility for some groups while trapping others in particular places. This trend is matched by new types of demand for more flexible communications services as new sectors and production processes restructure transportation and telecommunication needs to improve efficiency and competitiveness.

CONCLUSION: CITIES SHAPING INFRASTRUCTURE NETWORKS

These wider structural changes have helped to stimulate a new round of interest in the role of urban infrastructure networks as a vehicle for promoting local economic development. Cities have placed increasing emphasis on economic development strategies which develop a specific infrastructure slant including the development of teleports and advance communications infrastructures (see Chapter 9). Urban governments are attempting to lever in an advanced telecommunications infrastructure in an attempt to support wider economic development objectives and attract scarce mobile investment. The development of new types of transportation infrastructure is emerging as a key role in these initiatives.

The plans tend to go beyond the conventional focus on road-based infrastructure to include new and expanded airport developments, high-speed train and urban transit systems. The development of advanced logistic systems has been targeted by localities attempting to capture firm investment in JIT production strategies. Local authorities are also closely involved in experiments to demonstrate the potential of transport informatics through road pricing, autoroute guidance and passenger information systems.

There is a clear infrastructure 'push' to these initiatives to renew and improve the performance of established infrastructure systems while promoting the installation of new advanced telecommunications infrastructures. But because of liberalisation and privatisation there is considerable variation in the ability of localities to help underpin these economic development trajectories. There is little critical assessment of what these new infrastructure systems actually do for urban areas as it is simply assumed that new urban infrastructures are needed to support wider linkage into global economic structures and promote local competitive advantage.

There is a danger that the providers of infrastructure services are emerging as the 'new urban managers'. How these companies plan and configure their networks, the parameters against which they manage them and the basis on which customers are allowed to connect have profound implications for the social, economic and environmental performance of cities. However, it is not easy to assess the impact of these systems on particular networks or the functioning of the whole city. There are few studies on the role of telematics and urban infrastructure while 'the only information available has been provided by the companies marketing the system in question, and these more often pertain to theoretical performances than to observed effects' (Laterasse, 1992).

Although situations vary, urban policy in many countries has often had little connection with the provision and management of large technical networks and policy-makers are poorly equipped to develop a response and seem to be increasingly marginalised from the radical changes taking place in urban infrastructure management. But there may be potential for policy development. Perhaps the key feature of the new context is the huge diversity of the new infrastructure patchworks. The social, economic and environmental impacts are cast very unevenly between cities and nations, and different infrastructure sectors. Although it is difficult to predict how a particular city will be affected by these changes the issue for policy-makers is the degree of fit between public and private urban management strategies. How far do the strategies of infrastructure companies coincide with, and potentially strengthen, public policy? Or do they conflict with, and undermine public strategies?

In this situation, we would argue that new policy approaches are necessary at the urban level through which this new 'utility and transport landscape' can be influenced in ways conducive to the development of socially equitable, economically efficient and environmentally sustainable urban areas. Urban governments need to monitor closely the infrastructural developments within their territories, and the technological and political foundations for them. They need to use whatever influence they can muster to try and reshape market forces into regimes which meet their local needs rather than simply the profit margins of infrastructure developers. Innovative ways of using 'sticks' – wayleave rights, land ownership and planning powers – and 'carrots' – financial subsidies and local government infrastructure markets – need to be explored. The protection of marginal social consumers from 'social dumping' is a major concern that needs to be addressed by city governments working with political and consumer action groups.

At the same time, the physical issues to do with the laying of competing networks, the physical convergence of networks, and the supply of infrastructure to economically disadvantaged areas is also a pressing concern. Ultimately, the broadest infrastructural needs of a competitive urban location competing within global markets need to be in place. For this to happen, however, we would argue that new institutional innovations are necessary so that the concerns of city authorities are brought together with the profit-seeking concerns of infrastructure developers, rather than local politics being simply ignored. Efforts need to be made to collapse the institutional boundaries which tend to exist between urban policy-makers and those who increasingly control the destiny of infrastructure networks within cities. One final area of potential is joint action between networks of cities to push for more consideration of urban issues by infrastructure regulators at the national, and increasingly, supranational levels.

8

URBAN PHYSICAL FORM

INTRODUCTION

Current developments in telecommunications have important implications for the physical form of contemporary cities. In this chapter we examine the linkages between the economic, social and environmental issues reviewed in previous chapters and the changing physical form of the city. We show how tele-communications underpin urban restructuring processes that have profound implications for the form and structure of contemporary cities.

Much of the debate about telecommunications and urban form focuses on two alleged forms of spatial restructuring. The first is the role of telecommunications in dissolving the need for physical proximity between people and services leading to the inevitable dissolution of the city and a new form of home-centred life. The second pattern is based on the recentralisation of certain cities as tele-communications are utilised by the most powerful and important urban centres to reinforce their centrality as controllers of information flows. The tensions and contradictions between these two trends flow through much of the debate on urban form – telecommunication relations.

However, the focus on dichotomous models of the impact of telecommunica-tions upon patterns of urban form often fails adequately to grasp the complex and contradictory nature of the linkages. Instead, we will demonstrate that policy-makers and academics need to develop a more sophisticated analysis of the relations between electronic spaces and the physical form of urban places. Using a conceptual framework that acknowledges that these relations can take a number of different forms simultaneously, we want to show how complex patterns of centralising and decentralising tendencies are proceeding in parallel across cities.

In the rest of this chapter we focus on three key questions. First, what are the historical linkages between telecommunications and the early development of cities? Was it a facilitator of concentration or a motor of suburbanisation? Second, what are the linkages between telecommunications and contemporary patterns of urban physical form? How do different theoretical perspectives help us to understand the changing nature and outcomes of these relationships? Finally, how can we develop a more sophisticated framework to understand the relationship between telecommunications and urban form in the contemporary city? How can we integrate different types of social and economic changes with shifts in the physical forms of cities? Does the minimisation of space and time constraints provide new levels of mobility and access for some while creating disenfranchised ghettos of poverty for others? Are telecommunications involved in creating new potential for controlling access to urban places and electronic spaces in the city?

THE DEVELOPMENT OF THE CITY AND THE TELEPHONE

A useful way of examining the linkages between urban form and tele-communications is historically to review the role of the telephone in the development of the city. As we shall see, many utopian and futurist accounts of the impact of telecommunications on the contemporary city tend to ignore the role of the telephone in different stages of urban development. Here we briefly examine the role of telecommunications by taking a number of historical 'slices' through the development of urban form between the 1870s to the 1970s. Figure 8.1 provides a generalised illustration of the phases of urban development of a typical large North American city. This relates the development of the telephone and the bundle of transport and energy infrastructure networks to the development of particular patterns of urban form.

Cities developed, at least in part, to make communications easier. In the early industrial city of the mid-1800s communication was dependent on the physical movement of people, goods and services. In this pre-telephone and electricity era movement was either by foot, horse or river and all activities required close physical proximity to reduce travel times to manageable levels. Consequently, the spatial limit of the 'walking city' was two to three miles and perhaps double that for travel by horse. The concentration of activities in one area helped to generate 'agglomeration economies' in which close physical proximity allowed markets to develop, workers to reach employment and the coordination of urban governance.

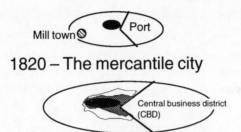

1820 – The mercantile city

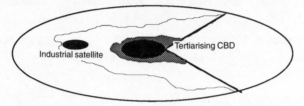

1870 – The competitive industrial city

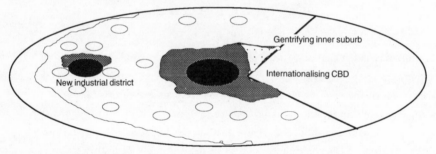

1920 – The corporate monopoly city

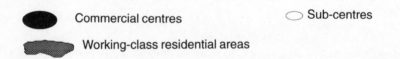

1970 – The state-managed Fordist city

Commercial centres   ⬭ Sub-centres

Working-class residential areas

Figure 8.1 Evolution of urban form of a North American city: 1820–1970
Source: Soja, 1989

The necessity of face-to-face communication meant that in spatial terms the walking city was densely settled with an intermingling of businesses, residences and factories. As we saw in Chapter 3, the development of dense cities allowed travel time constraints to be overcome by minimising distance constraints.

BOX 8.1 THE EARLY TELEPHONE AND THE STRUCTURE OF CITIES

As the telephone broke down old business neighbourhoods and made it possible to move to cheaper quarters, the telephone/tall-building combination offered an option of moving up instead of moving out. Before the telephone, businessmen needed to locate close to their business contacts. Every city had a furriers' neighbourhood, a hatters' neighbourhood, a wool neighbourhood, a fishmarket, an egg market, a financial district, a shippers' district, and many others. Businessmen would pay mightily for an office within the few blocks of their trade center; they did business by walking up and down the block and dropping in on the places where they might buy or sell. For lunch or coffee, they might stop by the corner restaurant or tavern where their colleagues congregated. Once the telephone was available, business could move to cheaper quarters and still keep in touch. A firm could move outward, as many businesses did, or move up to the tenth or twentieth story of one of the new tall buildings. Being up there without a telephone would have put an intolerable burden on communication.

The option of moving out from the core city and the resulting urban sprawl has been much discussed, but most observers have lost sight of the duality of the movement; the skyscraper slowed the spread. It helped keep many people downtown and intensified the downtown congestion....

The telephone is a facilitator used by people with opposite purposes.... It served communication needs despite either the obstacle of congested verticality or the obstacle of distance; the magnitude of the opposed effects may differ from time to time, and with it the net effect. At an early stage the telephone helped dissolve the solid knots of traditional business neighbourhoods and helped create the great new downtowns; but at a later stage, it helped disperse those downtowns to new suburban business and shopping centers....

Thus we find many relationships between the development of the telephone system and the quality of urban life; strikingly, the relationships change with time and with the level of telephone penetration. The same device at one stage contributed to the growth of the great downtowns and at a later stage to suburban migration. The same device, when it was scarce, served to accentuate the structure of differentiated neighborhoods. When it became a facility available to all, however, it reduced the role of the geographic neighborhood.

(Pool *et al.*, 1977; 127–157)

The development of new networked technologies including gas, electricity, transit systems and the telephone helped to overcome some of the physical limits to the size of the walking city. In Box 8.1, Sola Pool (1977) graphically describes how transit systems and the telephone underpinned important changes in the physical structure of the city. In the initial phases the new networks facilitated the development of further increases in the concentration of activities in the centre of cities. Businesses no longer had to rely on close physical proximity between headquarters and manufacturing to coordinate production and distribution processes. Instead, manufacturing processes could be located outside the city, using the telephone to maintain close contact with centrally located headquarters, which from the late 1800s were increasingly located in capitals or regional centres. As new transport and telephone networks overcame the constraints on travel time and distance, new urban patterns developed based on the creation of 'a large downtown containing a miscellany of commercial and marketing activities that needed to be accessible to a variety of clients and customers' (Pool, 1982; 452).

Although, initially at least, the telephone helped facilitate the growth of cities, a parallel trend was also developing 'first in a small way and then massively. That second trend was dispersion from the city to suburbia and exurbia' (Sola Pool, 1982; 453). Dispersal predated the telephone as it was initially based on more effective and efficient transit systems which helped minimise the travel time involved in commuting to the central city. While this trend was noted in the late 1800s, even in the 1940s assessments of the impact of telephones on the city continued to stress its centralising effect.

But from the early 1960s the effect was reversed and the predominant direction of movement of both business and residences was a movement away from the city cores. While the trend towards dispersal developed unevenly between countries, from around the middle of the twentieth century a more diverse pattern of urban form developed than the typical single hub of the early industrial city. Jean Gottmann (1990) characterised the new type of city as a form of 'megalopolis' with highly differentiated spatial pattern of centres and subcentres and complex interactions between them. There was increasing separation between home and work, the development of branch plants, shopping malls and specialist regional centres. Although this pattern developed most quickly in the USA – exemplified by the urban structure in Los Angeles from the early 1960s – similar trends followed later in Europe. Fundamentally, the new urban form was underpinned by the motor car and automobile. The mobility of the motor car made it possible to live and work in very different locations but the 'ability to pick up a telephone and get a message through without moving was just as essential as the car' (Pool, 1982; 454).

The development of these decentralising trends was based on relatively small but significant improvements in telecommunications technology. These included increasing, although not universal, levels of telephone ownership and the greater capacity and reliability of telephone services. This meant it was easier to maintain contacts with employers and more dispersed social and familial networks. There were also improvements in the quality of services in rural areas which allowed physically remote spaces to be utilised for living and production. The development of private telecommunications, more effective corporate telephone exchanges, higher capacity, and faster and more reliable forms of data communications encouraged large corporate users to consider new locations for production and retailing activities. As the space-shrinking potential of telecommunications became more advanced, there was increasing interest in its ability to dissolve the physical 'glue' that provided the initial rationale for the development of concentrated cities.

This review demonstrates the complex and changing nature of the linkages between the telephone and the city. More effective and reliable transport and telecommunications helped to create new spatial patterns, characterised by increasing separation of home and work and between different parts of production processes. At least for the first fifty years, the telephone facilitated the development of centralised cities with specialist functions located in urban downtowns.

Although there was evidence of dispersion effects from the turn of the century, these did not become significant until the 1960s when the effectiveness, efficiency and diffusion of telecommunications allowed activities to become more spatially dispersed. Pool was careful to acknowledge the dual nature of the spatial trends, emphasise the importance of national variations in changing urban form and not to predict the end of the city. However, most of the more recent attempts to predict the impact of telecommunications on urban form have tended to ignore the complex and contradictory roles of the networks and mainly to focus on their potential for decentralising or even dissolving the need for cities.

TELECOMMUNICATIONS AND THE CONTEMPORARY CITY

Telecommunications, Gottmann argues, dramatically alter the 'significance of distance in the organization of space. For the transmission of information across space, distance has become a secondary consideration. What matters most is the organisation of the network installed to transmit the artifact and its technique, not the physical facts of distance and extent of space' (Gottmann, 1990; 194). As we saw in Chapter 4, the development of a whole range of telemediated services mean that physical presence may no longer be necessary for banking, shopping, education, health services and perhaps even work. The annihilation of distance and time constraints could undermine the very rationale for the existence of the city by dissolving the need for physical proximity. Consequently, the debate about the relationship between contemporary patterns of urban form and telecommunications has focused on the tensions between two alternative views of the future of the city. As Gottmann asks, will telecommunications lead to a 'renaissance' of the contemporary city or do they signal the 'dissolution' of the city? (1990; 192). Put simply, do telecommunications reinforce the importance of cities as centres of production, government and consumption or instead do they provide a substitute for face-to-face contact and dissolve the need for physical proximity?

THE DISSOLUTION OF THE CITY?

A remarkable degree of commonality exists in futurist and utopian assumptions about how current advances in telecommunications will influence cities (see Chapter 3). As with the often related approaches deriving from technological determinism, the 'logic' of telecommunications and electronic mediation is interpreted as inevitably supporting urban decentralisation or even urban dissolution. As Andy Gillespie argues, 'in all . . . utopian visions, the decentralising impacts of communications technology are regarded as unproblematic and self evident' (Gillespie, 1992; 67).

Futurists then go on to argue that this will iron out geographical differences, totally redefine cities and make location virtually irrelevant. Telematics networks will be able not only to replace physical concentration but also to solve many of the problems which futurists and utopianists see as the results of concentration in industrial cities. To Marshall McLuhan, for example, the emergence of the 'global village' meant that the city 'as a form of major dimensions must inevitably dissolve like a fading shot in a movie' (McLuhan, 1964; 366; quoted in Gold, 1990; 23). In 1968, Melvin Webber, in his assertion that society had reached of the 'post-city age', predicted that 'for the first time in history, it might be possible to locate on a mountain top and to maintain intimate, real-time and realistic contact with business and other societies. All persons tapped into the global communications network would have ties approximating those used in a given metropolitan region' (Webber, 1968). Anthony Pascal predicts that 'with the passage of time [will come] spatial regularity; the urban system converges on, even if never quite attains, complete areal uniformity' (Pascal, 1987; 602).

Joseph Pelton notes that what he calls the 'global electronic machine' is 'making the global village a reality... our ability to shape information and knowledge is today creating global trade and culture. Tomorrow we will begin to form a global brain – a global consciousness' (Pelton, 1989; 9). Within this 'telepower economy', he predicts that 'cities will be increasingly freed from geography and defined not only by place but by intellect' (Pelton, 1992). Widely dispersed cities around the Pacific – the area of fastest growth in tele-communications – will emerge as a vast 'Pacific telecity', linked virtually into one functioning urban unit via telecommunications (Pelton, 1992; Cornish, 1993). Cities will therefore 'vanish' as their chief *raison d'être* – face-to-face contact – is substituted by electronic networks and spaces (Pascal, 1987; 59). New rural societies will emerge, as people exercise their new freedom to locate in small, attractive settlements that are better suited to their needs (Goldmark, 1972).

Telecommunications provide the means for overcoming time and physical space constraints on the location of activities. The space–time compression features of telecommunications mean that the 'inconvenience costs' of interaction amongst geographically separated points have declined consistently while the capacity for the exchange and use of information has increased dramatically. Travel time between pairs of points 'has been reduced significantly' (Fathy, 1991; 40). At a number of different levels telecommunications can now overcome the time and space constraints which forced activities to locate in close physical proximity within cities.

There are a two key aspects to the transition to a new form of decentralised and dispersed pattern of activities. The first aspect focuses on the 'on-line' integration of telecommunications with a range of personal services. The development of teleservices delivered electronically into the home – such as retailing and shopping, information, entertainment and banking services – could eliminate the need for physical presence (see Chapters 4 and 5). Instead of travelling to services the customer–producer relationship is mediated through electronic flows and there is no need for physical proximity with service providers.

The second aspect is the development of teleworking which could shift substantial areas of production from centralised specialist buildings back into the home, which theoretically could be located anywhere. Fathy argues that as telecommunications increase the ease and cost of communication they are 'likely to encourage people to live further away from their jobs or other places they visit frequently. Improvements in accessibility of activities encourage dispersion of both homes and activities. Telecommunications are speeding this process of decentralisation for jobs and services. Thus, city centres are no longer favoured over other places' (Fathy, 1991; 44). Consequently, cities will not necessarily be the only centres of economic activity as they are challenged by rural areas who are able to attract 'footloose' information services.

BOX 8.2 SPECULATION ABOUT THE FUTURE SIZE AND FORM OF URBAN AREAS

Production of Goods. Manufacturing requires workers to be present simultaneously, but increased automation and manufacturing's declining share of gross national product (GNP) mean that the influence of this urbanising force will continue to decline. The fraction of GNP allocated to artists and craftpersons may rise, but much of this production can be carried out locally, near home, implying a reduced demand for transport facilities.

Production of services. This is the greatest area for the potential decentralisation of activity through telecommunications and linked computer networks. The only limits on having all work take place at home are our ability to exercise managerial control (here piece-work pricing can help), and the human desire for a change of scenery. Otherwise, this is a powerful decentralising influence. Goodbye downtown office boom!

In fact, the external economies of agglomeration that Chinitz (1964) argued cogently accounted for the dramatic difference in size between New York City and Pittsburgh are represented primarily by services whose productivity can be enhanced by improved telecommunications. To the extent that the necessary trust of specialised financial and legal services, and the creative excitement of a new marketing strategy, can be conveyed electronically, these services can be provided to any location from agents scattered all over the world. A question remains: can the creative energy spawned by electronic interactions be equal to that through personal contact? The answer to that question should also suggest the future spatial structure of universities.

Wholesale and retail trade. Catalogue shopping is currently making major inroads on retail trade in wealthy neighbourhoods, and the availability of interactive video catalogues in most homes with smart terminals should greatly accelerate shopping from home. The limits to the use of electronic shopping are for those commodities where more than two senses are required, for those goods where the shopping process is lengthy (usually highly-valued goods), and for those occasions where shopping is also a social activity.

Otherwise, home shopping may imply further urban decentralisation of commercial activity since many 'catalogue operations' may serve merely as brokers, and deliveries may be direct from manufacturer to customer. The exception will be those goods and commodities, such as food, where same-day delivery is required and therefore nearby inventories must be maintained. Nevertheless, due to the scale economies in inventory maintenance, this may imply a concentrating tendency for warehouse activities within an urban area.

Social interaction. This may be the one human activity where all senses are employed and personal contacts are desired – a concentrating force. Nevertheless, preliminary searches may be conducted more efficiently through telecommunications systems, and in this case transportation and telecommunications activities may be complementary.

What all this seems to add up to is a net continuation of forces for the decentralisation of urban areas with more miles of transportation links which, however, each bear a smaller volume of traffic. The continuing focal points are for manufacturing and the warehousing of retail goods, but these activities do not require substantial employment. In fact the largest number of physical human interactions may arise for social and large-cost retail activities. The great unknown is to what extent the office worker (service provider) of the future can work from an electronically connected home. Can privacy be secured for sensitive negotiations by wire or air-wave, and to what extent can trust and managerial overseeing be exercised remotely? Conversely, can people stand spending so much time in their homes? The answer depends in part on where people live and the nature of their home environment.

Thus future urban areas may continue to sprawl, with many downtown offices vacated (tomorrow's trendy 'loft' apartments?) and service providers located either in the old central business district (CBD), where personal social interactions are facilitated, or dotted in rural hamlets on the periphery. In between would be a ring of industrial plants and warehouses that could draw their employees from either the CBD or the exurbs. This general pattern of decentralisation and dispersion should lessen even further the desirability of using mass transit systems, except perhaps for outward commuting by households located in the centre, to their manufacturing jobs in the ring, in the case of the most densely populated metropolitan areas. However, with electronically-wired households, fewer workers will commute to work, so that work-related commuting and congestion should diminish greatly, unless offset by increased travel for pleasure because of lower traffic volumes. The other offsetting feature will be the increased volume of commercial hauling of retail items delivered to households after electronic sales, but these deliveries could be widely distributed throughout the day....

Urban living may be somewhat less costly in the future, and when coupled with less congestion due to work-related commuting, city centres may appear to be quite attractive habitats for the electronic house-hold. Nevertheless, except for urban ring roads, traffic volumes on inner city streets may be reduced because of less commuting and commercial traffic.

What this analysis does not provide is an answer to the question of where the poor and disadvantaged will locate. With a more dispersed population and less commuting, begging and scavenging will be more difficult. Serving the disadvantaged and bringing them into the electronic age may be the biggest municipal challenge of the twenty-first century – in exchange for finding and filling fewer pot-holes!!

(Schuler, 1992; 297–309)

By increasing the flexibility of locational decisions, telecommunications help spatially to rearrange the distribution of work, retail, services, manufacturing and leisure activities. The extract by Schuler (see Box 8.2) examines how telecommunications are likely to support the further decentralisation of activities away from the city centre. The new city is likely to be characterised by the relaxation of home–work relationships, the dispersion of manufacturing, a reduction in face-to-face communication in certain occupations, and a decreasing number of journeys to work with potential to relieve urban congestion. This will result in a major modification to the city – particularly locational patterns – 'by transforming them from a basis of territoriality and proximity to "communities without propinquity" by breaking down their confinement' (Fathy, 1991; 45).

Taking this argument a stage further, telecommunications are centrally implicated in the development of virtual cities. The mixing of home and work and the provision of teleservices into the home could be used to create new organisational configurations no longer based in conventional cities but taking place in information or electronic space (see Chapter 5). Here, social reality is no longer based on real space but on a type of 'symbolic proximity' in which personal face-to-face interaction is replaced by remote contact through electronic networks. As services and work are provided electronically, new types of 'city' and 'neighbourhood' are created in electronic space and real physical streets as centres for transactions could become increasingly irrelevant. Wireless technologies increasingly blur spatial boundaries as communications are not themselves fixed but based on seamless networks of radio-based technologies. One result is the development of 'psychological neighbourhoods' set in the immediate physical environs but also including home-based electronic webs connecting people, spaces and functions in virtual structures. As activities are no longer bounded by the constraints of distance and time the capabilities of the electronic network structure people's conception of the city and space.

... OR AN URBAN RENAISSANCE?

Alternative theoretical perspectives have asked more fundamental questions about the relationship between telecommunications and the physical form of cities. Critical and dystopian perspectives do accept that by 'collapsing the relative distance between locations, albeit unequally and differentially, *communications technologies are necessarily implicated in the establishment of new spatial interrelationships and new forms of spatial organisation*' (Gillespie, 1992; 66; emphasis added). But, in contrast to utopian perspectives, they argue that '*it does not necessarily follow that the compact city has been made obsolete* and that settlements will disperse throughout the countryside' (Gottmann, 1990; 194; emphasis added). Critical approaches examine the use of telecommunications within existing trajectories of economic, social and spatial restructuring. An historical perspective is needed to examine how telecommunications contribute to different forms of spatial restructuring. The specific applications of telecommunications will largely be driven by the broader social and political context within which they are situated. Consequently, it cannot be assumed that telecommunications will inevitably result in a simple one-dimensional decentralising effect on urban form. Instead the broader spatial effects will depend on the specific context and process within which tele-communications are utilised.

Telecommunications can help to intensify control over urban space through deciding who has access to, and the use of, geographical space. The most advanced telecommunications are not universally available across space and access and control over the networks are usually mediated through powerful private telecommunication companies or proprietary telecommunications networks which are unevenly developed across space. Consequently, telecommunications can be used to open up new areas of space for exploitation while bypassing other areas which are then less able to benefit from economic growth. Because cities lie at the centre of most physical communication networks it is hardly surprising that telecommunications are now being used to reinforce and recreate new systems of using and controlling space. The continued concentration of specialised labour markets, business services, social and cultural facilities, information sources and transport infrastructures in the largest cities also operates to maintain their key role as business hubs (see Chapter 4). This analysis attempts to link the debate about urban form within contemporary changes in the economic restructuring of cities. Telecommunications are firmly embedded in these wider processes of economic and social change. At the most fundamental level they provide ways of overcoming time and space constraints – but those organisations which control access to flows of information and space within and between cities are in an extremely powerful position. They can determine who has access to space, on what basis, how space is reconstituted and which areas are cut off.

These perspectives emphasise more contradictory tendencies towards the recentralisation of communications in certain cities. Telecommunications are not able simply to compress space and time constraints evenly in every location. Geography is still important, not least because telecommunications are unevenly developed and physical transportation is still extremely important for the distribution of people, goods and services. This means that there is still friction in spatial and temporal terms which places important limits on levels of decentralisation and dispersal away from cities. More critical perspectives do accept that there may be 'deconcentration in some aspects, but basically transactional activities are not likely to be scattered throughout rural territory just because the technology is becoming able to overcome distance' (Gottmann, 1990; 198). Instead Gottmann argues that 'the information services are fast becoming an essential component, indeed the cornerstone, of transactional decision making and of urban centrality' (1990; 197).

BOX 8.3 TELECOMMUNICATIONS AND THE CENTRALITY OF CITIES

From their rather different starting points, both Castells (1989), in his explorations for 'the informational city', and Gottman (1983), in his analysis of 'the transactional city', are well aware of the complex nature of the relationship between communications technologies, economic change, and urban spatial outcomes. Castells, for example, contends that 'the newly emerging form, will not be determined by the structural requirements of new technologies seeking to fulfil their developmental potential, but will emerge from the interaction between its technological components and the historically determined process of the restructuring of capitalism' (p. 21).

Both authors conclude, however, on the basis of the evidence that they marshal, that modern communications technologies are serving to enhance rather than undermine the functional authority of the major city. Castells (1989), drawing on the empirical work of Moss (1986; 1988), contends that 'telecommunications is reinforcing the commanding role of major business concentrations around the world' (p. 1), whereas Gottman (1983) argues that 'urban settlements will not dissolve under the impact of this technology, although they may evolve and, indeed, are evolving. For transactional activities, the new technology is rather helping concentration in urban places; first, in large urban centres, already well established transactional cross-roads; and second, in a greater number of small centers of regional scale or highly specialised character' (p. 28).

Why is this so? Quite simply because there is mutually reinforcing interaction between the existence of information-intensive and communications-intensive activities and investment in the advanced infrastructure needed to support advanced telecommunications services. As Hall has recently argued, echoing the findings of Gottman, Moss and Castells, 'technical advance will be most rapid in the existing centres of informational exchange, above all the world cities, where demand produces a response in the form of rapid innovation and high-level service competition' (1991, p. 19).

One of the more commonly advanced myths is that the 'new electronic grids' are ubiquitously available. Although they are certainly capable of going anywhere, and although they undoubtedly span the globe, they remain inherently nodal. Indeed, the combination of increasingly specialised networks and more market-driven systems of regulation are serving to exacerbate rather than diminish this nodality.

We are, in effect, moving into an era of 'non-universal service' (Gillespie and Robins, 1991), in which cities are very well placed to capture a disproportionate share of the benefits associated with the use of advanced communications networks (Goddard, 1990). For rural areas, there are very real fears that they will become 'bypassed' by advanced networks, unable to demonstrate sufficient levels of spatially concentrated demand to justify the investmevnt needed to provided specialised services (Gillespie *et al.*, 1991; Hudson and Parker, 1990; Price Waterhouse, 1990).

The 'electronic highway' analogy has previously been criticised on the grounds of its failing to recognise the essentially private and proprietary nature of many of the advanced telecommunications networks (Gillespie and Hepworth, 1988). Even for those networks that provide switched services, and which can be regarded as publicly available, 'electronic railways' would be a more appropriate analogy than electronic highways, for it draws attention not only to network infrastructure but also to the need for place-specific access to the network through stations or other terminal facilities. For many rural areas, the new electronic communications networks can 'pass through' en route to somewhere else, without providing access points, as the Mercury Communication network in Britain well illustrates (Economic and Transport Planning Group, 1989; Gillespie and Williams, 1990).

Far from acting, then, as Berry (1973) foresaw, as a 'solvent which would dissolve the core-orientated city', advanced communications technologies are rather acting as a very powerful magnet, an 'electronic superglue' which is enhancing and extending the influence of the city within nodal communications networks. As Gottman (1983) has astutely observed, 'the centrality of the city is constantly being re-born' (p. 15).

(Gillespie, 1992; 67–78)

In Box 8.3, Andrew Gillespie examines the historical development of cities and shows how they developed as primary centres of control over forms of communication. Control over canal, rail, road and air transportation enabled cities to channel communications between cities at national and international levels – so extending their control over the use of geographical space. Although Gillespie acknowledges that telecommunications have the potential to decentralise activities it must be recognised that 'communications technologies have the parallel capability to "bind space together" more effectively, and, in so doing, to facilitate further expansion of the monopolies of information enjoyed by certain nodal points in the communications network' (1992; 71). The extract from Gillespie graphically illustrates that instead of leading to the decentralisation of the city telecommunications have reinforced and powerfully recreated the centrality of the city. Consequently, the unevenness of telecommunications infrastructure, private control over the networks and differential levels of transportation infrastructure, and the other advantages of cities conspire to reinforce the centrality and importance of large cities in many cases. Even though some activities will disperse away from cities, despite the physical distance they are in fact still under the control of higher order functions based in cities.

Telecommunications also provide new ways of overcoming the constraints to continued urban growth. While increased congestion, crime and costs of locating in cities may be overcome by decentralising some functions to lower cost areas, higher level functions will still be located in urban centres. Here, tele-communications can be used to manage transportation and utility networks to overcome congestion and help keep time and distance constraints to manageable levels. For instance, the development of Road Transportation Informatics (RTI) technologies provides ways of rationing access to valuable road space, enabling those groups that can afford access to road pricing or electronic route guidance with ways around the high costs of traffic congestion (see Chapter 7).

Clearly, then, there are different ways of looking at the relationship between urban form and telecommunications. Utopians and technological determinists point to the dissolution of the city while dystopians urge a more sophisticated approach which examines the context within which telecommunications are used to reformulate the control exercised by cities over communications networks and the use of space. In this context telecommunications have a range of complex and contradictory effects – facilitating the decentralisation of some services while reinforcing the power and control of global cities on the hub on tele-communication networks.

MAPPING CITY–TELECOMMUNICATIONS RELATIONS

There are, however, serious problems in relying on this simple dichotomous model of city–telecommunication interactions. While the tensions between the centralising and decentralising roles of telecommunications provide a useful device for examining the future of the city, it fails to capture the complex and contradictory nature of the linkages. Previous chapters have analysed a wide range of economic, social and environmental issues raised by telecommunications, but it is difficult to insert these debates into the simple dichotomous model. In fact, it is clear that a range of different relationships exist between the physical form of urban places and the development of electronic spaces. We highlight four key aspects of city–telecommunications relations. We have termed these 'physical and developmental synergies', 'substitution effects', 'generation effects' and 'enhancement effects', shown diagrammatically in Figure 8.2. This typology usefully shows the dynamic interplay between cities and telecommunications that we have explored through this book and has profound implications for the physical form of the contemporary city.

PHYSICAL AND DEVELOPMENTAL SYNERGIES

Cities and telecommunications tend to develop synergistically rather than in opposition. Cities are the hubs and engines of communications; telecommunications are some of the most important networks strung out within and between them that handle the growing demands for intra- and inter-urban communication. Despite the rhetoric that telecommunications will somehow mean the end to cities, it is clear that, both within and between cities, the patterns of intense telecommunications development correspond very closely with the routes and areas where physical transportation movements and developments in urban places are most concentrated. Cities dominate the development and use of tele-communications networks (see Chapter 4). The relationships between tele-communications and the physical dynamics of cities seem to be largely complementary.

Here there are synergies between the parallel effects of physical and electronic changes in cities. Telecommunications and physical changes in cities push in similar

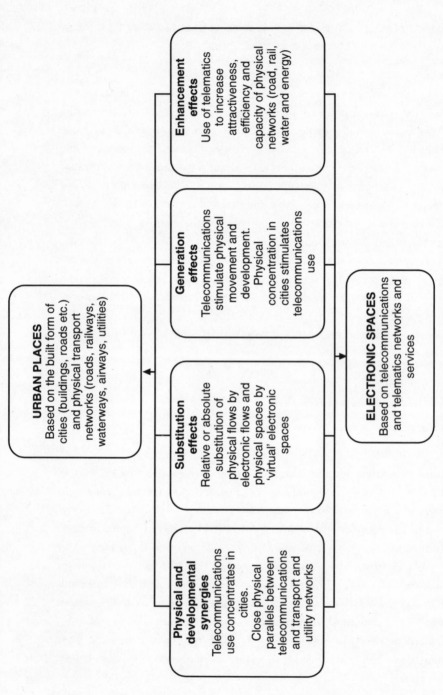

Figure 8.2 Mapping telecommunications–city interactions:
a typology of the relations between urban places and electronic spaces

directions. For example, the development of multicentred, fragmented cities or home-centred social life is supported by both the development of tele-communications networks in electronic space and the development of transport networks and suburbs in urban places. It also seems that current innovation in architecture and planning combines with the use of telematics to support the privatisation of urban public space, a process paralleled by shifts in the regulation of telecommunications and urban electronic spaces (see Chapter 5). The parallels between the 'fortress effect' in post-modern architecture and the exclusionary nature of private telematics networks are excellent examples of such synergy.

Physical synergies are also demonstrated by the fact that the world financial centres of New York, London and Tokyo are the hubs which dominate global telecommunications investments and global electronic flows at the same time as being the hubs of global airline networks, national rail and road systems, and the centres where the largest transport demands are concentrated within cities. Movement and telecommunications seem to be growing together in mutually reinforcing loops which further concentrate transport and telecommunication investment on to 'hubs' within the larger cities. The shift towards market-based telecommunications development is leading to greater differences between the telecommunications infrastructures and services available at these hubs and the more peripheral and rural areas. The patterns and flows of transport and telecommunication movement are clearly interrelated, part of a wider shift towards a society which is intensely based on rapid movement and circulation of information, symbols and people and commodities (Lash and Urry, 1994).

At a smaller scale, there are very close physical parallels and synergies between the development and routeing of telecommunications networks within and between cities and the patterns of other infrastructures (see Chapter 7). Because of the costs of developing new telecommunications networks, all efforts are made to string optic fibres through water, gas and sewage ducts; between cities existing railway, road and waterway routes are often used. In the UK, for example, Mercury, the second largest telecommunication company, laid their City of London network within an old network of hydraulic ducts that still underlies the main financial district. Between the main cities of the UK, companies eager to enter the liberalised telecommunication markets are stringing out competing optic fibre networks along the railway lines (Mercury), along the tops of electricity pylons (Energis) and even along the banks of the old canals that were the main transport networks during the industrial revolution two hundred years before (British Waterways). Once again, the physical synergies between tele-communications and physical networks mean that both tend to concentrate in the same city centres and the same corridors between them.

Finally, social patterns of access and exclusion to electronic spaces reflect very closely the wider processes of social and economic polarisation currently underway in most capitalist cities. The areas of ghettoisation, growing poverty, high structural unemployment and rising crime are also the sparse regions on the network maps, the blank zones on the roll-out plans of telecommunications companies, and the districts where access even to basic information and communications services is problematic. By contrast, the affluent segments of cities are the centres of most rapid development of infrastructure and services and the beneficiaries of the market-led processes of telecommunications development. Again, this trend reflects the use of electronic spaces to maintain and exercise social power and the wider structural patterns of exclusion (see Chapter 5).

SUBSTITUTION EFFECTS

As we have seen, substitution controversies run right through all the areas considered by this book. The growing teleworking debate, for example, centres on the degree to which telecommuting to work can replace the cost, stress and environmental effects of physical commuting (see Chapter 6). Other business debates consider the amount of expensive city centre office space that can be substituted by people working in new patterns, linked up within electronic spaces while being physically mobile or separated (see Chapter 4). Similarly, one can question the extent to which virtual reality or videoconferencing technologies linked over telecommunications can allow people to substitute for physical conferences and meetings. Or can 'back offices' be deployed and developed in cheap locations to service small headquarters, so substituting for the traditional typing pool?

Many other substitution questions have been explored in this book. For example, on the social front, can virtual communities and bulletin boards, electronic mail and computer conferencing substitute for the physical associations generated by spatial propinquity within urban districts? Will the growth of computer-mediated entertainment replace the physical theatres and cinemas of city centres? Are the days of face-to-face banking threatened because of the growth of telebanking? Would it even be possible to construct a new convivial sense of urbanism within the confines of electronic spaces to substitute for the increasingly unsatisfactory, unsafe and decaying physical reality of many cities? What, on the other hand, are the implications of social groups spending so much time in 'cyberspace' that they cease to interact with urban places in any meaningful sense?

While these discussions are important, the possible substitution of electronic flows and spaces for physical flows and places is only one element of a complex interrelationship between the two which highlights the myth of the idea of simple substitution to be a myth.

GENERATION EFFECTS

A third set of linkages between urban places and electronic spaces can be termed 'generation' effects, representing the generation of 'knock-on' effects of developments. On the transportation front, evidence actually points in the reverse direction to the myth of simple substitution. Three key areas of telecommunications–transport innovation currently suggest that telecommunications either generate more transportation than they substitute for, or allow rising transport demands to be accommodated and managed.

First, instantaneous and reliable telecommunications links allow for the generation of many extra transport trips because it becomes possible to coordinate transport flows much more effectively. Phone calls, faxes, and electronic mail set up arrangements through electronic space and are used to coordinate journeys through physical space. Many new social trips and transport movements between manufacturing firms become possible, because telecommunications allow people and firms to synchronise themselves in such a way as to make the transport predictable, useful and effective. For example, the use of networks of Electronic Data Interchange (EDI) and Just-in-Time (JIT) management methods between networks of manufacturers radically increase physical transport shipments.

A second area of innovation centres on mobile telematics technologies, which actually convert the 'dead' time of travel into 'live' working time. Corporate workers remain working in real time as part of a functioning organisation, irrespective of the fact that they are travelling. As Chapters 2 and 7 show, mobile data, fax, image, voice and graphics communications are merging with mobile computers to provide powerful mobile office technologies. These are rapidly being applied as part of the re-engineering of corporate business practices. Flexible and mobile use of electronic spaces within and between cities therefore supports the flexible and mobile use of urban places and the corridors between them.

Another set of generation effects centres on the use of electronic service innovations to create many extra transactions that would not have happened previously because of travel and time limitations. A good example is the use of phone banking or Automatic Teller Machines (ATMs) for personal banking transactions. Global telephone and e-mail networks generate entirely new forms of social and community interaction which would be deterred or impossible in a physical sense by space–time confines.

Advances in telecommunications technologies and application also help to induce radical generation effects on the physical form of cities and their economies. As we have shown, the invention and application of the telephone allowed offices to move from factories and to centralise in business districts. The skyscraper itself – that symbol of the industrial city – was only made possible because the telephone allowed its occupants to communicate without the need for thousands of messengers (Pool, 1977). More recently, as we saw in Chapter 4, advances in telematics have allowed multinational service companies to centralise their activities within the world's financial capitals. The real-time control capabilities of corporate telematics networks allow giant, global organisations to be controlled with unprecedented precision from single corporate headquarters. This was a major factor behind the explosive physical and economic growth in London, Tokyo and New York in the 1980s (see Chapter 4).

In the reverse direction, in line with the physical synergies outlined above, the layouts and patterning of activities within and between cities obviously has a powerful generation effect on the traffic in electronic spaces (see Chapter 6). Often, the patterns of generation between urban places and electronic spaces and flows are complex and mutually reinforcing. The recent explosion in mobile voice and data communications, for example, is allowing people to operate usefully within electronic spaces while also moving through urban places. As Chapter 4 shows, this leads to faster and more flexible work arrangements based on ever higher intensities of both physical and electronic communications, which are linked synergistically and work in parallel – in turn reflecting the physical synergies between the two outlined above.

ENHANCEMENT EFFECTS

We have termed the final set of interrelationships between physical and electronic spaces 'enhancement' effects. These embody the use of electronic spaces and networks to enhance the capability, efficiency and attractiveness of physical networks such as roads, railways, airline networks, energy and water systems. In all of these areas, as Chapters 6 and 7 show, increasing demands for physical movements are being met by enhancement of the capabilities of existing physical networks through the use of the control capabilities of telematics. Financial, environmental and political imperatives often preclude new physical constructions. This is being done by using telematics to construct parallel electronic spaces through which the management and monitoring of physical flows can be improved in real time. Thus, with Road Transport Informatics (RTI), for example, 'dumb' road space is wired up and tuned into a 'smart highway', which can then be minutely controlled to maximise the traffic loads. Many debates about the so-called 'intelligent city' centre on the need to create these enhancement effects by paralleling the existing physical fabric of cities with complex control systems constructed in electronic space.

CONCLUSIONS: WHAT FUTURE FOR THE CITY?

There are obviously no simple answers to the questions we posed at the start of this chapter. We cannot definitively argue that telecommunications will signal either the rebirth or the death of the contemporary city. Instead we argue that it would be dangerous to assume that the direction of change is simply one-dimensional, leading inevitably to decentralisation or even recentralisation of cities. We need to be aware of the complexity of change, the problems of inertia and the difficulties of interpreting the direction of change. We also need to be aware of the different trends between sectors. Clearly, while decentralisation is affecting routinised services, centralisation tends to occur in higher order control and decision-making functions. But perhaps most fundamentally, we need to be alert to the dangers of simplistic interpretations which neutrally forecast new spatial changes on the basis of technological potentials and the apparent capabilities of new telecommunications technologies. There are highly contradictory tendencies pushing for further decentralisation in some cases while reconcentrating other activities. We need to accept that cities will not necessarily 'dissolve under the impact of this technology, although they may evolve, and are indeed evolving' (Gottmann, 1990; 201).

The relationship between telecommunications and the physical form of the city is certainly much more complex than the utopian and deterministic vision of the dissolution of cities to a new decentred era of electronic cottages. Telecommunications interact with cities in a whole variety of ways – reinforcing the centrality of world cities, facilitating the decentralisation of some activities, shifting physical flows to electronic flows and mediated access to physical space such as roads and buildings, thus removing the barriers to further urban growth.

In this conclusion we attempt to go beyond these simple dichotomous concepts. Instead we utilise the insights developed from our more sophisticated framework of city telecommunications interactions to examine the potential implications for the urban form of the city. Figure 8.3 charts the likely implications of telecommunications on a North American city. Drawing upon the insights developed by Pool's work on the historical city and the telephone, the decentralisation and recentralisation debate which has framed most conventional analyses and our own framework of the multitude of linkages between the city and telecommunications, we attempt to highlight the main issues for the current stage of urban development.

Urban centrality is still important for high-level managerial functions that cannot be transferred into flows. These may include specialised education institutions, recreation areas, specific production centres, special face-to-face delivery activities such as hospitals or boutiques and high-level corporate decision-making functions. There is likely to be continued decentralisation around edge of cities but within the context of extended metropolitan regions for locating back offices and teleservice centres. There may be some teleworking but this is likely

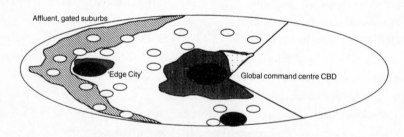

1995 – the post-Fordist global metropolis

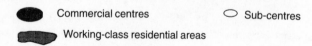

Figure 8.3 The post-Fordist global metropolis
Source: Adapted from Soja, 1989; Davis, 1992

to be focused around major metropolitan centres. It may take off in remote rural centres if the costs or comparative advantage outweigh existing forms of office organisation, but only for routine activities where there is rarely any need for physical interactions in the city. When urban centres are economically dis-advantaged by traffic congestion, telecommunications can facilitate dispersal to urban fringes but at same time RTI can help overcome limits to urban growth by rationing road space.

Despite the complexity and contradictory nature of these interactions what is central to an understanding of telecommunications and urban form is the control and power aspects of space and time compression. Telecommunications provide new potentials for exploiting space in new and innovative ways. The new advanced telecommunications infrastructures are not simply ubiquitous and available everywhere at the same costs. Instead, they are unevenly developed; not all users have access to the network and this access varies in cost and quality. Control over telecommunications networks has a central role in the management of different types of geographical and virtual space. For instance, the example of teleworking illustrates how radical new demands are placed on the home as the boundaries between public and private space become increasingly blurred. Companies can use teleworking to offload the costs of employment on to the employee while the new advanced communications infrastructure could be used for the delivery of home-based teleservices. The role of telecommunications in smart or intelligent buildings is also closely related to the control of space, for example by changing work processes and increased surveillance, monitoring and control of workspace.

Within cities telecommunications create a range of new potentials for the control of access to and use of physical urban space. Increasing use of Closed Circuit Television (CCTV) systems in city centres can help displace physical presence with more reliable twenty-four-hour telepresence. This allows closer scrutiny of movement, but it also seems likely that these systems push crime into new places which are unable to afford sophisticated surveillance. New road transport informatic technologies are based on overcoming the space and time constraints of urban congestion. Rather than using telecommunications to decentralise activities, another option is to ration access to road space at premium times through road pricing or electronic guidance systems. RTI allows some groups to overcome the physical constraints on the city by using tele-communication to buy mobility at premium rates.

As we saw in Chapter 5, the increased development of teleservices for basic functions such as banking, shopping and other information services may help to underpin the withdrawal of the physical presence of these activities from cities.

Teleservice providers will be able to target their services at premium markets, switching out access to marginal and low-income households. In turn these groups may have increasing difficulty in obtaining access to physical services as these are withdrawn from non-profitable and marginal locations. There is increasing potential for deciding which groups have access to both urban places and electronic service provision through the capabilities telematics bring for controlling space.

Clearly, the city is being redefined and redrawn in both physical and electronic space. In the informational city the boundaries between home and work, public and private, electronic and physical are becoming increasingly blurred. Telecommunications do not simply substitute or displace space – they redefine how space is perceived, used and controlled. Crucially, they facilitate increasing control over space for powerful groups while creating new physical and electronic ghettos for marginal, low-income and disenfranchised households. The city becomes much more fragmented and polarised as physical and electronic space is used in new ways. Rather than seeing the 'end' of the city, these processes create a complex new patchwork of different types of spaces – some real and others configured electronically.

United Nations Centre for Regional Development 20th. Anniversary Programme
INTERNATIONAL CONFERENCE
ON
Mega-Infrastructure '91 in Gifu
—Airports, Seaports, Telematics, togistics—

9

URBAN PLANNING, POLICY AND GOVERNANCE

Lifelines Linking Gifu and the World

国際連合地域開発センター20周年記念事業

国際メガインフラ会議ぎふ '91
空港・港湾・情報通信・商流・物流を語る

オープンフォーラム

岐阜と世界をむすぶ生命線

91年11月9日(土)午後2時〜　会場/岐阜羽島駅前「日健総本社　国際会議場」

野中ともよ　（ジャーナリスト）
梶原　拓　（岐阜県知事）
加藤　晃　（岐阜大学学長）
安田　梅吉　（岐阜商工会議所顧問）
鈴木　秀郎　（㈱セイノー情報サービス社長）
　他、海外専門家

専門家会議
平成3年11月11日(月)〜平成3年11月13日(水)
テーマ/「地域開発におけるメガインフラ整備の効果と対応」

問い合せ先/(財)岐阜県企画設計センター　☎(0582)72-1111(代)
主催/国際連合地域開発センター設立20周年記念事業組織委員会、(財)岐阜県企画設計センター
協力/岐阜大学土木工学科土木計画学研究室、名古屋大学土木工学科土木計画学研究室
(株)日健総本社、(財)岐阜県国際交流センター
後援/岐阜県、羽島市

Why should we care about this new kind of architectural and urban design issue – [the design of cyberspace]? It matters because the emerging civic structures and spatial arrangements of the digital era will profoundly affect our access to economic opportunities and public services, the character and content of public discourse, the forms of cultural activity, the enaction of power, and the experiences that give shape and texture to our daily routines.

(Mitchell, 1995; 5)

INTRODUCTION: CITY POLICY-MAKERS AS 'SOCIAL SHAPERS' OF TELEMATICS

In the last six chapters, we have built up a comprehensive overview of city–telecommunications relations. All of the issues raised in these discussions clearly impinge directly on to the ways in which contemporary cities are planned, managed and governed. Urban planners, policy-makers and governors are on the 'front line' in dealing with the effects of economic globalisation and restructuring; social and spatial polarisation and its associated knock-on effects on unemployment, poverty and crime; the crisis in urban environments and transport and infrastructure; and the physical restructuring of cities.

But what do these telematics-based shifts actually mean for those institutions and individuals responsible for the management, planning and governance of cities? How can those responsible for planning, managing and governing cities come to terms with threats and opportunities posed by telematics? All six chapters beg the question: given these changes, what can localities and policy-makers *actually do* about the disparate and complex telematics-based shifts underway in their cities? Are there genuine powers available at the local level for reducing the damaging effects of telematics on cities and socially constructing more beneficial applications? How can city policy-makers address the problems and take advantages of the policy opportunities which present themselves? What are the best approaches to take in making telecommunications policies for cities? Related to this is the question: how are telematics being used within urban governance and public service delivery?

This chapter explores these questions. The first point to stress is that, starting from a position of very low levels of knowledge of, and influence over, telecommunications, there is much evidence that city authorities are increasingly trying to shape how they develop within their areas, often for the first time (Graham, 1994). This is being driven by different approaches to urban governance as well as a proliferation of new proactive strategies attempting to shape the economic, social, physical and environmental development of cities (Mayer, 1995). The growing importance of these strategies is linked with the erosion of the power of nation states. This has led local and regional governments to assume much more active roles in trying to shape urban development – particularly urban economic development. It has also led to a shift from national, public telecommunications to complex global patchworks of telecommunications marketplaces. This offers both threats and opportunities to urban policy-makers who are attempting to intervene in telecommunications (Graham, 1993).

Municipal governments and urban development agencies are currently engaged in an increasingly pervasive competitive struggle to attract investment and secure nodal positions for their cities on corporate global networks (Harvey, 1989; Amin and Thrift, 1992). 'City revitalization has become a prerequisite strategy for cities keen to develop marketable and strategic functions in the wider network of global city economies' (Imrie and Thomas, 1993; 87). In the wake of urban economic restructuring and globalisation, there is now a remarkable homogeneity in the shift from what David Harvey calls 'managerial' service-oriented forms of urban governance to new strands of 'urban entrepreneurialism' (Harvey, 1989; Preteceille, 1990). Related to this has been a renewed 'localism' (Lovering, 1995) and the emergence of new types of local public–private alliances and growth coalitions geared towards securing the competitiveness of individual cities in the global economy (Harding, 1991). The construction of favourable urban images, or the reconstruction of unfavourable ones, is a central component of these new policies (Harvey, 1989).

At the same time, financial and political shifts in urban governance are reducing and refocusing the roles of urban governments in social, consumption and welfare services (Mayer, 1995). In most cities, universal welfare services are being restructured as 'safety nets' for those in most desperate need and voluntary agencies are taking a more significant role. Wider processes of restructuring are leading to a greater emphasis on the privatisation of urban services, business-style management methods and stress on efficiency and targeting rather than social redistribution and territorial justice.

In all these shifts, telematics are a growing focus – both as new potential tools for economic and social development strategies and as a support to the wider restructuring in the delivery of urban services and in urban governance. Thus, we can identify telecommunications and telematics both as a central set of technologies involved in the wider changes underway in cities and as an increasingly important focus of the policy responses to these diverse changes. However, we argue that for this policy action even to start to address the many issues and problems we have raised – whether at the urban, national or supranational level – it will be necessary fully to understand the increasingly tele-mediated nature of urban life and the new, often bewildering restructuring processes that are currently reshaping cities. In this chapter, we attempt to assess what can be done progressively to influence urban telecommunications innovation. By 'progressively', we mean the use of telematics to underpin genuine new economic growth for cities and to ameliorate tendencies towards social and spatial polarisation within and between cities. We also critically explore how telematics are being used to reshape the ways in which cities are managed and governed. We do this by reviewing a wide range of 'leading edge' examples of planning, management and government initiatives which are attempting to address the issues we have raised through explicit intervention in tele-communications and telematics.

We have already seen some examples of embryonic attempts by city governments and policy-makers to begin shaping telecommunications in cities – to start intervening in the development of urban electronic spaces as well as the urban places which have traditionally dominated urban policy. In Chapter 2 we saw how four Californian cities were developing Public Information Utilities (PIUs) in different ways. Chapter 3 described the range of optimistic and often utopian 'wired city' projects developed during the 1970s and early 1980s. The more recent trend towards virtual urban communities was discussed in Chapter 5. New environmental strategies using telematics for environmental monitoring and to support teleworking were discussed in Chapters 7 and 8 and the proliferation of Road Transport Informatics (RTI) initiatives were described in Chapter 8. This chapter will build on these examples by focusing in particular on economic, social and physical strategies, and on telematics initiatives within urban government.

THE NEED FOR URBAN TELEMATICS POLICY INNOVATION

The theoretical perspective built up in Chapter 3 teaches us that space exists within which urban planners, policy-makers and city managers can 'socially construct' telecommunications and telematics initiatives in a wide variety of ways which are geared to different needs and interests. Indeed, their often intimate knowledge of urban needs and contexts can give them advantages over either the crudity and polarising effects of market forces or the distant imperatives of national policies in this regard. Geoff Mulgan argues that cities need vigorously to pursue telematics policies to complement other plans aimed at strengthening the physical aspects of urban life:

Alongside new policies for housing, transport and education it is becoming clear that the new vision of the city will also emphasise its nature as a means of communications, a place where people can meet, talk, and share experiences, where they think and drink together. . . . City councils are much better placed for the kinds of holistic approach that makes the ecology of a city work than either national governments or untrammelled markets.

(Mulgan, 1989; 19)

We would argue that the precise results of these policies depend on the contingent political, social and institutional processes of technological construction in each case. Telematics and information technologies are nothing if not flexible. Many different types of technologies can be used and organised in many different ways. Different assumptions, objectives, perceptions and 'visions' amongst those socially constructing urban telematics plans are likely to lead to very different projects. The effects of these policies on cities and urban governance can, therefore, vary dramatically. Not surprisingly, this means that a wide variety of urban telematics projects are actually now underway, reflecting the diversity of processes and patterns of urban governance and planning across cities.

The growing importance of urban planners, managers and governments in shaping telematics developments in their cities is increasingly being recognised by telecommunications and urban policy-makers at national and supranational levels. In Japan, for example, a wide range of centrally driven telematics-based policy programmes have been developed as springboards for projects at the urban level (Edgington, 1989). The National Information Infrastructure (NII) plans in the USA increasingly recognise the need for the involvement of the cities and communities it will link up. The European Union, too, is becoming much more sensitive to specifically urban telematics issues. Until recently, it pursued a 'blue skies', 'technology push' approach to telematics development, pouring billions of dollars into a range of separate technical research programmes on integrated broadband communications, medical technologies, advanced multimedia or Road Transport Informatics, and the construction of telecommunications infrastructure in lagging regions. The assumption tended to be that because the technology existed, it would diffuse to have wide (and beneficial) application through European society. Now, a more balanced 'technology push' and 'demand pull' approach is being developed which increasingly brings together technologists with representatives of users and city policy-makers to develop telematics applications that are more appropriate to real social and economic needs in real cities. The new 'telematics sites' programme is oriented towards supporting a wide range of telematics experiments that are integrated within particular cities.

Successful intervention at the urban level in the new, complex and fast-moving area of telecommunications and telematics is both risky and far from easy. There are four problems here. First, the hampering effect of the paradigm challenge in urban planning and policy has to be faced (discussed in detail in Chapter 2). Once this has been done, a second danger must be confronted – the risk of adopting simplistic versions of technological determinism of futurism as the basic policy model. Often, policy-makers seem to be seduced by the lustre and glamour of the technology and hype up their policies as some quick 'technical fix' for the problems of their city (Gillespie, 1991).

It is instructive to note that the language of the current wave of urban telematics projects is actually very similar to that used by cities and technology companies when describing the anticipated promises of cable in the 1960s and 1970s (Strover, 1988). It was assumed that broadband cable networks would link all homes and bring electronic democracy, unlimited varieties of interactive services, special local channels and new economic bases for cities (Strover, 1988). In fact, these promises have not been met: in the USA cable has emerged as a plethora of one-way entertainment channels offering reruns and commercial pay-per services. Promised local services for cities have often never materialised or been withdrawn (Strover, 1988). In the 1980s similar hopes for an urban 'techno fix' were declared for the teleport (Moss, 1987; 544). Currently, the hype centres on National Information Infrastructures, the so-called 'information superhighway', multimedia telematics and the supposed need for broadband optic fibre networks to every home – the list of instant urban technological panaceas continues.

It is not surprising that the many unrealistic approaches to urban tele-communications policy often lead to spectacular failures. As the cable example above demonstrated, urban technology and telematics projects are often unsuccessful. Technological and regulatory change in telematics often outstrips the ability of local policy-makers to innovate and adapt to changing circumstances, resulting in technological white elephants. Often, it could be argued that resources would be better used on non-technological projects (Gibbs, 1993). Urban telematics policies are often over-ambitious, ill-informed and poorly thought through, reflecting the fact that 'urban policy makers and technologists both tend to share a common trait of believing that they have more control over the [policy] environment than they actually do' (Newton, 1991; 97). Often, initiatives are unsustainable because they fail to meet the actual needs of users and rely too much on long-term subsidy and a blindly optimistic 'technology push' idea (see Qvortrup et al., 1987; Qvortrup, 1988; Cronberg et al., 1991). Good quality evaluation of these failures is rare, so often the same pitfalls are fallen into again and again.

Given these problems, a healthy scepticism to technological hype, techno-logical determinism and futuristic urban visions is necessary. They provide a very poor basis for successful urban telematics policy-making. More sophisticated views of city–telecommunications relations are necessary. Policy-making needs to be less ambitious and more realistic. We would agree with Michael Keating when he argues that, 'given their limited resources, local governments must concentrate on low-cost initiatives within their capacity' (Keating, 1991; 200). This should not extend to major investments in technological infrastructure; rather, supporting the application of telematics services should be the main policy focus.

The need, then, is for social innovation rather than just technical innovation involving telematics in cities. This inevitably raises broader normative questions about the sorts of cities we want and the ways in which telematics can help us to create those cities. As William Mitchell (1995; 5) argues, 'the most crucial task before us is not one of putting in place the digital plumbing of broadband communications links and associated electronic appliances (which we will certainly get anyway), nor even of producing electronically deliverable "content", but rather one of imagining and creating digitally mediated environments for the kinds of lives that we will want to lead'. Rather than concentrating on the abstract and glamorous capabilities of the latest hardware and software, attention needs to focus on the social organisation of telematics systems – what Lars Qvortrup calls 'orgware' (Qvortrup, 1988; see Cronberg et al., 1991). How, in other words, can sustainable, successful and truly beneficial telematics applications be socially constructed in real-world situations and in real cities? The main question is: how do you translate this technological potential, or indeed do you want to – into sustainable applications that actually meet the day-to-day needs and demands of a largely urban society at large? To quote Dave Spooner (1992), a former technology policy officer for Manchester City Council in the UK:

The problem is not one of technical development but the lack of appropriate applications. There are hundreds of technical telematics products that have enormous potential for economic [and social] development [in cities], yet they mostly lie on laboratory benches waiting for someone with an application – a classic 'solution looking for a problem'.

Spooner notes that large telecommunications companies like BT tend either to be engineer driven or are increasingly geared towards meeting the needs of large corporate customers for private networks. They are much less able to 'think across product lines or between technical and social/economic disciplines' in ways suited to actually meeting the needs of cities and communities on a holistic basis. It is on this basis that the contribution of urban planners, governors and managers is potentially so important.

Third, there are fundamental tensions between cities as territories and telecommunications as networks (Negrier, 1990). People, firms and organisations outside the jurisdictional boundaries of city authorities can often benefit as equally as those within the city boundary from a new urban telecommunications project. While the satellite facilities of a teleport may be based in the municipal area that supported it, its services may actually enhance the competitive position of firms in other areas who link into it through optic fibre links. All they have to do is to link into the project electronically. This makes it difficult for local policy-makers completely to control the new computer communications or teleport projects that they finance and support. In principal, the benefits of access can easily be spread well beyond the 'local' to virtually any spatial scale – even globally. On the other hand, the overwhelming political imperative in cities to create projects which are perceived to be based in and benefit the (often arbitrary) 'home' area of the jurisdiction, may create problems from the point of view of network development. The flows of information, capital, services and labour that may be based on new networks 'transcend the narrow confines of local (and indeed national) boundaries and are . . . impossible to monitor and control at the level of individual cities' (Robins and Hepworth, 1988; 167).

The final problem is that the policy space available to city authorities and governments in which to address telecommunications issues is heavily shaped by approaches both to telecommunications policy and to urban development policy at the national level (see Dutton *et al.*, 1987; Graham, 1994, 1995a). Despite globalisation and reducing importance for the national state, conditions for urban telematics initiatives are still highly contingent on national level support. Nations vary significantly in the support they give to urban telematics initiatives and projects. For example, while France has a comprehensive range of state-backed strategies for such initiatives and a national regulatory approach that stresses urban and regional issues, in Britain central government regulates telecommunications as a free market and in a completely 'spaceless' manner – as though on a 'pinhead'. No formal role is allotted to local authorities in influencing how telecommunications develop within their jurisdictions. As Dutton *et al.* argue: 'in some areas of communications policy in Britain, there is remarkably little concern for the role of local authorities. For example, in the cable area (a natural arena for local control given the nature of the medium), local authorities have virtually no role at all' (Dutton *et al.*, 1987; 469).

TELEMATICS POLICIES FOR URBAN DEVELOPMENT AND PLANNING

The old-style planner talked about physical zoning, the balance of employment, housing and open space and traffic flows. The new style planner has to consider the configuration of electronic systems and Local Area Networks (LANs) and the provision of bandwidth to each urban area. The town planner dealt with the stocks and flows of vehicles. Today's public authorities have to face the stocks and flows of information.

(Howkins, 1987; 427)

Telecommunications are becoming a new component in urban and regional development planning. [The] desire is to use telecommunications as a structuring element in cities and regions and to incorporate telecommunications in economic and social development.

(Machart, 1994)

This stress on caution and social innovation puts us in a position to explore some examples of the urban development and planning policies based around telematics that have so far occurred in advanced capitalist cities. There are four main policy areas here: national policy programmes for urban telematics development; teleports and competitive economic policies; inter-urban networking initiatives; and experiments with electronic public spaces.

NATIONAL URBAN TELEMATICS DEVELOPMENT
STRATEGIES

In certain nation states, notably Japan, France, the Netherlands and Singapore, comprehensive strategies blending urban, industrial and telecommunications policies have been used over the past fifteen years in an attempt to create new and futuristic cities and urban spaces. The Japanese case is examined in Box 9.1.

BOX 9.1 TELEMATICS FOR FUTURISTIC URBAN DEVELOPMENT: THE CASE OF JAPAN

In Japan, early experiments in the 1970s with wired cities have been superseded by a widening range of futuristic urban telematics projects. Here, ambitious and often grandiose telematics strategies for old and new communities have been developed (see Edgington, 1989). These are a central element of the Japanese approaches to national physical and development planning, telecommunications policies and industrial policies, which centre on engineering Japanese society and economy into the 'information age' (see Newstead, 1989). A strong emphasis exists on public intervention to address the failures of private sector investment in rolling out universally accessible and integrated telematics infra- structures on a national basis. Driven by inter-ministerial as well as inter- city rivalry, such initiatives range from social and community experiments to urban environmental and physical–economic renewal strategies. The national rivalry between the Ministry of International Trade and Industry (MITI) and the Ministry of Posts and Telecommunications (MPT) has been particularly important.

In the 1970s, a large new town for public research and development was developed at Tsukuba near Tokyo. Over 25 towns and cities in the peripheral regions of Japan were also designated as 'Technopolis' projects – targets for the decentralisation of high technology industry from Japan's congested core. Local telecommunications, rail and airport infrastructures were improved to help support this planned decentralisa- tion. Host computers, Local Area Networks (LANs) and on-line business support services are also common in Technopolis towns.

In addition, a wide range of social and community telematics experi- ments have been pursued; MITI has built around 20 'new media commu- nities' across Japan to build up access in smaller cities and towns to local cable and videotex services. The MPT, meanwhile, has developed its 'Teletopia' programme to assist public information services in sixty- three model cities (Edgington, 1989). Further complicating the situation, the Ministry of Construction has developed its 'Intelligent City' pro- gramme to encourage the hard-wiring of cities with optic fibre systems and advanced urban management networks based on LANs and WANs (Terasaka *et al.*, 1988).

Despite this multitude of initiatives, it remains unclear exactly how successful these highly ambitious plan-led initiatives have been. Behind the hype and lustre of the technology, and the 'information age' rhetoric, there have been problems. Consumer demand for the home-based, on-line services that the new infrastructures were intended to support has been very difficult to encourage (Mulgan, 1991; see Chapter 5). Filling the many new infrastructures with sustainable applications that actually address social and economic problems has been a perennial problem. While the innovation efforts in Japan represent the most important national approach to urban and regional planning based on telematics, it can be criticised for taking too much of a 'technology push' approach geared to injecting hard infrastructures into cities and areas of cities. Much less effort has been put into socially constructing appropriate applications for these infrastructures. This, and the watering-down effect of so many liberally scattered projects, means that the lesson is that 'each city and region should develop a tightly targeted set of technology objectives' (Edgington, 1989; 25).

Often, these policies are aimed at the construction of whole new communities oriented around the latest telematics services and technologies. The first generation of these were the so-called 'wired city' experiments that were supported by most western nations in the 1970s and early 1980s. These were new communities linked up to advanced and interactive cable networks – for example, Biarritz in France, Hi-Ovis in Japan, and the QUBE system in Columbus, Ohio (Dutton *et al.*, 1987). However, these had a mixed success as often they relied too much on technology push approaches and had insufficient technologies for their ambitions.

A second example of nationally supported urban telematics development comes from France. Here a strong tradition of state-led urban and tele-communication planning resulted in national programmes for urban telematics projects in the 1980s. The most important of these were: the *Plân Câble* – an early plan for a national optic fibre network to every home; a national programme of French-style *téléport* developments; and national support for urban management systems based on telematics. All these projects drew on a broader national strategy in the 1980s for transforming French society through indigenous telematics-based innovation within a national telecommunications monopoly. This was a response to the perceived threat of national subordination to foreign TNCs in telematics industries. One result of this has been the free distribution of over 6 million videotex terminals (known as Minitel/Télétel) to households – a unique national infrastructure upon which 20,000 consumer information, retail and municipal services are now thriving.

But the many problems with the French approach, and a shift towards liberalisation in planning and telecommunications · policies, has led to an abandonment of most of these policies. The *Plân Câble* failed because of a lack of consumer demand; private firms now run the (unprofitable) cable networks. The thirteen *téléports*, which were designated within parts of cities known as *zones de télécommunications avancées* (advanced telecommunications zones) have proved to be largely symbolic – a prop to the declining power of the national tele-communications monopoly. Rather than supporting competitive telecommunica-tions provision for cities, as the first teleports in North America were originally concerned to do, they were actually a national attempt to placate calls from some cities for more liberalisation and competition in telecommunications.

With the progression of liberalisation, new and more innovative models are developing in France, encouraged by a continued commitment to national and regional planning by the national agency supporting such projects called the *Observatoire des Télécommunications dans La Ville* (the urban telecommunications observatory). For example, attempts to reduce traffic congestion are leading to the construction of a planned network of neighbourhood telework centres around Paris. Such projects mean that the French integration of telecommunications and telematics into mainstream approaches to urban and regional planning is rarely matched in places where decentralised planning operates.

Singapore presents perhaps the most widely known example of comprehensive national planning for urban development based on telematics (see Corey, 1991; 1993; Motiwalla *et al.*, 1993), driven by an ambitious national strategy to transform the island from a manufacturing centre into a service-based 'intelligent island'. These developments are also supported by a system of highly centralised political and social control. The whole of Singapore is being 'wired' into an integrated, carefully planned set of telematics networks. A national Information Technology Plan, finished in 1986, produced a blueprint for upgrading the island's telecommunications infrastructure, developing IT personnel, 'promoting an IT culture', finding new applications for IT, and encouraging 'creativity and entrepreneurship' in IT (Corey, 1993; 52). Through the centralised power and control of the national IT committee, and massive state investment, whole sectors of the local social and economic fabric have, as a result, been pushed 'on line' – from retailing through education, public administration and the crucial port and trading sectors (through the TRADENET system). A major new industry of small

IT entrepreneurs has been built up. In March 1992 a new plan – IT2000 – was published to consolidate these changes.

While the economic growth and modernisation has been impressive, the strategies have made the tight social and political control of the island increasingly difficult to maintain. In short, Singapore has been a unique laboratory in national and urban telematics planning, but one which can actually offer few direct lessons to urban policy-makers in other more typical advanced capitalist nations (Corey, 1993).

METROPOLITAN PLANNING FOR TELEMATICS: THE 'TELEPORT' PHENOMENON AND COMPETITIVE ECONOMIC STRATEGIES

More relevant here, particularly as the global shift towards market-based development of telecommunications gains ground, is the increasing number of urban telematics strategies in nations where national physical and telecommunications planning does not exist, as in the USA, the UK and most of western Europe. Following the developments in Singapore, and in France and Japan during the 1980s, the acquisition of better telecommunications infrastructures than competing cities, or at least the generation of the perception of a better infrastructure, is seen as an increasingly important policy objective for city authorities in these countries (Batty, 1990b). This concern is borne out by a recent survey of the top 500 European companies where 'the quality of telecommunications' came out as the second most important location factor (58 per cent), only behind 'easy access to markets, customers and clients' (quoted in Cornford et al., 1991). The increased unevenness of telecommunications development that we traced in Chapter 4, caused by the shift to market-based regulation, adds a further stimulus to urban policies in this area.

The result of these shifts is rapid current growth in urban strategies aimed at using telecommunications to improve the economic competitiveness of individual cities. At the same time, there is a tendency for debates about 'urban regeneration' to centre on the need for property-led initiatives, and the perceived social, cultural, economic and environmental needs of corporate business élites (Imrie and Thomas, 1993). The much-vaunted 'public–private' partnerships behind these initiatives often represent what Derek Shearer calls the 'edifice complex'. Here, he argues, progress is equated with 'the construction of high-rise office towers, sports stadiums, convention centres, and cultural megapalaces, but ignores the basic needs of most city residents' (Shearer, 1989; 289).

Teleports – satellite links associated with property developments and links to local telecommunications networks – are one such body of initiatives. The thirty or so operational teleports in western cities actually involve a variety of different kinds of initiatives (IBEX, 1991). As we noted in Chapter 1, teleports effectively consist of nodes for advanced national and international telecommunications services, implanted into a part of a city and linked usually to local tele-communications networks for distributing access to the services locally. Often teleports form the centrepiece of ambitious urban redevelopment plans, with new office, industrial and high-status housing property developed around the facility. This is the case in Amsterdam, New York and Cologne.

Through their catalytic function, and their potent marketing potential (Cornford et al., 1991), municipal authorities see teleports as potential centres of excellence and innovation in business telematics. Teleports aim to emerge as centres for the diffusion of innovation into the wider urban economy, through the creation of 'hot spots' of telematics demand in these key sectors, as well as providing new linkages between the urban and the global economy. This may improve economic competitiveness and the chances of attracting inward investment as a result through linking these new services to the needs of key sectors of the city economy. A common approach is to tailor the teleport to the specialised needs of a key local economic sector, whether it be media industries (Cologne Media Park), broadcasting (London Docklands teleport), financial services (New York and Edinburgh teleports), textiles (Roubaix teleport), high technology research and development (Sophia Antipolis, France) or maritime, port and logistics industries (Le Havre maritime city initiative and Bremen teleport).

Municipal and central government involvement in teleports also varies considerably. The one feature they have in common is their attempt to use advanced telecommunications and telematics networks and services to enhance the development prospects of cities, or parts of cities, in the national and international urban system. This ambition to catalyse the economic regeneration of cities through telematics is epitomised by the views of the Mayor of Metz, a declining industrial city in France which has developed a teleport. He declares that 'from a disabled region we have succeeded in building a centre of excellence in telecommunications' (quoted in Bakis, 1995). Each city is driven by the fear that it will be excluded from some global teleport network in the future, a fear encouraged by the lobbying efforts of the World Teleport Association (WTA).

Hasagawa (1990; 91) predicts, for example, that by 'the year 1995, the world could be divided into two groups; developed cities connected to the teleport network, and those areas which are neglected'.

In Osaka and Tokyo, the teleports form the centrepieces of massive land reclamation and metropolitan property development projects of between 700 and 1,000 hectares each (Itoh, 1988). These are being fuelled by intercity rivalry, following the liberalisation of Japanese telecommunications and finance industries. Tokyo 'Teleport Town', the new settlement on the reclaimed land, is being built at a cost of over $12 billion. The anticipated telecommunications systems underpinning it are shown in Figure 9.1.

The nearby city of Kawasaki is taking these ideas for telematics-based urban renewal even further. It is pursuing a £100m master plan of eighteen 'intelligent plazas' – smart buildings linked via optic fibre and cable services and geared towards advanced research and development and global service organisations (Batty, 1987). Here the confidence in futuristic telematics-based solutions for urban planning problems reaches its zenith with the aim to create the 'world's first on-line city' (Batty, 1987). The areas of the new city are themed around different industries: a 'techno-venture' park for new telematics applications, a 'techno-community' for older industries, a 'technopia' for the existing city centre and a 'technoport' for the existing sea port (Batty, 1991; see Figure 9.2). Economic, social, cultural and public administration telematics applications are to be fed to all elements within the new city and 'each plaza will be modelled on the idea of a neighbourhood office which will contain an extensive array of information technology . . . the plan is for each office to contain up to 1000 workers' (Batty, 1991; 155).

These developments tend to be geared directly to the needs of TNCs and their associated transnational corporate classes, with 'world trade centres', five-star hotels, restaurants, leisure spaces and expensive residential areas. Often, they are at the vanguard of 'intelligent building' technology and are infused with the latest security and surveillance systems to help exclude those social and economic groups that are felt not fit to belong in these spaces. Thus, teleports seem likely to support socio-spatial polarisation within cities. They usually represent part of a broader shift towards the subsidisation of the interests of multinational land and property developers, multinational services corporations and technology firms within concentrated urban 'renewal' programmes (Hepworth, 1989; 198).

In declining industrial cities, the teleport function is often as much symbolic as substantive. A recent study by Richardson et al. (1994) of Amsterdam, Cologne, Edinburgh and Paris teleports found that the actual advantages of

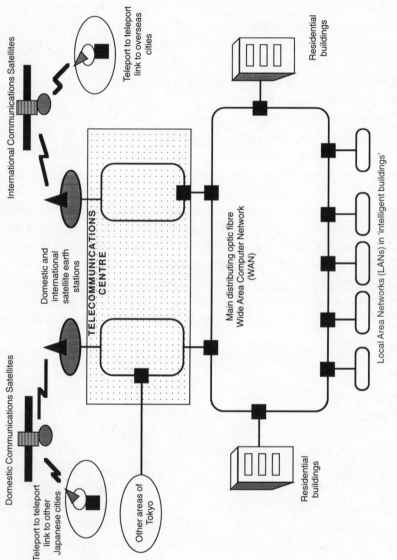

International Communications Satellites

Teleport to teleport link to overseas cities

Domestic and international satellite earth stations

TELECOMMUNICATIONS CENTRE

Residential buildings

Main distributing optic fibre Wide Area Computer Network (WAN)

Other areas of Tokyo

Domestic Communications Satellites

Teleport to teleport link to other Japanese cities

Residential buildings

Local Area Networks (LANs) in 'intelligent buildings'

Figure 9.1 A diagram of the planned telecommunications systems for Tokyo Teleport Town

Source: Adapted from Tokyo Teleport Center Inc, undated

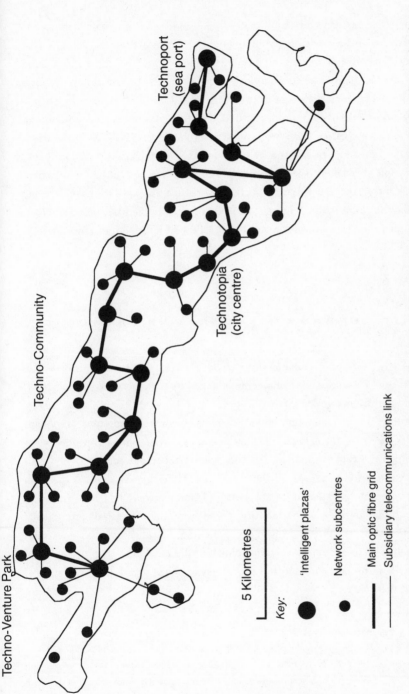

Technoport (sea port)

Technotopia (city centre)

Techno-Community

Techno-Venture Park

Key:

● 'Intelligent plazas'

• Network subcentres

— Main optic fibre grid

— Subsidiary telecommunications link

5 Kilometres

Figure 9.2 Map showing the telematics-based urban plan for the transformation of Kawasaki City, Japan, into a network of 'intelligent plazas'

Source: Adapted from Batty, 1991

teleport sites in terms of their telecommunications facilities with comparably placed urban areas are quite limited. Often the demand for the services that are developed ahead of demand in a 'technology push' fashion fails to build up. In fact, in many cases, city authorities can be accused of heralding teleports as 'quick technological fixes' to urban decline and dereliction, thereby (so the theory goes) stimulating property development, enhanced land values and economic revitalisation. Such urban boosterism for teleports builds on the wider claims by futurologists and the WTA about their importance. Fjordor Ruzic, for example, predicted that 'in future world consisting of National Information Systems integrated via teleports, such networks may be able to make a strong contribution to international peace, as well as to satisfy the basic demand within all countries for free expression and the free flow of information' (Ruzic, 1989; 115).

INTER-URBAN COLLABORATIVE NETWORKS

Increasingly, the purely competitive urban strategies like teleports which characterised the 1980s are being complemented by the emergence of collaborative interurban networks (Dawson, 1992; Parkinson, 1992; Robson, 1992). Rather than simply competing within 'markets' for investments in a 'zero-sum' game, or working through the 'hierarchies' of the state, city governments are searching for alternative models of urban policy (Stoker and Young, 1993; 180). In the words of Eric Lavocat of the French IDATE agency, 'a new set of urban policies are emerging based on a network of "inter-town solidarity"' (Lavocat, 1989). It has even been argued that the development of parallel territorial telematics networks may even begin to overcome the asymmetry of power and information between localities and multinational capital that directly fuels the development of parochialism and entrepreneurialism. Castells, for example, suggests that:

On-line information systems linking local governments across the world could provided a fundamental tool for countering the strategies of flows-based organisations, which would then lose their advantage, deriving from their control of asymmetric information flows. Information technologies could provide the flexible instrument to reverse the logic of domination of the space of flows built by the processes of socioeconomic restructuring.

(Castells, 1989; 353)

The last decade has witnessed a proliferation of policy initiatives which involve horizontal coordination and collaboration between different cities operating to achieve mutually complementary benefits. These emerging policies stress the importance of inter-municipal cooperation, resource pooling, information exchange and mutual commitment to shared goals across spatial boundaries. Rather than emphasising the need for cities to compete relentlessly within the global political economy, as part of the drive towards 'urban entrepreneurialism' (Harvey, 1989), these policies stress the role of networked rather than market-based forms of inter-urban organisation. While not replacing competition, this shift may represent the first significant steps in the reorientation of urban politics towards a mixture of competition with collaboration (Dawson, 1992). So far, inter-urban networks have developed to fulfil three functions: as a method of information exchange on best practice for policy development; as a stimulant to economic cooperation between cities; and as a method of improving the effectiveness of lobbying to national and European government for resources (Parkinson, 1992). The shift towards interurban collaboration is especially strong within the European Union, where many policy programmes are encouraging these developments.

Interestingly, many of these policy initiatives are centring on the development and use of telematics networks between cities as the technological basis for new types of inter-municipal policy development. Clearly, telematics networks and services offer much potential to this exploration of network-based, inter-urban policies. Telematics are space-transcending technologies that are perfectly suited to reducing the costs of transaction involved between widely dispersed sites such as disparate networks of cities. In principle, inter-municipal telematics networks could offer similar benefits of responsiveness, flexibility and real-time information-sharing between many widely dispersed sites that are already used in the corporate sector – a potentially powerful tool in the exchange of policy experience, the development of economic cooperation and the coordination of lobbying. Telematics offer much potential for supporting inter-city as well as intra-city applications which are of benefit to cities.

Intra-regional, national and transborder telematics-based urban networking centre on the goals of economic cooperation and information exchange. Such networks are most highly developed in France, where telematics are being centrally addressed within mainstream urban and regional planning activities, a process encouraged by the supportive urban and telematics policies of the national government (Tucny, 1993). The extreme municipal fragmentation of French local government presents a particularly strong demand for the emerging ethos of 'intercommunality' in France, which is becoming embodied in the use of inter-communal telematics networks. The first national expression of this trend was Le Réseau Villes Moyennes (RVM; the medium-sized town network), set up in 1987 and described by Marcou in Box 9.2 (Marcou, 1990).

BOX 9.2 LE RÉSEAU VILLES MOYENNES (RVM) – AN EXAMPLE OF NATIONAL COLLABORATIVE URBAN NETWORK

Day in, day out, town administrators from Montpellier to Valenciennes overcome a host of problems.... The medium-sized town network Réseau Villes Moyennes – RVM for short – organises direct information exchange between towns aided by a computer system. France – a country with countless subnational authorities and 36,000 municipalities – was the natural breeding ground for a network of this kind. RVM links up medium-sized towns, with populations between 20,000 and 100,000, using two interlocking tools: a professional messaging service enabling the daily interchange of detailed, up-to-date information in the form of questions and answers; and an exchange fund or projects directory co-produced by the member towns of the network and comprising files in which the towns describe their method of organisation and manage-ment, together with their achievements. RVM was thus born of a need to resolve the countless problems arising in daily municipal management and a firmly entrenched habit: that of finding out, when faced with a problem, what solutions have been designed and tested elsewhere. The running and promotion of the network for the authorities of participating towns are handled by a 'local RVM correspondent' who passes on the information in printed form.

The network operates according to two principles. The first is balance between information production and consumption: each member town undertakes to answer the questions put by its partners and to produce a minimum number of files to feed into the exchange fund, thus ensuring that the content matches users' needs. The second principle is that each town accepts full responsibility for the information it gives out, which, incidentally, relates only to its own activities or relations with third parties.

RVM is a generalist network, open to all administrative departments and all areas of interest. The information transmitted is resolutely operational and deals in particular with the management of municipal amenities (nurseries, swimming pools, science and technology colleges), town planning, social action, internal management (finance, personnel) and third-party relations (contracts). Since the RVM was set up in 1987, thousands of questions and answers usable by all partners have been exchanged: all new members of the network have these results at their disposal. RVM also compiles the answers given; it is thus able to produce a summary answer to any specific question. This answer will be sent out at the same time to all network members. Such answers are then stored in the public records of the messaging service.

(Marcou, 1990; 34)

Others telematics systems in France link cities together that are developing technopolis and teleport-style high technology development initiatives and broadband cable networks (Graham, 1995b). Broadband cable links are being established between the four cities which surround Montpellier in the Côte d'Azur, as a stimulus to the interaction of technopoles and teleport initiatives. These 'new solidarities' between cities are based on what the IDATE development agency call 'complementary poles of excellence' in the different city economies (Lavocat, 1989). The theory is that, through combining the complementarities of a network of cities, the whole region will be able to compete more effectively for inward investment within the Single European Market, as well as creating new synergistic economic processes of development. This applies to the efforts of Roubaix in NE France to use its teleport as the basis for strengthening transborder economic linkages with adjacent textile-dominated industrial zones in Belgium.

International inter-urban telematics networking initiatives are focused as much on improving the lobbying for national and European resources as for information exchange and stimulating economic collaboration. A major stimulant to the development of these networks has been the European Commission's own initiatives, based on its view that 'widening the horizons of regional and local actors is one of the keys to successful economic and social development' (CEC, 1991; 201). Two current initiatives where trans-European inter-urban telematics networks are being explored are the the Eurocities/Telecities network (Figure 9.3) and the European Urban Observatory initiative (Figure 9.4).

Eurocities is an informal club representing about sixty 'second cities' from across the European Community (Parkinson, 1992). The activities of the network are diverse, ranging across lobbying of the EC, influencing EC policy decisions, exchanging best practice and supporting economic links (Dawson, 1992). Eurocities are currently developing an electronic messaging and information system as the infrastructural basis for its future programmes. This 'telecities' initiative is being supported directly by the Telecommunications Directorate General in the EC, which increasingly stresses the importance of trans-European networking between cities and the vital role of urban policy experiments with telematics. The telecities network (Figure 9.3), which will be based on the Geonet global communications system, aims to use telematics as a tool for actually

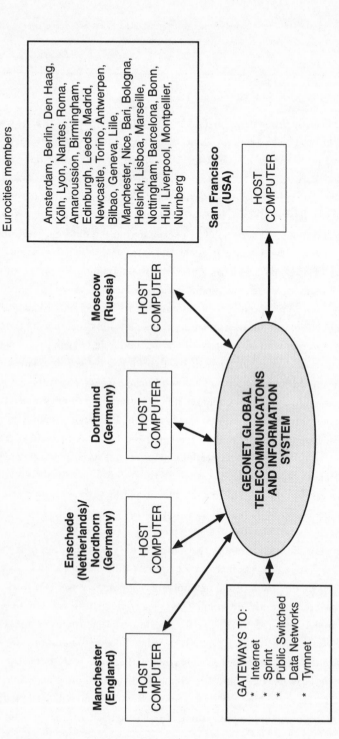

Eurocities members

Amsterdam, Berlin, Den Haag,
Köln, Lyon, Nantes, Roma,
Amaroussion, Birmingham,
Edinburgh, Leeds, Madrid,
Newcastle, Torino, Antwerpen,
Bilbao, Geneva, Lille,
Manchester, Nice, Bari, Bologna,
Helsinki, Lisboa, Marseille,
Nottingham, Barcelona, Bonn,
Hull, Liverpool, Montpellier,
Nürnberg

Manchester (England)

HOST COMPUTER

Enschede (Netherlands)/ Nordhorn (Germany)

HOST COMPUTER

Dortmund (Germany)

HOST COMPUTER

Moscow (Russia)

HOST COMPUTER

San Francisco (USA)

HOST COMPUTER

GEONET GLOBAL TELECOMMUNICATONS AND INFORMATION SYSTEM

GATEWAYS TO:
* Internet
* Sprint
* Public Switched Data Networks
* Tymnet

Figure 9.3 The technological basis for the 'Telecities' initiative: municipal host computers on the global Geonet system

Source: Adapted from Telecities, 1994

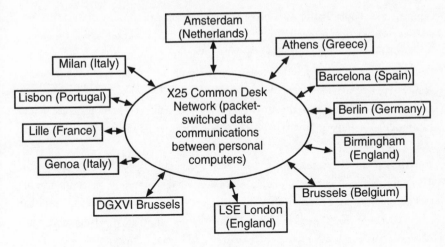

Figure 9.4 The European Urban Observatory initiative
Source: Van Meerten, 1994

improving the telematics policies underway in the member cities for economic and social development.

The European Urban Observatory initiative (EUO) similarly aims to explore new ways for supporting urban decision and policy-making (Van Meerten, 1994). It acts as an up-to-date and real-time decision support system to its member cities, using data communications to distribute information on common problems and pooled knowledge. The system is to include information on economic development, social cohesion, urban environment, local government, transport and mobility, urban infrastructure, and urban development planning in European cities.

URBAN POLICY EXPERIMENTS IN 'ELECTRONIC PUBLIC SPACES'

We saw in Chapter 5 how many urban commentators are seeing the growth of virtual urban communities as a result of people searching electronic spaces for the conviviality and public sphere that seems increasingly lacking in the physical spaces of contemporary cities. We also pointed out that many are advocating new policies for constructing 'electronic public spaces', not just as a boost to the public, civic aspects of cities and to help to eliminate social isolation, but also to combat the rampant social polarisation surrounding telematics in cities. Many of these policies address the particular needs of groups excluded from market-based telematics developments in cities: disabled people, women, ethnic minorities, technologically illiterate groups and people in poverty or on low incomes. They can be seen as part of a wider struggle between concepts of public access to the new telematics networks and the emerging 'private club' of commercialised and commodified telematics services where access is restricted strictly by ability to pay (Mulgan, 1991).

Urban policy attempts at creating electronic public spaces are growing fast. These initiatives centre on cable, videotex and computer communications technologies. A wide range of community access initiatives involving cable has recently been developed by city authorities in the USA. Using the First Amendment and local regulatory powers, cable access programming involving channels dedicated to community TV production are now fairly common. These channels can be considered as 'electronic public spaces', which offer a counter to the commercial imperatives of the marketplace (Aufderheide, 1992). These policies, however, have resulted from long battles against the centralisation of control over cable and the overwhelming power of the market imperative in shaping their development. Only 10–15 per cent of networks have such facilities. Far from being an automatic benefit of the supposed diversity of cable services, they have only developed where 'citizens carve out public spaces with ingenuity, against the odds, and rarely noticed in the national media' (Aufderheide,1992; 53).

Strikingly similar ideas fuel a wide variety of videotex experiments in France linked with the ubiquitous Minitel system. The initiatives of the municipality of Marne la Vallée, near Paris, is typical (Weckerle, 1991). Aiming to structure a whole new set of 'modern public spaces' based on publicly accessible information and communication services on Minitel, the city authority has made efforts to develop a diversity of local social applications on the system, aimed at widening participation in telematics as much as possible. The hope is to support new approaches to local democracy, through Minitel-based networks between citizens, centres of education, social organisations and municipal departments (Weckerle, 1991). Equally significant are a fast-growing range of urban computer–communications systems in North America and Europe known as Public Information Utilities (PIUs), Freenets and Hosts. These are outlined in Box 9.3.

BOX 9.3 PUBLIC INFORMATION UTILITIES (PIUS),
THE FREENET MOVEMENT AND HOST COMPUTERS

Following some radical technology experiments in the 1980s, such as the Berkeley Community Memory Project (see Athanasiou, 1985) community-oriented telematics initiatives or 'electronic public spaces' are increasingly moving into the mainstream of urban public policy in the United States. A wide constituency of activist groups, such as the National Public TeleComputing Network (NPTN) and Computer Professionals for Social Responsibility (CPSR), are, in combination with local activist and special interest groups, proving increasingly powerful in lobbying for national support for local, public telematics networks within Bill Clinton's National Information Infrastructure programme.

There are three main policy models here. First, current efforts in Santa Monica and other Californian cities to develop Public Information Utilities (PIUs) between municipal departments and citizens, allowing computer communications on key local issues, are one model (see Guthrie and Dutton, 1992; Chapter 3). The Santa Monica Public Electronic Network (PEN), for example, allows for electronic town meetings between citizens across the city and elected representatives.

Second, electronic 'Freenets' use commercial and municipal sponsorship to develop freely accessible electronic civic telematics networks. These services, which offer electronic mail, conferencing, information services, bulletin boards, and, often, wider Internet access, are becoming increasingly common in North American cities (Winner, 1993); there were over twenty-five at the end of 1994 and eight others are being developed in Western Europe. Special equipment is often provided for people with various forms of disabilities to use Freenet services. Many Freenets are actually set up with structures that are analogous to the different physical elements of cities themselves. Figure 9.5 shows the main menus of the Cleveland Freenet system, one of the first systems to start up.

Growing policy efforts are also being made to support the access of low-income groups to Internet services. For example, the Public Utilities Commission in Ohio recently ruled that $20 million of subsidy should be made for building computer access centres in low-income communities and schools. Freenet-style projects are increasingly common outside the USA. In Amsterdam, the Netherlands, for example, the Digital City project provides free access to the Dutch parts of the Internet to all the city's citizens. By November 1994, some 7,000 users were registered, including most civic organisations in the city.

Finally, there is a wide range of simple bulletin board services (BBS), most delivered by computer enthusiasts from their own personal computers. It is estimated that there are 300 BBS services in Los Angeles alone. SF Net, a BBS service with 3,000 regular users in San Francisco, has set up 20 coin-operated terminals in cafés across the city. These are aimed at people without normal access to personal computers and modems.

Other electronic public spaces are being developed in the UK based

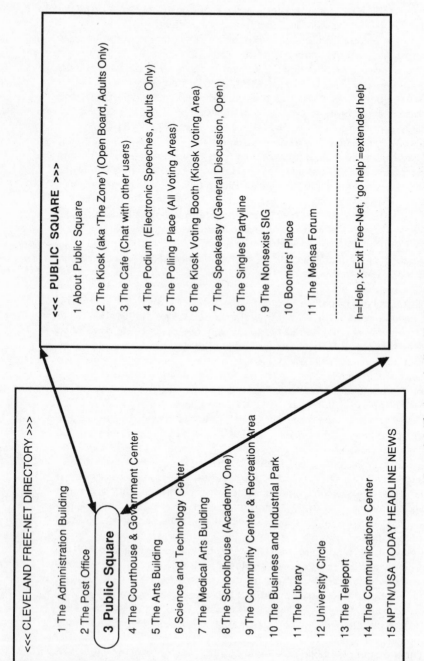

Figure 9.5 The 'menu' offered by the Freenet system in Cleveland, Ohio, showing the analogies made with the physical elements of cities

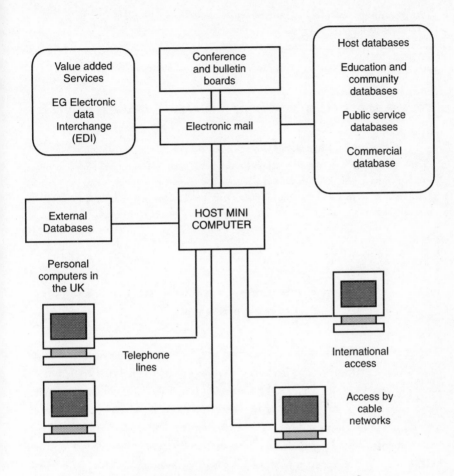

Figure 9.6 The Manchester Host Computer Communications System
Source: Adapted from Manchester City Council, 1991

on a network of municipally controlled Host computers. The Manchester Host (Figure 9.6) offers a wide range of electronic mail, bulletin boards and database services to registered users through computers attached to phone lines. While its services are charged – unlike those of Freenet – excluded groups are being supported through a network of 'Electronic Village Halls' in the city. These are physical centres where training and Host services are supported for 'communities of interest' – for example, the Bangladeshi community, women's organisations, disabled groups and old people – and also distinct geographical communities.

Increasingly, the various urban experiments in developing electronic spaces are themselves being interlinked via the Internet. Special networks linking electronic communities and city-led World Wide Web

experiments have recently emerged which provide an integrated interface between these initiatives on a global, real-time basis.

TELEMATICS AND URBAN GOVERNANCE

In most UK towns and cities, the information economy does not centre on on-line industries,
international head offices, high tech industry – or, on teleports, intelligent buildings and 'silicon'
factory estates. Rather, it pivots around public sector institutions, created during the post-war
golden years of Keynesianism and the Welfare State. Local authorities, as such, are probably the
dominant sector of the information economy in most areas of the United Kingdom.

(Hepworth, 1991a; 172)

TELEMATICS STRATEGIES IN URBAN GOVERNMENT

Urban government in advanced industrial nations is becoming ever-more reliant
on telematics infrastructures and services (Taylor and Williams, 1990). Indeed,
as Mark Hepworth suggests, local government is a major player in the burgeoning
'information economy' within most western nations in which telematics are a
central set of infrastructures. The management and organisation of urban
government is, fundamentally, an 'information business'; this leads Hepworth
(1991) to coin the term 'municipal information economy'. Urban government
relies on the effective use and communication of information about citizens, land
parcels, buildings, infrastructure networks and service delivery networks to
function successfully. This means that networks of interlinked computers and
computerised equipment offer radically new capabilities for managing and
improving the organisational fabric of municipal government. This has three
dimensions which stem from the three types of telematics services currently
under development:

- communications services (such as electronic mail between personal
  computers);
- information services (such as remote databases and videotex systems);
- transactional services (such as electronic fund transfer, teleshopping and
  electronic data interchange).

The welfare-oriented service structures built up in local government under the post-war 'Keynesian boom' are being forced to restructure rapidly right across the developed world (Pinch, 1989; Pickvance and Preteceille, 1991; 4). The traditional public administration oriented culture concerned with the delivery of collective services as 'public goods' to all the citizens of a municipality is declining (Taylor and Williams, 1990). The current emphasis in urban government is increasingly on facilitating the delivery of public services to targeted groups of local citizens, using dwindling local resources as efficiently as possible and only supporting social welfare through a final 'safety net' rather than through uniform service provision (Pickvance and Preteceille, 1991; 5).

Under this new set of political and economic pressures, the bureaucratic structures characteristic of post-war municipal government are being restructured into business-style service organisations. Often, many previously public services are being privatised or contracted out, under pressure from neo-liberal national governments. Such new structures often have more in common with firms (or 'quasi firms') than with the traditional bureaucratic hierarchies of collectivised public administration (Hepworth et al., 1989). Financial crisis at both the national and local level mean that economic criteria of efficiency and price are often replacing the social justice arguments of welfare in determining the nature and quality of local government service provision, under the onslaught of 'new right' legislation in many western nations.

The general shift to market-based and more cost-conscious urban government structures and approaches means that local government needs to control and manage very complex and fast-moving information flows. This management is founded increasingly on telematics-based information and communications systems. Without the application of the full range of telematics innovations – computerised information systems and local and wide area computer and data networks – many of these changes in urban government revolution would simply be impossible. The survival of local authorities 'depends on mobilising information resources to achieve higher levels of competitiveness, efficiency and flexibility' (Hepworth et al., 1989).

Taylor and Williams (1990) go so far as to represent the local government revolution as a transition from the principles of 'public administration' to the new principles of the 'information polity'. The new policies of targeted provision, managed networks of service providers, and newly responsive approaches to customers or citizens succinctly capture the cultural and political transformations which together make up the radical changes in local government. Without sophisticated information systems, including Geographical Information Systems

(GISs), distributed via municipal telematics networks (see Figure 9.7), the emerging real-time methods for managing local government service provision in such a market-oriented fashion would rapidly falter. The crucial role of telematics is shown by the fact that the markets for telematics in urban government are amongst the largest. The municipal market for telematics systems in the UK alone is set to reach £1,000 million by the year 2000 (Hepworth *et al.*, 1989). State governments in the United States spent $20 billion on IT in 1989 (Dutton *et al.*, 1994).

Increasingly, these dedicated Wide Area Networks (WANs) are being used to integrate data and voice communications, support electronic transactions with suppliers and citizens, and underpin the electronic delivery of municipal services. These applications extend right across the spectrum of urban government functions, from library networks, housing services, education and traffic

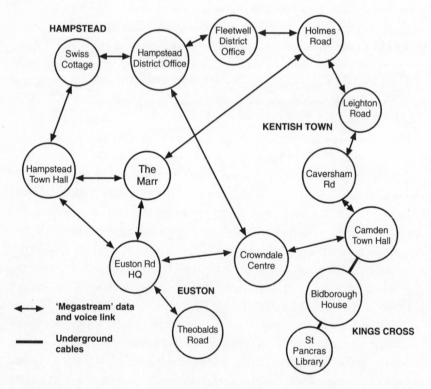

Figure 9.7 An example of an integrated municipal Wide Area Network (WAN) – the London Borough of Camden

Source: Adapted from Camden Council, 1990

management, to social services, recreation management, refuse collection, emergency services and consumer relations. Telematics provide opportunities for better:

- planning and control over service operations;
- productivity measurement, monitoring and control to improve service quality;
- innovation in the 'street level interface' with local citizens;
- better methods of resolving complicated issues (Kraemer and King, 1987; 24).

We should add that transactional telematics system such as Electronic Data Interchange (EDI) are also being widely explored as new efficient systems for organising the links between urban governments and their many suppliers of goods and services. The government of the Balaeric Islands in Spain, for example, hopes to have 90 per cent of its suppliers on-line by 1997. In France, a nationally supported EDI programme for municipal tendering is underway.

However, equally, telematics can be used to support the transformation of public goods into private goods. The control capabilities of telematics can shift the historical definition between what is public and what is private. They can help to open up the traditional roles of the local state in providing the education, welfare, health and infrastructure services that private capital could not deliver, to private, profit-seeking firms. Robins and Hepworth (1988; 166) note that 'through these technologies it becomes possible and viable for the private sector to reappropriate and recapitalise what had become public goods. The national health service, state education, or public service broadcasting all become threatened as it becomes (technologically) possible to market these services as information commodities in the emerging network marketplace.'

This 'commodification' can apply to many urban government services. The control capabilities of telematics can be used to set up new finance systems for taxation and service charging (as with the commodification of library services). In the United States, 'some federal, state and local government agencies view electronic [service] delivery as an opportunity to recover costs or actually generate net revenues' for services that were previously free. This is especially the case for electronic kiosks and phone-based service support lines (Office of Technology Assessment (OTA), 1993). Similar trends have been observed in Britain, especially with the commodification of local government information services such as land searches, planning and socioeconomic information

(Hepworth *et al.*, 1989). Telematics can also be used to exclude non-payers from newly commodified park, transport, education, culture and leisure and other municipal services. This inevitably results in sharpened patterns of social exclusion as people become treated as consumers who are valued according to wealth, rather than citizens with universal entitlements and basic levels of service support. In California, Jim Davis (1994) argues that 'as government services have been reduced, the poor are most affected. The transformation of information into a commodity item over the past few decades has paralleled the defunding of public libraries, museums, schools, and other programs that delivered information and skills to people regardless of ability to pay. Once the barrier of an admission price is raised, those with no money are effectively excluded.' As we saw with the 'pay-per' revolution in Chapter 5 and electronic road pricing in Chapter 7, there are many examples where telematics are being used in this way.

Telematics can also underpin competitive tendering between firms for public service provision, processes which lie at the centre of American debates about 'reinventing government' and British debates about 'market testing' and 'compulsory competitive tendering' (Bellamy and Taylor, 1994; Dutton *et al.*, 1994). Related to this, they can also support mass redundancies and downsizing in government workforces and the contracting out of routine data processing to back offices in peripheral areas or less developed countries. They can underpin more regressive service structures based on highly intrusive surveillance systems, social exclusion, ghettoisation and funding cuts (Davis, 1994). It is clear, then, that the effects of telematics in urban government depend largely on how they are socially and politically constructed: 'the democratising potential of telematics . . . has to be fought for' (Healey *et al.*, 1995).

THE ELECTRONIC DELIVERY OF PUBLIC SERVICES

As well as opening up the potential for commodification and privatisation of public services, telematics provide new opportunities for changing the ways in which those services are delivered to citizens within cities. The information, communications and transactional capabilities of telematics are being explored throughout western nations as a new means of delivering urban public services (OTA, 1993). Increasingly, electronic information kiosks, videotex terminals, automatic teller machines and smart cards are being used to cut the costs and improve the effectiveness of dealing with the millions of day-to-day information requests, communications, and transactions between urban government and citizens. In many American states, for example, government is following the private sector with advanced telematics systems development. These support the electronic delivery of benefits and replace the physical offices of certain government services with electronically mediated kiosks (OTA, 1993). The Public Information Utilities described in Chapter 3 are part of this broader set of innovations.

However, such innovations can be shaped in many ways, and the trade-off between risks and benefits remains unclear. On the one hand, they may simply support the substitution of the physical apparatus of urban government with electronic spaces – a way of cutting costs and improving efficiency. In New York, the city's INet optic fibre system is already used for conducting remote videoconferences between prison inmates and their legal aid lawyers. Touch screens and kiosks in shopping malls are rapidly emerging in the United States, but with a counter trend towards the withdrawal of public bureaucracies to fewer back offices (Bellamy and Taylor, 1994). In California and Hawaii, for example, multimedia kiosks in shopping malls offer advice on anything from HIV and local services, through job listings to transactional capabilities such as offering local fishing or driving licences. But such initiatives may lead to major job-shedding in local government. One company, markets these kiosks under the heading 'fewer workers . . .' (North Communications, undated). A Californian policy-maker who uses them admits that 'if hundreds of kiosks are answering routine questions . . . the bureaucracy can function with fewer workers'. An industry spokesman predicts that 'they could eventually replace $6–$8 [per hour] workers' (North Communications, undated).

Other emerging examples of the possible substitution of electronic for physical space can be drawn from France. Here the ubiquitous Minitel system is supporting many information, communications and transaction applications linking citizens and municipalities: a theatre booking systems operates in Metz; special graffiti removal hotlines have been set up in Bordeaux; part of the Nantes job market has gone on-line; and babysitting brokerage operates in Blagnac. The emergence of hundreds of specialised telephone help lines is also part of this trend.

Dutton (1993) notes the 'long range, strategic vision' in US urban government 'of employing multi-media kiosks and other information technologies to extend the reach of public facilities that increasingly find their physical facilities imposing constraints on the services they could provide'. Colin Muid (1992; 79), head of the UK Government Centre for Information Systems, argues that 'potentially, services to citizens can be made available directly into the home and at whatever time is convenient to them. There is, as yet, no public service equivalent to First Direct bank. . . . Such possibilities may occur to private sector firms who win market test competition and take over services from government.'

It seems likely that these processes of change will parallel those in banking and retailing noted in Chapters 4 and 5: electronic, home-based services could substitute for physical networks of offices. As well as the likely job losses, they bring with them issues of equity and privacy. How, for example, will people without phones or computer literacy fare in electronic service delivery, especially when their physical access to services may be lost through office restructuring? As with shifts in banking and retailing, these processes may advance the interests of socially privileged, mobile and technologically literate groups while compounding the many disadvantages already faced by marginal groups of 'information have nots' (OTA, 1993; Dutton, 1993). As Dutton (1993) argues 'living in an information society, the educated public often takes exposure to information technology for granted. Yet information technology is invisible to many within the inner city. . . . In an era of so-called information overload, few managers or professionals can imagine a situation in which there is truly a lack of essential information, but this is precisely the case in the inner city.' Thus, huge demands for information within disadvantaged areas of cities often go unmet, a problem that can be compounded by regressive shifts in urban public support services.

It is clear that special projects are required to help to overcome existing inequity and that new changes will need to take account of these problems to avoid making them worse. Many examples of such services already exist. For example, the InfoLine service, a phone-based support service for disadvantaged communities in Los Angeles, handles 230,000 calls per year. Training, education, capital support and help with setting up on-line electronic public spaces are also necessary, geared to the specific needs of disadvantaged groups and areas. Without such projects, marginal inner cities or social groups may simply end up as the victims of the withdrawal of the physical presence of services while being unable to participate in the growing electronic spaces of public services. Changes in urban governance may simply compound the wider processes towards social and spatial polarisation within western cities highlighted in Chapters 4 and 5. There are even risks that local government may follow many private sectors firms and become less local, as it employs computer and data services from private sector firms in non-local regions.

There are also risks that privacy and civil liberties will be undermined by the use of telematics-based public services. In the United States, many food stamp programmes and public medical support services now operate through smart cards and Electronic Benefits Transfer (EBT). The embedded microchips within the cards are 'topped up' and 'spent' electronically. This supports whole electronic financial networks linking recipients, food retailers and government agencies (OTA, 1993). These are easier to verify and less easy to defraud than the old paper-based systems (Hausken and Bruening, 1994). Many municipal smart cards experiments are now underway in France aimed at integrating the tracking of all transactions and exchanges between citizens, transport companies and municipalities. Paris, for example is exploring the use of smart cards to transfer all cash transactions into the form of prepaid, electronic money, so saving the time and resources spent on transactions. Advances in smart cards and the much more capable optical card technologies seem likely to lead to much wider application in urban government (Hausken and Bruening, 1994).

While these systems can improve the flexibility, responsiveness and efficiency of services, they can also threaten civil liberties and privacy and lead to socially regressive practices of surveillance and exclusion. For example, the intimate health information sometimes contained on medical and social service smart cards – such as genetic, sexual and drug use information – could be accessed by agencies judging the suitability of individuals for credit, education or other services. Subjective assessments of an individual's 'demeanor, character, and mental state' (Hausken and Bruening, 1994) are sometimes put on to these records and can be further used in this way. Other electronic systems can be used for such covert social surveillance. Some kiosks are now being developed which check the alcohol content of a user's breath and add this information to that already held on the individual by the municipality (Dutton et al., 1994). The information collected in kiosks or on smart cards may be insecure and open to unscrupulous use or reselling in the network marketplace to the personal information bureaus mentioned in Chapter 5. Above all, these trends may simply support the development of more sophisticated systems of social exclusion and control, a regressive response to the combined processes of public sector privatisation, financial cuts and social and spatial polarisation in cities. Jim Davis comments on the growing use of Automatic Teller Machines (ATMs) to deliver food stamps and local government services in the United States:

While proposals to deliver welfare benefits electronically, via ATM cards, has some decided benefits for welfare recipients, including increased flexibility and security, it also poses serious risks. When food 'stamps' are delivered electronically, for example, the potential for tracking purchases and comparing them with other welfare data becomes a possibility.

(Davis, 1994)

On the other hand, telematics may be used radically to improve the delivery, user-friendliness and quality of public services to remote locations and people with poor mobility. For example, the kiosks mentioned above may improve the quality and timeliness of information delivery and make citizen–government transactions, in several languages if necessary, easier on both sides. The Santa Monica Public Electronic Network (PEN), discussed in Box 9.3 and Chapter 3, 'is widely perceived by city personnel to have enhanced the responsiveness of the city' to citizens (Dutton, 1993). An interactive cable service in Iowa City in the United States offers a wide range of seventy types of civic information to people's homes (Bankston, 1993). A teleshopping experiment in Gateshead, England, has long supported the accessing of basic foodstuffs by groups unable to leave their homes (Ducatel, 1994).

The time–space flexibility of telematics may underpin truly beneficial systems for information exchange, communication and transaction between urban government and its citizens. These may be more appropriate to the wide-ranging needs of the diverse cultural, geographical, ethnic, social and gender groups that make up cities. People with poor physical mobility potentially have the most to gain from the use of electronic services to overcome space and time barriers.

The resource-sharing potential of telematics may also support radically new innovations in distance learning and education, adult training, community building and democratisation in access to information, services and skills. Local cable and telematics systems, for example, are being used in several American cities and in Birmingham, UK, to provide special educational services to pre-school children. In New York, a distance learning and videoconferencing network, based on a Nynex optic fibre infrastructure, offers the city's schools remote access to its cultural and educational institutions.

Telematics may also be used to support the decentralisation of local government to locations more closely linked with the distribution of population centres – so bringing 'power to the people' (Baddely and Dawes, 1987). They may extend political participation through innovations in 'electronic democracy'. In Albuquerque, New Mexico, for example, a freephone system is being used to support state-wide balloting by phone. New York Telephone (1993; 9) argue that 'several communities around the nation are on the edge of electronic citizenship, meaning widespread access to public information and government resources, fostering increased community involvement and cultural enrichment'.

Finally, telematics may help to overcome the barriers which stem from social inequality. Electronic innovations in service delivery may, like the electronic public spaces discussed above, actually act to counter the trends towards heightened inequality in cities. More than most of the areas covered in this book, it seems, it is apparent that urban governments' use of telematics is open to social construction and political shaping.

CONCLUSIONS: TELEMATICS AND NEW VISIONS FOR URBAN POLICY AND GOVERNANCE

These burgeoning telematics-based applications in urban government and service bring us to the key conclusion of this chapter – that the uses to which telematics are put by urban planners, managers and governors are open to social and political construction (Bellamy and Taylor, 1994). They are not technologically determined; nor are they the result of the simple 'working through' of the political–economic imperatives of capitalism. Rather, the policy processes driving the many projects and initiatives reviewed here are the complex results of social and political interaction of different, usually dominant, interests. There is little doubt, though, that this goes on against the backcloth of broader political–economic changes in capitalist society: the reorientation of all levels of the state away from Keynesianism to market liberalism, the growth of inter-city economic competition and the increase in unemployment, marginalisation and social polarisation.

This inevitably biases the processes at work as urban governments and policy-makers struggle within the shifting wider context. But it is clear that considerable space remains at the urban level within which innovative telematics applications can be socially constructed which have more equitable and progressive results than either the polarising logic of market forces or the remote and distant decisions of national or supranational states.

This is not to cast urban and regional policy-makers and their policies as panaceas in the search for more equitable, democratic and sustainable cities. As we have seen, there remain many problems confronting urban policy intervention in telematics. The initiatives developed are often relatively insignificant compared to the broader forces we have examined in the rest of the book. Urban policy-makers are as prone as anybody to being seduced by notions of futurism and technological determinism. Their actions can often be criticised as being socially regressive, over-ambitious and based on dubious assumptions about the genuine power of municipal level policy-making over the global political economic forces of capitalism. Many mistakes result and projects can have unintended side effects.

Finally, telematics are being used in some cases as support for privatisation, reducing social services, and enhancing the degree of surveillance and control that dominant institutions have over the socially powerless.

Conversely, much promising policy innovation and exploration is emerging in which genuine social telematics innovation in cities is producing real benefits. There are many emerging examples where appropriate, low-cost applications of telematics are helping to open up new policy avenues based on inter-urban collaboration as well as competition, social empowerment and equalisation as well as social polarisation. Genuine attempts are being made to address the social, economic and environmental problems of cities without falling into the traps of techno-hype and the language of the quick technical fix.

The embryonic nature of this wide front of policy innovation makes it difficult to estimate the broader significance of these policies and their possible effects. The obvious diversity and rate of change make generalisation hazardous. The wider hope is that urban telematics policies will have a significant role to play within these processes whereby new, progressive urban visions can be developed that address the context of globalisation, fragmentation and polarisation that has been analysed in the rest of this book (and within which, ironically, telematics are themselves heavily implicated).

There are many difficult questions here (see Healey et al., 1995). How can cities and city regions respond institutionally to these instabilities and problems? How can urban politics be remade in ways which fight the growing unevenness and fragmentation of urban social and economic life? What does the city and its politics mean within the global shifts now at work? While it would require another book to explore these issues, we simply argue that telematics, when socially constructed in appropriate ways, can offer a new and potentially powerful set of policy tools for supporting innovation which may at least help in the first few steps towards possible solutions. Telematics have clear potential as policy tools for bringing together the urban fragments and adding much needed coherence and capacity to urban policy-making (see Healey et al., 1995). But they offer no hope on their own and must not be seen as some technical fix. They can only play a relatively small role as technological tools within the much wider process of social, institutional and political change that is the key prerequisite to such a shift. Without this, telematics can be little but an irrelevant distraction.

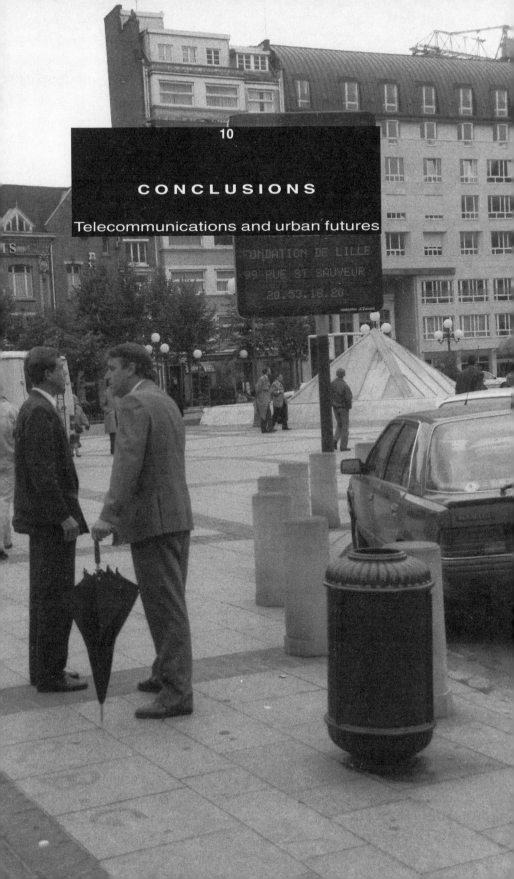

10

CONCLUSIONS

Telecommunications and urban futures

The network is the urban site before us, an invitation to design and construct the 'City of Bits' (capital of the twenty first century). . . . This will be a city unrooted to any definite spot on the surface of the earth, shaped by connectivity and bandwidth constraints rather than by accessibility and land values, largely asynchronous in its operation, and inhabited by disembodied and fragmented subjects who exist as collections of aliases and agents. Its places will be constructed virtually by software instead of physically from stones and timbers, and they will be connected by logical linkages rather than by doors, passageways, and streets.

(Mitchell, 1995; 24)

INTRODUCTION

This book has been motivated by the widening perception that we are losing our ability to view and understand the contemporary city: how cities operate within a globalised context as systems linking economic, social, cultural, geographical and political development. We argue that the increasing telemediation of urban life is a key factor in the apparent growth of difficulties faced by urban commentators or policy-makers, as they try fully to understand the dynamics of contemporary urban development.

Our long journey through the complex terrain of city–telecommunications relations is now complete. We have built up a picture in unprecedented detail of how all aspects of the development, management and planning of cities increasingly relate to the pervasive application of telecommunications and telematics. Our perspective has been broad and comprehensive – some might say over-ambitious. We have critically reviewed the theoretical approaches that can be taken to city–telecommunications relations. Using this, we have constructed a basic framework blending the insights of political economy and social constructivism to build up the argument that contemporary cities can only be understood as parallel constructions within both urban place and electronic space. Without understanding both, and the many interaction points between them, we believe that we will never be able to approach or understand the totality of the current transformation underway in advanced capitalist cities. As Law and Bijker (1992) suggest, 'all relations should be seen as both social and technical. . . . Groups and organizations are held in place by mixed social and technical means'. Following this, we would argue that the relations that shape cities and urban development are now, more often than not, 'held in place' by both social and telecommunications-based means.

This book suggests that the understanding of the contemporary city requires that one should grasp the complex interactions between urban places – as fixed sites which 'hold down' social, economic and cultural life – and electronic spaces, with their diverse flows of information, capital, services, labour and media which flit through urban places on their instantaneous paths across geographical space.

A NEW TYPE OF URBAN WORLD, NOT A POST-URBAN WORLD...

Clearly, the growth of electronic spaces is not somehow leading to the dissolution of cities, as so often argued by futurists and utopianists. Urban functions are not being completely substituted by dematerialised activities operating entirely within electronic spaces. In fact, this view is naive, shortsighted and dangerous; it perpetuates simplistic ideas about cities and telecommunications and undermines the potential for critical and sophisticated policy debates. We live in a fundamentally urban civilisation: cities as places still matter and will continue to matter. Urban places remain the unique arenas which bring together the webs of relations and 'externalities' that sustain global capitalism. They are of fundamental importance as the terrain for social and cultural life; they house the vast majority of our population; and they seem likely to remain the key economic, social, physical, cultural and political concentrations of advanced capitalist society. Much of what goes on in cities cannot be telemediated: 'only the most hopelessly nerded-out technogeeks could be persuaded to trade the joys of direct human interaction for solitary play with their lap tops in darkened rooms' (Mitchell, 1995). Urban places and electronic spaces can be seen to influence and shape each other, to be recursively linked; it is this recursive interaction which will define the future of cities.

In fact, this interaction is shifting us to a new type of urban world rather than a post-urban world and there will be complex terrains of winners and losers in this process. To discredit further the utopian dreams of some new technological rural idyll, what is emerging is a *more totally urbanised* world, where rural spaces and lifestyles are being drawn into an urban realm because of the time–space transcending capabilities of telecommunications and fast transportation networks. This represents not the death of cities and the renaissance of genuinely rural ways of life, but the emergence of a 'super-urban' and 'super-industrial' capitalist society operating via global networks. Healey *et al.* (1995) note that 'firms and households are dispersing across urban regions incorporating rural relations and cultures into the array of possibilities evolved in an urban context; villages become neighbourhoods within an urban region'.

There is also little evidence that transport demand and physical flows of people and goods are actually slackening to any significant extent as a result of the use of telecommunications as substitution – even though telecommunications demand is burgeoning. Demand for both transportation and telecommunications within and between cities is growing rapidly. This suggests that the trend is towards both a movement and communications intensive society based on growing flows of goods, people, services, information, data and images. Both, in fact, tend to feed off each other in positive ways. The main focus of innovations in teleworking is around the major cities, not in widely dispersed electronic cottages. Most teleworkers are still reliant on transport links to and from work at some stage during their work routines. What is happening is the use of combinations of transport and telematics innovations to make work processes more flexible in time and space. While this may reconfigure transport patterns within and between cities, there seems little sign yet that it will begin to undermine the practice of travel.

On the other hand, though, we do not argue that cities are unaffected by the remarkable extension of electronic spaces; their pervasive growth critically affects all aspects of urban development. New conceptual treatments of the 'city' and the 'urban' are required, raising many questions as to how cities can sustain the transition to telemediation in a progressive way. Of key importance here is the inherent logic of polarisation, which seems to be locked into current processes of economic and social development in cities. This polarisation is both reflected in, and supported and reinforced by, the development of electronic spaces. Fewer city economies seem set to do well; patterns of economic health become more starkly uneven at all spatial scales; and processes of change seem to reinforce the privilege and power of social élites while marginalising, excluding and controlling larger and larger proportions of the population of cities.

Clearly these shifts represent a transformation of society as much as a transformation of cities. Important questions are raised about what words like 'city' and 'urban' can actually mean in a world where urbanity seems almost total; where globalisation and global networks seem to undermine the notion that what goes on in cities is radically different from the processes elsewhere; and where electronic spaces provide the 'sites' for a growing proportion of urban life.

THE CITY AS AN AMALGAM OF URBAN PLACES AND ELECTRONIC SPACES

This book has shown how the contemporary city is, more than ever before, an amalgam whereby the fixed and tangible aspects of familiar urban life interact continuously with the electronic and the intangible. We have explored many examples across all aspects of urban life whereby superimposed systems combining presence in urban places with interactions in electronic spaces are being constructed – witness the electronic financial markets, the back-office networks, the global media flows, the 'intelligent building', 'intelligent city' and 'smart home' debates, the surveillance networks, the transport–telecom inter-actions, the virtual communities and the civic electronic spaces. Fixed construc-tions of places and buildings in urban places linked into electronic networks and 'spaces' seem to define contemporary urbanism. Telematics – the supports for electronic spaces – are woven increasingly into the built environments of cities. They are also filling the corridors between them as key infrastructures to underpin the shift to global urban and infrastructural networks. The social and cultural aspects of urban life also operate increasingly through constructions which meld the built environments of cities with uses and applications of telecommunications.

Together, the diverse electronic spaces that we have explored amount to a hidden and parallel universe of buzzing electronic networks. Largely free of space and time constraints, these interact with and impinge on the tangible and familiar dynamics of urban life on a twenty-four-hour-a-day basis and at all geographical scales. Thus, a car, rail, plane or bus journey, and the physical flows of water, commodities, manufactured goods and energy are supported by a parallel electronic 'networld'. These monitor, shape and control the physical flows underway on a real-time basis. Traffic snarl-ups are now the launch pads for countless electronic conversations and interactions. The apparently lifeless world of the office block hides an 'intelligent building' – a hub in an electronic universe of twenty-four-hour-a-day global flows of capital, services and labour power. The daily life of an urban resident leaves a continuous set of 'digital images' as it is mapped out by a wide array of surveillance systems – closed circuit TV cameras, electronic transaction systems, road transport informatics and the like. The fortressing of affluent neighbourhoods relies on old-fashioned walls and gates linked into sophisticated electronic surveillance systems. The most ordinary suburbs of most cities now act as hubs in the growing electronic cacophony of global image and media flows and the ongoing participation of people in virtual communities, often on a global basis. Urban policies and strategies are increasingly directed to try to shape both urban places and electronic spaces. This shadowy world of electronic spaces exists through the instantaneous flows of electrons and photons within cities and across planetary metropolitan networks, which, unseen, underpin virtually all that we experience in our daily lives.

Not surprisingly, this world of electronic spaces is as diverse and complex as the landscapes and life of the actual cities. Like the geographical landscapes of cities, there are many segmentations, divisions and social struggles underway over the definition and shaping of electronic space. There are 'information black holes' and 'electronic ghettos' where the poor remain confined to the traditional marginalised life of the physically confined, and there are intense concentrations of infrastructure in city centres and élite suburbs supporting the corporate classes and transnational corporations. As with geographical landscapes, the results can be 'read' as reflections of complex processes whereby social, ethnic, gender and power relations play out against the backdrop of the globalising political economy of capitalism.

The proliferation of electronic spaces seem to be heavily involved as we move from the standardised, rigid, hierarchical and rhythmic world of the industrial age to new and much more fluid societal processes. Telecommunications and electronic spaces bring profoundly new relationships between cities and the fundamental matrices of space, time and power within a globalising capitalist society. We feel that they are implicated as central underpinnings to the changing nature of capitalism and the capitalist city – not as technologies 'impacting' autonomously on cities but as social constructions that are being used to explore new ways of controlling and organising space, time and social processes in a crisis-ridden urban world.

These complex interactions between urban places and electronic spaces challenge the simplicity of many widely held assumptions about telecommunications, notably those stemming from technological determinism and futurism, which are widely infused into research on telecommunications and cities. In much urban telecommunications research, we diagnose the use of many over-simplified and 'autonomous' notions of technology, unjustified assumptions and a dearth of real empirical analysis. As we have seen, neither the urban studies nor the urban policy communities have made much progress in understanding these dynamics with the sophistication they demand. Most approaches still only accommodate urban places and the tangible social and economic dynamics of cities. Mechanical, Euclidean and Cartesian notions of urban development, distance, time and space are often still privileged over 'electronic' ones, which accommodate the interrelationships between electronic spaces and other forms of space and time. Remarkably few urban commentators even mention telecommunications and telematics; when they do, the treatment tends to be rather simplistic. Old-fashioned ideas of the development of cities that are separated in 'Euclidian' space and shaped by the physical 'friction of distance' surrounding them often still implicitly dominate and are used to try and accommodate the effects of telecommunications. Concepts that capture the free and flexible meshing of times and spaces together into global networks, and the often bewildering processes of time–space compression which are associated with telematics, are still poorly developed.

OVERCOMING THE MYTHS OF DETERMINISM: CONTINGENCY WITH BIAS

To improve our understanding of the recursive interactions between cities and telecommunications, we must leave behind the myths of determinism, both technological and social. The assumption that telecommunications impact in some simple, universal and linear way on cities is still common; the stress that they simply reflect some abstract capitalist political economy is still prevalent in some critical literatures. Both, we argue, are unhelpful. An integrated perspective of cities and telecommunications teaches us that social action shapes telecommunications applications in cities in diverse and contingent ways, even if this goes on against the backcloth of broader political economic trends. New telecommunications technologies bring new options and capabilities within which urban processes can be shaped. Much of what we have seen in this book represents a range of efforts to address problems and crises in this way. But the complex interactions between the social and technological lead to diverse effects, some intended, some unintended, and to the emergence of new problems. Social conflict and struggle between unevenly equipped groups and organisations is a key feature of the processes at work. As Graham Murdock argues:

The history of communications is not a history of machines but a history of the way the new media help to reconfigure systems of power and networks of social relations. Communications technologies are certainly produced within particular centres of power and deployed with particular purposes in mind but, once in play, they often have unintended and contradictory consequences. They are, therefore, most usefully viewed not as technologies of control or of freedom, but as the site of continual struggles over interpretation and use. At the heart of these struggles, lies the shifting boundary between the public and private spheres.

(Murdock, 1993; 536–537)

These struggles inevitably bias the design and application of telecommunications in cities. Transnational corporations gain access to optic fibre and private networks; people in disadvantaged ghettos are lucky to access a pay phone. But this bias does not shape and determine all the applications and developments of the technology in all cities. Once technologies are available, political and social struggle and actions can redirect their application and change their effects – just as political and social influences can redirect the shaping of urban politics and the built environments of the urban places. Thus, similar technologies can be used in very different ways and have very different effects. The effects of telecommunications on cities also seem to vary considerably across time and space. They do not conform to the simplistic and often naive models of technological forecasters where cities were assumed to be 'impacted' universally, linearly and directly by telecommunications (Gökalp, 1988; Gillespie, 1992). In other words, the implications of telecommunications for cities are indeterminate. They are not predefined either through some abstract technological 'logic' or by some mechanistic political economy imprinting itself on to society. These effects tend to be evolutionary rather than revolutionary. Like all technologies, telecommunications and telematics have influences only through their involvement in the wider process of social, political–economic and geographical transformation. As Geoff Mulgan suggests, 'information technologies continue to be most revolutionary not in creating the new out of nothing but rather in restructuring the way old things are done' (Mulgan, 1991; 13).

This leads to complex and apparently contradictory situations. The same technologies can be applied to empower and assist disadvantaged groups as well as to disenfranchise or exploit them. There are, as we saw in Chapters 5 and 9, many examples of disabled people, women and ethnic minorities benefiting substantially from telematics. Telematics can be used to strengthen the public, local and civic dimensions of cities as well as to support social fragmentation and atomisation. They can help improve urban public transport or further the domination of cars in cities. They can assist in the search for sustainable models of urban development or help maintain the growth of highly unsustainable cities. And profoundly different political styles of urban government can all develop, each using telematics in particular ways to support their approach. The clear lesson is that telecommunications and telematics are nothing if not flexible.

The effects of telecommunications within cities are therefore complex and ambivalent. They allow urban infrastructures and transport systems actually to extend their capacities, so removing barriers to further urban growth and concentration. Telecommunications are supporting environmentally damaging increases in transport flows as well as promising assistance in the drive towards sustainable cities through telecommuting and reduced transport flows. Telecommunications assist the globalisation of the economy in which city economies are being fragmented as nodes on the global telecommunications networks of transnational corporations. They also provide new policy tools of urban management at the local level through which the public, civic face of cities can be strengthened and local economic and social cohesion supported. On the social front, telecommunications can help overcome isolation, disadvantage and disability, as well as furthering the degree to which the 'information poor' are marginalised and exploited. On the one hand, new telematics services promise a global playground to the affluent élites who are 'switched into' broadband cable TV and telematics systems and whose lives are saturated with these technologies at home and work. A short distance away, however, there are invariably ghettos of the so-called 'information poor' who fail to benefit even from the supposedly 'universal' basic telephone service – that most rudimentary point of access to the so-called 'information society'. Many households in inner cities without the basic phone are actually surrounded in the physical sense by lattices of sophisticated optic fibre networks supporting corporate telecommunications flows from which they are totally excluded. The nature of telematics means that physical proximity has very little to do with electronic proximity or access.

TELECOMMUNICATIONS AND URBAN FUTURES

This leads us to the future of urban places and electronic spaces. Debates about the future of cities and urban society, while gaining ground, are currently somewhat stifled. Futurism is often discredited; the fall-out from the failures of many futuristic and modernist urban plans of the 1960s and 1970s continues. Faith in science and technology as redeeming forces has long since withered with the collapse of modernist assumptions about progress, knowledge and technical rationality. The rate and complexity of change in contemporary cities makes even more daunting the task of the extrapolation of these processes and the prediction of urban futures (Healey *et al.*, 1995). Given the crisis of urban social polarisation, the many questions over the future of urban economies, the continuing environmental crisis in cities, and the radical shifts underway in urban policy and planning, it is not surprising that looking into the future often seems something of a luxury from current standpoints. The 'paradigm challenge' brought by the proliferation of electronic spaces further compounds these problems. The inertia and dominance of anachronistic ideas about cities are deeply rooted; the barriers facing sophisticated analysis and intervention in urban telecommunications are daunting. And these problems make it even more difficult to be normative and suggest what kinds of cities and electronic spaces we want and how these urban ideals may be brought to fruition. Of course the answers to such questions presume that there is such a group which can be easily identified from the many disparate and conflicting bodies that make up cities. Perhaps debate should start here by exploring how this 'we' might first be identified.

Thankfully, these issues lie beyond the scope of this book. While aiming to be comprehensive, this book represents only a small part of this much wider project: to reconfigure urban studies and urban policy-making in ways which directly reflect the increasingly tele-mediated nature of contemporary cities and so help reinvigorate debates about urban futures. The overwhelming conclusion of this book is that a concerted research drive is urgently needed if telecommunications are to take their appropriate place at the centre of current conceptions and understanding about the development, planning and management of cities. Critical urban research now needs to turn to the complex interactions between electronic spaces and urban places.

GUIDE TO FURTHER READING

3 APPROACHING TELECOMMUNICATIONS AND THE CITY: COMPETING PERSPECTIVES

The four edited books by John Brotchie and colleagues (Brotchie *et al.*, 1985, 1987, 1991) and the OECD, (1992) provide the best collections exploring the urban impacts of telecommunications. A related area of work centres on the 'long waves' of urban and regional development is best explored by Hall and Preston (1988). Good examples of futurist and utopian treatments can be found in the pages of *The Futurist* journal. See also Maisonrouge (1984), and Mason and Jennings (1982). The most influential approaches have been by Toffler (1981) and Martin (1981). Not surprisingly, many promotional brochures and videos produced by telecommunications and information technology companies take a similarly optimistic stance. A more considered approach to forecasting Europe's technological and urban futures is provided by Masser *et al.* (1992).

The most important book taking the political economy stance is *The Informational City* by Manuel Castells (1989). Shorter pieces demonstrating a more critical approach include Gillespie and Robins (1989), (1991); Robins and Gillespie (1992); Graham (1994) and the full version of the Robins and Hepworth article included here (Robins and Hepworth, 1988). Broader critical collections on telecommunications and society include Slack and Fejes (1987); Webster and Robins (1986); and Mosco and Wasko (1988). For more cultural approaches, see Postman (1992); Carey (1989) and Bender and Druckrey (1994).

The Social Construction of Technology approach has rarely treated specifically urban telecommunications developments. Dutton and Guthrie (1991) and Guthrie and Dutton (1992) (included above) explicitly use SCOT to study the

Public Information Utilities in Californian cities. The book by Dutton *et al.* (1987) comparing the 'wired city' initiatives developing across western nations takes a broadly SCOT based approach. *The Social Impact of the Telephone*, edited by Pool (1977), also includes a range of articles demonstrating the historically contingent nature of telecommunication effects on cities (see Chapter 8).

4 URBAN ECONOMIES

A historical feel for the role of telecommunications in urban economic development can be found in Pred (1977), Meier (1962) and Pool (1977). The excellent book by Beniger traces the 'control crises' that led to the historical origins of information technologies (Beniger, 1986). The best work on the broad development of the 'information economy' and the 'information city' is by Mark Hepworth (1986, 1987, 1989, 1991). Hepworth (1989) discusses the development of new industrial districts, which are also well covered in Hall and Jacques (1989) and Antonelli (1988).

Literature covering the globalisation of city economies and the role of corporate telematics networks includes Knight and Gappert (1989), Bakis *et al.* (1993), Amin and Thrift (1992, 1995), and Dematteis (1988, 1994). Howells and Wood (1993) cover manufacturing and telematics. The general book by Kellerman (1993) provides a useful overview of geographical aspects of telecommunications. The critical stance on telematics and globalisation is well represented by Irwin and Merenda (1989) and Schiller and Fregaso (1991). Similar critiques of the wisdom of opening up cities to such networks are developed by Gillespie (1991), Gibbs (1993), and – from the point of view of less developed nations – Samarajiva and Shield (1990).

Research which traces the links between the information economy, the growing power of corporations, and the uneven development of cities and regions can be found in Reich (1992) and in the work of scholars at the Centre for Urban and Regional Development Studies at Newcastle University (see, for example, Robins, 1992; Gillespie and Hepworth, 1988; Gillespie and Robins, 1989, 1991; and Gillespie and Williams, 1988). The most influential critical perspective on urban economic restructuring, the role of telematics, and the new relations between economic activity and location, however, has been put forward by Castells (1989).

5 THE SOCIAL AND CULTURAL LIFE OF THE CITY

The best sources on the globalisation of urban culture are the books by Bird *et al.* (1993), Knox (1993), Featherstone (1990), Lash and Urry (1994), Watson and Gibson (1995) and Amin (1994). The journals *Theory, Culture and Society* and *Media, Culture and Society* also explore these relationships. On the unevenness of access to technologies the best work is by Peter Golding and Graham Murdock (Golding, 1990; Golding and Murdock, 1986; Murdock and Golding, 1989). More general approaches to telematics and social development are collected in Featherstone (1990). The two books by Miles (1988) and van Rijn and Williams (1988) provide the best treatments yet of general home telematics issues. Silverstone *et al.* (1992) look at technologies within households.

Issues surrounding the 'pay-per revolution' are best explored from a critical perspective by Mosco and Wasko (1988). Social surveillance of all kinds is well covered by David Lyon's book (1993) and, in the specific case of Los Angeles, by Mike Davis (1990). Good quality discussion of virtual urban communities is rare. Featherstone and Burrows (1995) offer the best collection. Harasim (1993) addresses the global dimensions; Poster (1990) takes a post-modern perspective; and Rheingold (1994) is fairly promotional. Calhoun (1986) provides by far the best treatment of the relationships between telematics and social integration in cities.

6 URBAN ENVIRONMENTS

The urban environmental management research and policy agendas are covered in Breheny (1992) – particularly useful on urban form debates. CEC (1990) reviews the issues facing European cities. Douglas (1983) and Elkin *et al.* (1991) review the nature of environmental problems in contemporary cities while Cooper and Ekins (1993) provide a more scientific and engineering perspective. None of these texts specifically mentions the role of telecommunications.

There are a number of sources on the direct environmental effects of telecommunications. The Department of the Environment (1988) provides guidance on the local amenity issues associated with the development of telecommunications network and infrastructure in the UK while Longhini (1984) examines these issues in US cities. For a direct comparison of the resource use of telecommunications see Meyer (1977) and Tuppen (1992). The British Telecom report (1992b) provides an environmental audit of a major telecommunications supplier. Coolidge *et al.* (1982) examines the environmental impacts of telematics technologies while Young (1994) provides a useful review of the environmental problems associated with the production of microchips.

The literature on the environmental applications of telecommunications is very diffuse. The key source on the use of computer networks for the exchange of environmental information is Young (1994) while Rittner (1992) provides a guide to on-line environmental information. For a review of the technology and growing markets for remote environmental monitoring see Atkins (1991) and Bogue (1992). There are few case studies of remote environmental monitoring but Environmental Resources Ltd (1991) provides a useful guide to systems in European cities and problems of introducing a network in the UK. Clarke (1986) provides a handbook on environmental monitoring, Moeller (1992) provides an introduction to environmental health covering all the main sectors. There are a number of sources on the environmental role of smart metering technologies – see Dauncey (1990), Rosenfield *et al.* (1986), and Sioshansi and Davis (1989). On electronic road pricing the best source is Hepworth and Ducatel (1992), but see also Pickup *et al.* (1990). This literature does not focus on the direct environmental implications of road pricing.

The most comprehensive literature on the environmental role of tele-communications is on transport–telecommunications trade-offs. Start with the classic book by Nilles *et al.* (1976). Kraemer and King (1982) provide a comprehensive review of the literature in the early 1980s. Compare the relatively pessimistic view of the potential energy savings of teleworking in the UK (British Telecom, 1992) with the optimistic findings of a California case study (State of California, 1990). The classic articles which really emphasise the contradictory nature of the relationship between telecoms and transport are Mokhtarian (1990) and Salomon (1986).

The most useful literature on the role of telecommunications in the management of transport networks and urban utility systems is examined in Chapter 7. Although Dupuy (1992), Laterasse (1992), Hepworth and Ducatel (1992), Giannopoulos and Gillespie (1993) do not specifically focus on environmental issues – they give a good sense of the direction from which telecommunications are being fitted over older networks. They also highlight the increased control these applications give and the potential this could provide for environmental management. The literature on intelligent buildings is still very diffuse but in the first instance try the article by Gann (1990) and for applications in a domestic context see Miles (1988).

7 URBAN INFRASTRUCTURE AND TRANSPORTATION

There is not a very clearly defined social science literature on the linkages between telecommunications and urban infrastructure. Most work tends to be dominated by engineering and technical interests, but with a bit of digging around it is possible to pull together some useful sources.

There are a number of books on cities and infrastructure that are worth consulting. For an historical review of the relationship between cities and infrastructure see Tarr and Dupuy (1988). Although Grubler (1989) does not specifically focus on cities he provides an elegant overview of the long run historical development of infrastructure systems. The links between telecommunications and urban infrastructure are covered in OECD (1992) – where you will find the Dupuy article – also see Laterrasse (1992). The best sources for transport informatics are Giannopoulos and Gillespie (1993) and Hepworth and Ducatel (1992). On the relationship between transportation and telecommunications, the two best sources are Salomon (1986) and Mokhtarian (1990). They provide useful reviews of the substitution and enhancement effect with lots of empirical examples and further sources of reading.

There are also journals specialising in cities and networked technologies – try *Flux* (published twice a year alternating between French and English), the Australian *Urban Futures* and the *Journal of Urban Technology* from the USA. Although there is increasing social science interest in the relationship between cities and technical networks you will still find it useful to monitor computing, telecommunications and transportation journals and magazines produced by utilities and telematic suppliers.

8 URBAN PHYSICAL FORM

The most useful sources on the historical relationship between the telephone and the development of the city are in the three books by Pool (1977, 1982 and 1990) and the reprint of an early paper by Gottmann (1990). All these provide a useful overview of the complex and contradictory nature of the impacts of the telephone on the city. Although now somewhat dated they provide an extremely good antidote to the utopian and deterministic forecasts that we stand on the edge of urban dissolution.

There is a large literature on the relationship between telecommunications and the contemporary city. Longhini (1984) provides a useful review of the physical planning issues raised by the new urban telecommunications infrastructure. Clarke (1991) reviews debates in the architectural profession about the relationship between smart offices and theories of design. The intelligent building debate is reviewed in Gann (1992), but for a technical overview see Greig (1987) and Oades (1990). Gurstein (1991) provides an extremely good critical review of the spatial and psychological issues associated with home working, but see also Holcomb (1991).

There are a number of sources for the contemporary debate about the role of telecommunications in the spatial restructuring of cities. The book by Fathy (1991) provides an extensive review of the decentralising effects of tele-communications. For shorter, more manageable reviews see the introduction in Brothchie et al. (1985), Coates (1982) and Schuler (1992). A well-referenced chapter by Gillespie (1992) provides a very useful antidote to visions of a decentralised city. Mitchell (1995) provides a fascinating analysis of the challenges raised by the development of virtual cities and their implications for conventional ways of viewing the physical city.

9 URBAN PLANNING, POLICY AND GOVERNANCE

See Corey (1987) and Mandlebaum (1986) for discussion of the paradigm problem facing urban policy when considering modern, telecommunications-based cities. A large literature on urban development strategies and their use of telecommunications and telematics now exists. See Strover (1988) and Schmandt *et al.* (1990) for American experience; Hepworth (1990; 1991b) on European examples; Graham (1991; 1992a) and Graham and Dominy (1991) on the UK; Newstead (1989) and Edgington (1989) for Japan; and Corey (1993) on Singapore. The articles by Graham (1991, 1994) and the books by Brotchie *et al.* (1991) and the OECD (1992) include many worldwide examples of urban telematics strategies. For discussion of wired city experiments in cities, see Dutton *et al.* (1987). The two books by Noothoven van Goor and Lefcoe (1986) and Duncan and Ayers (1988) give an upbeat impression of the potential of teleports to revitalise cities. Interurban networks and their use of telematics is summarised in Graham (1995a,b). The experiments in electronic public space at the urban level can be explored by visiting the World Wide Web home for community networks (http://nearnet.gnn.com/wic/free.20.html) and the City.Net initiative (http://www.city.net). Mitchell (1995) offers a good summary of the policy issues here from an architect's perspective.

Local Government and telematics issues and applications are best summarised by the recent US Office of Technology Assessment (1993) report, and by Taylor and Williams (1990), Kraemer and King (1987) and Hepworth (1991a). Baddeley and Dawes (1987) treat a particular case of decentralisation; Cronberg *et al.* (1991) trace some interesting examples in Denmark.

BIBLIOGRAPHY

Abler, R. (1974) 'The geography of communications', in M. Eliot Hurst (ed.), *Transportation Geography*, New York: McGraw
Hill, 327–346.

—— (1975) 'Effects of space adjusting
technologies on the human geography of the
future', in R. Abler, D. Janelle, A. Philbrick
and J. Sommer (eds), *Human Geography in a
Shrinking World*, North Scituate, MA: Dux-
burg Press.

—— (1977) 'The telephone and the evolution of the American metropolitan system',
in I. de Sola Pool (ed.), *The Social Impact of
the Telephone*, London: MIT Press, 318–341.

Adam, B. (1990) *Time and Social Theory*, Cambridge: Polity.

Adamiak, M.G., Roberts, D.C. and Ketz, S.D.
(1990) 'A microprocessor-based system for
the implementation of variable spot pricing
of electricity', *IEEE Computer Applications in
Power*, 3(4), 43–48.

Adams, J. (1991) *Energy Management Application
Study*, PC 2000, Martlesham Heath, Suffolk:
British Telecom Laboratories.

ADUML (1991) *Plan de Développement Urbain de
la Communication*, Agence de Développement D'urbanisme de la Métropole Lilloise.

Agre, P. (1994) 'Orwell was off by 499 channels
and what to do about it', mimeo.

Aiello, J. (1994) 'Computer-based work monitoring: electronic surveillance and its
effects', *Journal of Applied Social Psychology*
23(7), 499–507.

Aird, W.W. (1988) 'The radical impact of
telecommunications – as an emerging agent
of design change, *Architecture* 77(2),
112–116.

Aldrich, M. (1982) *Videotex: Key to the Wired
City*, London: Quiller.

Allen, J. and Pryke, M. (1994) 'The production
of service space', *Environment and Planning
D: Society and Space* 12, 453–475.

Alles, P. and Esparza, S. (1994) 'Telecommunications and the large city–small city divide:
evidence from Indiana cities', *Professional
Geographer* 46(3), 307–316.

Alovic, T. (1993) *Corporate Networks: The Strategic
Use of Telecommunications*, Norwood Ma:
Artech House.

Ambrose, P. (1994) *Urban Process and Power*,
London: Routledge.

Amin, A. (1994) *Post-Fordism: A Reader*, Oxford:
Blackwell.

Amin, A. and Thrift, N. (1992) 'Neo-
Marshallian nodes of global networks', *International Journal of Urban and Regional Research*
16(4), 571–587.

—— (1995) 'Globalisation, institutional
'thickness' and the local economy', in P.
Healey, S. Cameron, S. Davoudi, S. Graham

and A. Madani Pour (eds), *Managing Cities: The New Urban Context*, London: Wiley, 91–108.

Anderson, B. (1983) *Imagined Communities*, London: Verso.

Antonelli , C. (1988) *New Information Technologies and Industrial Change: The Italian Case*, London: Kluwer.

Appadurai, A. (1990) 'Disjuncture and difference in the global cultural economy', *Theory, Culture and Society* 7, 295–310.

Arnbak, J. (1993) 'The European (r)evolution of wireless digital networks', *IEEE Communications Magazine*, December, 74–80.

Arrow, K. (1980) 'The economics of information', in L. Destorzos and J. Moses (eds), *The Computer Age: A Twenty Year View*, Cambridge, Ma: MIT Press.

Athanasiou, T. (1985) 'High-tech alternativism: the case of the community memory project', in P. Golding (ed.), *Making Waves: The Politics of Communications*, London: Radical Science Collective / Free Association, 37–51.

Atkins, W.S. Management Consultants (1991) *Markets for Environmental Monitoring Instrumentation*, London: Department Of Trade and Industry, HMSO.

Atkinson, R. (1995) 'Technological change, service employment and the future of cities', mimeo, Washington: Office of Technology Assessment.

Aufderheide, P. (1992) 'Cable television and the public interest', *Journal of Communication* 42(1), 52–65.

Ausubel, J.H. (1989) 'Regularities in technological development: an environmental view', in J.H. Ausubel and H.E. Sladovich, *Technology and the Environment*, Washington D.C.: National Academy Press.

Ausubel, J.H. and Herman, R. (eds) (1988) *Cities and their Vital Systems: Infrastructure Past, Present, and Future*, Washington D.C.: National Academy Press.

Baddeley, S. and Dawes, N. (1987) 'Information technology support for devolution', *Local Government Studies*, July/August, 1–16.

Bakis, H. (1995) 'Territories and telecommunications – shift of the problematics: from the "structuring" effect to "potential for interaction"'. Paper given at the conference on 'Telecom Tectonics', Lansing, Michigan, March.

Bakis, H., Abler, R. and Roche, E. (1993) *Corporate Networks, International Telecommunications and Interdependence*, London: Belhaven.

Bankston, R. (1993) 'Instant access to city hall: an examination of how local governments use interactive video to reach citizens', *Multimedia Communications: Forging the Link*, Conference Report, Chapter 75.

Bannister, N. (1993) 'Banks step up war on cash with plastic card that can be topped up by phone', *The Guardian*, 9 December.

——— (1994a) 'Networks tap into low wages', *The Guardian*, 15 October, 40.

——— (1994b) 'Go-slow on the European multimedia superhighway', *The Guardian*, 26 October.

Bannister, N. and Atkinson, D. (1995) 'Banks profit from hi-tech job losses', *The Guardian*, 3 April, 3.

Bartolucci, A. and Morini, A. (1992) 'The future uses of intelligent homes', *Proceedings of Housing Technology and Socio-Economic Change*, World Congress on Housing, Birmingham, UK, 495–503.

Batten, D. (1994) 'Network cities: flexible urban configurations for the 21st century', mimeo.

Batty, M. (1987) 'The intelligent plaza is only the beginning', *The Guardian*, 17 September, 19.

——— (1990a) 'Invisible cities', *Environment and Planning B: Planning and Design*, 17, 127–130.

——— (1990b) 'Intelligent cities: using

information networks to gain competitive advantage', *Environment and Planning B: Planning and Design* 17(2), 247–256.

—— (1991) 'Urban information networks: the evolution and planning of computer–communications infrastructure', in J. Brotchie, M. Batty, P. Hall and P. Newton (eds), *Cities of the 21st Century*, London: Longman, 139–158.

Batty, M. and Barr, B. (1994) 'The electronic frontier: exploring and mapping cyberspace', *Futures* 26(7), 699–712.

Beard, S. (1994) 'The futures market', *The Observer Life*, 20–21.

Beaumont, J.R. and Keys, P. (1982) *Future Cities: Spatial Analysis of Energy Issues*, Chichester: Research Studies Press.

Beckouche, P. and Veltz, P. (1988) 'Nouvelle économie, nouveau térritoire', supplement to the June *Datar Letter*.

Bell, D. (1973) *The Coming of Post Industrial Society*, New York: Basic.

Bell, E. (1994) 'Nation shall network unto nation', *The Observer*, 5 June, 5.

Bellamy, C. and Taylor, J. (1994) 'Introduction: exploiting I.T. in public administration – towards the information polity?', *Public Administration*, 7(2), Spring, 1–12.

Bender, G. and Druckrey, T. (1994) *Culture on the Brink: Ideologies of Technology*, Seattle: Bay Press.

Beniger, J. (1986) *The Control Revolution: Technological and Economic Origins of the Information Society*, Cambridge Mass: Harvard University Press.

Bernard, C. (1994) 'De la conception à la réalisation', *France Télécom News*, Septembre, 3–12.

Bernardini, O. and Galli, R. (1993) 'Dematerialization: long term trends in the intensity of use of materials and energy', *Futures*, May, 431–448.

Berrie, T. and Berrie, T. (1993) 'Utility management, ownership and accountability in the 1990s', *Utilities Policy*, January, 81–85.

Berry, B.J.L. (1973) *The Human Consequences of Urbanisation*, New York: St Martin's Press.

Bianchini, F. (1989) 'The crisis of urban public social life in Britain: origins of the problem and possible responses', *Planning Practice And Research* 5(3), 4–8.

—— (1990) 'Reimagining the city', *Working Paper No. 18*, Liverpool: Centre For Urban Studies, University Of Liverpool.

Biehl, D. (1986) *The Contribution of Infrastructure to Regional Development*, Final Report, Brussels: Infrastructure Study Group, CEC.

Bijker, W., Hughes, T. and Pinch, T. (eds), (1987) *The Social Construction Of Technological Systems*, Cambridge: MIT Press.

Bijker, W. and Law, J. (1992) *Shaping Technology, Building Society: Studies In Sociotechnical Change*, London: MIT Press.

Biocca, F. (1992) 'Communication within virtual reality: creating a space for research', *Journal Of Communication* 42(4), 5–21.

Bird, J., Curtis, B., Putnam, T., Robertson, G. and Tickner, L. (1993) *Mapping The Futures: Local Cultures, Global Change*, London: Routledge.

Bleeker, S. (1994) 'Towards the virtual corporation', *The Futurist*, March–April, 11–14.

Boer, B. (1990) 'Big deal: closing the trade gap', *Telecom World*, September, 10–12.

Boettinger, H. (1989) 'And that was the future . . . telecommunications: from future-determined to future-determining', *Futures*, June, 277–290.

Boghani, A., Kimble, E. and Spencer, E. (1991) *Can Telecommunications Help Solve America's Transportation Problems*, Cambridge Ma: Arthur D. Little Inc.

Bogue, R. (1992) 'Europe senses opportunities', *Physics World*, 31–35.

Bowie, N. (1990) 'Equity and access to information technology', in Institute for Information Studies, *The Annual Review*, Institute for Information Studies, 131–167.

Boyer, C. (1993) 'The city of illusion: New York's public places', in P. Knox (ed.), *The Restless Urban Landscape*, Englewood Cliffs: Prentice Hall, 111–126.

Brain, D. and Page, A. (1991) *Review of Current Experiences and Prospects for Teleworking*, Brussels: European Commission.

Brasier, M. (1989) 'Merrill Lynch flees costly Wall Street', *The Guardian*, 29 June.

Breheny, M.J (ed.) (1992) *Sustainable Development and Urban Form*, London: Pion.

Brewer, H. (1989) 'Diversification attempts by electric utilities: a comparison of potential vs. achieved diversification', *Energy Policy*, June, 228–234.

British Broadcasting Corporation (1993) 'Caught on camera', Close Up North Television Documentary, Newcastle.

British Telecom (1991) *BT Environmental Policy Statement*, March.

—— (1992a) *BT and the Environment*, Environmental Performance Report 1992, London: British Telecom.

—— (1992b) *A Study of the Environmental Impact of Teleworking*, a Report by BT Research Laboratories, London: British Telecom.

Brody, H. (1993) 'Information highway: the home front', *Technology Review*, August–September, 31–40.

Brotchie, J., Batty, M., Hall P. and Newton, P. (eds) (1991) *Cities of the 21st Century*, London: Halsted.

Brotchie, J., Hall, P. and Newton, P. (eds) (1987) *The Spatial Impact of Technological Change*, London: Croom Helm.

Brotchie, J., Newton, P., Hall, P. and Nijkamp, P. (eds) (1985) *The Future of Urban Form: The Impact of New Technology*, London: Croom Helm and Nichols.

Brown, A. (1994) 'The tedium is the technology', *The Independent*, 28 January, 13.

Brown, L. (1994) 'The seven deadly sins of the information age', *Intermedia*, June/July,

22(3).

Bruce, A. (1993) 'Prospects for local economic development: a practitioners view', *Local Government Studies* 19(3), 319–340.

Brunn, S. and Leinbach, T. (eds) (1991) *Collapsing Space and Time: Geographic Aspects of Communications and Information*, London: Harper.

Buckingham, L., Culf, A. and Goldenberg, S. (1993) 'The battle for global vision', *The Guardian*, 23 October, 23.

Budd, L. (1994) 'The growth of global strategic alliances in different financial centres', paper presented at the conference Cities, Enterprises and Society on the Eve of the 21st Century, Lille, March.

Budd, L. and Whinster, S. (eds) (1992) *Global Finance and Urban Living*, London: Routledge.

Buijs, S. (1994) 'Urban networks', *Flux*, January–March, 51–58.

Burrows, R. (1995) 'Cyberpunk as social theory', mimeo.

Business Week (1994) 'Digital juggernaut', 13 June, 36–39.

Calhoun, C. (1986) 'Computer technology, large-scale social integration and the local community', *Urban Affairs Quarterly* 22(2), 329–349.

Callon, M., Courtai, J., Turner, W. and Baulin, S. (1983) 'From translation to problematic networks: an introduction to co-word analysis', *Social Science Information* 22, 191–235.

Camden Council (1990) 'Proposed integrated voice and data network', mimeo.

Cancelieri, A. (1992) *Habitat du futur*, Paris: Documentation Française.

Capello, R. (1989) 'Telecommunications and the spatial organisation of production', Newcastle Studies of the Information Economy, Working Paper No. 10.

Capello, R. and Gillespie, A. (1993) 'Transport, communication and spatial organisation: future trends and conceptual frameworks', in G. Giannopoulos and A. Gillespie, *Transport and Communications in the New Europe*, London: Belhaven, 24–58.

Carey, J. (1989) *Communication as Culture: Essays on Media and Society*, London: Routledge.

Caso, O. and Tacken, M. (1992) 'Deconcentration of work, effects on the design of dwellings', *Proceedings of Housing Technology and Socio-Economic Change*, World Congress on Housing, Birmingham, UK, 504–520.

Cassirer, H.R. (1990) 'Dissent – communications vs. the environment?, *Intermedia* 18(1), 10–12.

Castells, M. (1985) 'High technology, economic restructuring and the urban–regional process in the United States', in M. Castells (ed.), *High Technology, Space and Society*, London: Sage, 11–39.

—— (1989) *The Informational City: Information Technology, Economic Restructuring and the Urban–Regional Process*, Oxford: Blackwell.

Castells, M. and Hall, P. (1994) *Technopoles of the World: The Making of 21st Century Industrial Complexes*, London: Routledge.

Caulkin, S. (1994a) 'Why city must be slicker', *The Observer*, 13 February, 8.

—— (1994b) 'Engineering problems', *The Observer*, 28 August, 6.

CEC (Commission of the European Communities) (1990) *Green Paper on the Urban Environment*, Com(90) 218, Brussels: CEC.

—— (1991) *Europe 2000: Outlook for the Development of the Community's Territory*, Brussels: CEC.

—— (1992a) *Perspectives on Advanced Communications in Europe*, Volume I, Summary, Brussels: CEC.

—— (1992b) *Exploratory Investigation of Employment Trends in Rural Areas*, Brussels: CEC.

CEED (1992) 'The environmental impact of teleworking', *CEED Bulletin* 38, March–April, 10–11.

Channel 4 (1994) *Once Upon a Time in Cyberville*

(programme transcript), London: Channel 4.

Chevin, D. (1991) 'All the right connections', *Building*, 19 July, 46–48.

Chinitz, B. (1964) 'Introduction: city and suburb', in B. Chinitz (ed.), *City and Suburb: the Economics of Metropolitan Growth*, Englewood Cliffs, NJ: Prentice Hall, 3–50.

Chittick, D.R. (1992) 'Technology's impact on the environment: both problem and solution, *AT&T Technical Journal*, March/April, 2–4.

Christopherson, S. (1992) 'Market rules and territorial outcomes: the case of the United States', *International Journal of Urban and Regional Research* 17(2), 274–288.

—— (1994) 'The fortress city: privatized spaces, consumer citizenship', in A. Amin (ed.), *Post Fordism: A Reader*, Oxford: Blackwell, 409–427.

Clark, M., Burall, P. and Roberts, P. (1993) 'A sustainable economy', in A. Blowers (ed.), *Planning for a Sustainable Environment*, London: Earthscan.

Clarke, R. (1986) *The Handbook of Ecological Monitoring*, Oxford: Gems/Upep Publication, Clarendon Press.

Clarke, T. (1991) 'Machine dreams', *Marxism Today*, June, 40.

Coates, J.F. (1982) 'New technologies and their urban impact', in G. Gappert and R. Knight (eds), *Cities in the 21st Century*, London: Sage.

Confederation of British Industry (1989) *Transport in London: The Capital at Risk*, London, CBI.

Cooke, P. (1983) *Theories of Planning and Spatial Development*, London: Hutchinson.

—— (1988a) 'Flexible integration, scope economies and strategic alliances: social and spatial mediations', *Environment and Planning D: Society and Space* 6, 281–300.

—— (1988b) 'Modernity, postmodernity and the city', *Theory, Culture And Society*, 5, 475–492.

Cooke, P. and Morgan, K. (1991) *The Network Paradigm: New Departures in Corporate and Regional Development*, RIP Report No. 8, University of Cardiff.

Cooke, P., Moulaert, F., Swyngedouw, E., Weinstein, O. and Wells, P. (1992) *Towards Global Localization*, London: UCL Press.

Cooke, P. and Wells, P. (1991) 'Uneasy alliances: the spatial development of computing and communications markets', *Regional Studies* 25(4), 345–354.

Coolidge, A.B., Coates, J.F., Hitchcock, H.H. and Gorman, T. (1982) *Environmental Consequences of Telematics: Telecommunication, Computation, and Information Technologies*, EPA Report, 6000/8-81, April (13).

Cooper, I. and Ekins, P. (1993) *Cities and Sustainability*, AFRC-SERC Clean Technology Unit and ESRC, London.

Corey, K. (1987) 'The status of the transactional metropolitan paradigm', in R. Knight and G. Gappert (eds), *Cities of the 21st Century*, London: Sage.

—— (1991) 'The role of information technology in the planning and development of Singapore', in S. Brunn and T. Leinbach (eds), *Collapsing Space and Time: Geographic Aspects of Communications and Information*, London: Harper, 217–231.

—— (1993) 'Using telecommunications and information technology in planning an information-age city: Singapore', in H. Bakis, R. Abler and E. Roche (eds), *Corporate Networks, International Telecommunications and Interdependence*, London: Belhaven, 49–76.

Cornelius, S. (1994) 'GIS in the Environment', in D.R. Green, D. Rox, and J. Cadoux-Hudson (eds), *Geographic Information 1994*, London.

Cornford, J. and Gillespie, A. (1992) 'The coming of the wired city? The recent development of cable in Britain', *Town Planning Review*, 63(3), 243–264.

—— (1993) 'Cable systems, telephony and local economic development in the UK' *Telecommunications Policy*, November, 589–602.

Cornford, J., Gillespie, A. and Robins, K. (1991) 'Telecommunications and the competitive advantage of cities in the European urban system', paper presented at the Communications and Economic Development Conference, Manchester, May.

Cornford, J., Graham, S. and Marvin, S. (1994) 'Towards phones for all? Universal service in a liberalised environment', Programme on Information and Communications Technology, Newcastle Working Paper, No. 14.

Cornish, E. (1993) 'Man and megamachine', *The Futurist*, March-April, 38–39.

Corr, F. and Hunter, J. (1992), 'Worldwide communications and information systems', *IEEE Communications Magazine*, October, 58–62.

Cowe, R. (1994) 'Milk rounds and small shops face extinction', *The Guardian*, 9 June.

Cox, K. (1993) 'The local and the global in the new urban politics: a critical view', *Environment and Planning D: Society and Space* 11, 433–448.

Cox, K. and Mair, A. (1988) 'Locality and community in the politics of local economic development', *Annals of the Association of American Geographers*, 78 (2), 307–325.

Cramer, J. and Zegveld, W.C.L. (1991) 'The future role of technology in environmental management', *Futures*, June, 23 (5), 451–468.

Cronberg, T., Duelund, P., Jensen, O. and Qvortrup, L. (eds) (1991) *Danish Experiments: Social Constructions of Technology*, Copenhagen: New Society Social Science Monographs.

Curry, M. (1995) 'GIS and the inevitability of ethical consistency', in J. Pickles (ed.), *Ground Truth: The Social Implications of Geographic Information Systems*, London:

Guildford Press, 68–87.

Curtis, T. and Means, K. (1991) 'Market segmentation and the IBN policy debate', in M. Elton (ed.), *Integrated Broadband Networks: The Public Policy Issues*, Amsterdam: Elsevier, 23–38.

Dabinett, G. and Graham, S. (1994), 'Telematics and industrial change in Sheffield, UK', *Regional Studies* 28(6), 605–617.

Daniels, P. (1988) 'Producer services and the post-industrial space-economy', in D. Massey and J. Allen (eds), *Uneven Redevelopment*, London: Hodder and Stoughton, 107–123.

Danziger, J., Dutton, W., Kling, R. and Kraemer, K. (1982) *Computer in Politics*, New York: Columbia University Press.

Darby, G. (1994) 'Can digital kiosks for travellers bring digital services to the local loop and make a city, a village, smart? A development strategy', *Pacific Telecommunications Council, 16th Conference Proceedings*, 230–233, Travellers Services.

Dauncey, G. (1990) 'The role of new metering technologies in combating the greenhouse effect', *Proceedings of Sixth International Conference on Metering Apparatus and Tariffs for Electricity Supply*, Institution Of Electrical Engineers, Conference Publication No. 317, London.

Davies, S. (1994) 'They've got an eye on you!', *The Independent*, 2 November.

Davis, J. (1993) 'Cyberspace and social struggle', *Computer Underground Digest*, 28 November, 5 (89).

Davis, J.S., Nelson, A.C. and Dueker, K.J. (1994) 'The new 'burbs – the exurbs and their implications for planning policy, *Journal of the American Planning Association*, Winter, 45–58.

Davis, M. (1990) *City of Quartz: Excavating the Future in Los Angeles*, London: Verso.

———(1992) 'Beyond Blade Runner: urban control, the ecology of fear', *Open Magazine*, Westfield, New Jersey.

———(1993) 'Who killed LA? A political autopsy', *New Left Review* 199, 29–54.

Dawson J. (1992) 'European city networks: experiments in trans-national urban collaboration', *The Planner*, 10 January, 7–9.

Dawson, T. (1994), 'Framing the villains', *New Statesman and Society*, 28 January, 12–13.

Dear, M. (1993) 'In the city, time becomes invisible: land use planning and the emergent postmodern urbanism', mimeo.

———(1995) 'Prolegomena to a post modern urbanism', in P. Healey, S. Cameron, S. Davoudi, S. Graham and A. Madani Pour (eds), *Managing Cities: The New Urban Context*, London: Wiley, 27–44.

De Gournay, C. (1988) 'Telephone networks in France and Great Britain', in J. Tarr and G. Dupuy (eds), *Technology and the Rise of the Networked City in Europe and North America*, Philadelphia: Temple, 322–338.

Delebarre, M. (1992) 'Information technology: an opportunity for cities', in OECD, *Cities and New Technologies*, Paris: OECD 15–16.

Dematteis, G. (1988) 'The Weak Metropolis', in L. Mazza (ed.), *World Cities and the Future of the Metropolis*, Milan: Electra.

———(1994) 'Global networks, local cities', *Flux* 15, 17–24.

Department of Energy (1994) *Energy, Emissions and the Social Consequences of Telecommuting. Energy Efficiency in the US Economy*, Technical Report One, Doe/Po–0026, US Government.

Department of the Environment (1988) *Telecommunications*, Planning Policy Guidance, No. 8, London: HMSO.

De Roo, P. (1994) 'Chapitre 1 La Métropolité, in A. Sallez (ed.), *Les Villes, Lieux D'Europe*, Mouchy: Datar, 9–17.

Detienne, K. (1993) 'Big brother or friendly coach? Computer monitoring in the 21st century', *The Futurist*, September–October, 33–37.

Deutsch, K. (1966) 'On social communications

and the metropolis', in A. Smith (ed.), *Communication and Culture*, London: Holt, Rinehart and Winston, 386–396.

Devins, D. and Hughes, G. (1995) 'Down the information superhighway to urban information inequality?', mimeo.

Dicken, P. (1992) *Global Shift: The Internationalization of Economic Activity*, London: Paul Chapman.

Dickey, J. (1985) 'Urban impacts of information technology', in J. Brotchie, P. Newton, P. Hall and P. Nijkamp (eds), (1985) *The Future of Urban Form: The Impact of New Technology*, London: Croom Helm and Nichols, 175–187.

Dillon, D. (1994) 'Fortress America', *Planning*, June, 8–12.

Dimcock, M. (1933) *British Public Utilities and National Development*, London: George Allen and Unwin.

Dizard, W. (1982) *The Coming Information Age*, London: Longman.

Dordick, H., Bradley, H. and Narris, B. (1988) *The Emerging Network Marketplace*, Norwood Nj: Ablex.

Douglas, I. (1983) *The Urban Environment*, London: Edward Arnold.

Downing, J. (1989) 'Computers for political change: PeaceNet and public data access, *Journal of Communication* 39(3), 154–162.

Downing, J., Fasano, R., Friedland, P., McCullough, M., Mizrahi, T. and Shapiro, J. (eds) (1991) *Computers for Social Change and Community Organizing*; London: Hawath Press.

Doyle, L. (1994) 'Identity cards arriving by stealth in Europe', *The Independent*, Thursday, 3 September, 1.

Druckrey, T. (1994) 'Introduction', in G. Bender and T. Druckrey (eds), *Culture on the Brink: Ideologies of Technology*, Seattle: Bay Press, 1–12.

DTI (1993) 'Edward Leigh announces entry of electricity companies into telecoms', press release, 25 May.

Ducatel, K. (1990a) 'Rethinking retail capital', *International Journal of Urban and Regional Research* 14(2), 207–221.

——(1990b) 'Future shop', *Transnet Technical Innovation Supplement*, Winter.

——(1992) 'Future shop', *Technical Innovation*, Winter, 1–5.

——(1994) 'Transactional telematics and the city', *Local Government Studies* 20(1), 60–77.

Duncan, K. and Ayers, J. (1988) (eds) *Teleports and Regional Economic Development*, New Holland: Elsevier.

Dunford, M. and Perrons, D. (1983) *The Arena of Capital*, London: Macmillan.

Dunn, P. and Leeson, L. (1993) 'The art of change in Docklands', in J. Bird, B. Curtis, T. Putnam, G. Robertson and L. Tickner (1993) *Mapping the Futures: Local Cultures, Global Change*, London: Routledge, 136–149.

Dupuy, G. (1992) 'New information technology and utility management', in OECD, *Cities and New Technologies*, Paris: OECD, 51–76.

Durham, P. (1994) CCTV–Newcastle City Centre, personal communication.

Dutton, W. (1993) 'Electronic services delivery and the inner city: the risk of benign neglect', mimeo.

Dutton, W., Blumler, J. and Kraemer, K. (eds) (1987) *Wired Cities: Shaping the Future of Communications*, Washington: Communications Library.

Dutton, W. and Guthrie, K. (1991) 'An ecology of games: the political construction of Santa Monica's Public Electronic Network', *Informatization in the Public Sector* 1, 279–301.

Dutton, W., Guthrie, K., O'Connell, J and Wymer, J. (1991) *State and Local Innovations in Electronic Services: The Case of Western and North Western United States*, Report prepared for the Office Of Technology Assessment, Congress of the United States.

Dutton, W., Taylor, J., Bellamy, C., Raab, C.

and Peltu, M. (1994) 'Electronic service delivery: themes and issues for the public sector', ESRC Programme on Information and Communications Technologies Policy, Research Paper 28.

Dziegielewski, B. and Beaumann, D.D. (1992) 'The benefits of managing urban water demands', Environment 34(9), 7–41.

Economic and Transport Planning Group (1989) 'Telecommunications in rural England', report to the Rural Development Commission and OFTEL, Rural Research Series No. 2, Salisbury: Rural Development Commission.

Ecler, P. (1994) 'Privacy on parade: your secrets for sale!', The Futurist, July–August, 38–43.

Edge, D. (1988) 'The social shaping of technology', Edinburgh PICT Working Paper No. 1.

Edgington, D.W. (1989) 'New strategies for technology development in Japanese cities and regions', Town Planning Review 60(1), 1–27.

Elam, J., Edwards, D. and Mason, R. (1989) 'Now US cities compete through information technology: securing an urban advantage', The Information Society 6, 153–178.

Elkin, T., Mclaren, D. and Hillman, M. (1991) Reviving the City: Towards Sustainable Urban Development, London: Friends Of The Earth.

Emberley, P. (1989) 'Places and stories: the challenge of technology', Social Research 56, 741–785.

Environmental Resources Limited (1991) An Enhanced Urban Air Quality Monitoring Network – Feasibility Study, Department of the Environment, Air Quality Division, February, A7–A11.

Eubanks, G. (1994) 'Moving towards a networked society', Business and Technology Magazine, March, 42.

European Union (1994) High Level Group on European Information Society, Report to the European Council, 25 June.

Fala, I. (1994) 'Nineteen-ninety four', The Independent, 7 November, 23.

Fathy, T. (1991) Telecity: Information Technology and its Impact on City Form, London: Praeger.

Featherstone, M. (1990) 'Global culture: an introduction', in M. Featherstone (ed.), Global Culture: Nationalism, Globalization and Modernity, London: Sage, 1–14.

Featherstone, M. and Burrows, R. (1995) Cyberpunk / Cyberspace / Cyberbodies, London: Sage.

Financial Times (1991) World Telecommunications Survey, 7 October.

———(1992) Telecommunications in Business, survey supplement, 18 June.

———(1994) Mobile Communications Survey, 5 September.

Finnigan, R., Salaman, G. and Thompson, K. (eds) (1987) Information Technology: Social Issues – A Reader, Sevenoaks: Hodder and Stoughton.

Fischer, C. (1992) America Calling: A Social History of the Telephone to 1940, Oxford: University of California Press.

Fishman, R. (1990) 'Metropolis unbound: the new city of the twentieth century', Flux 1, 43–56.

Florida, R. and Kenney, M. (1993) 'The new age of capitalism: immovation-mediated production', Futures, July/August, 637–651.

Forester, T. (1989) 'The myth of the electronic cottage', in T. Forester (ed.), Computers in the Human Context: Information Technology, Productivity and People, Oxford: Blackwell, 213–227.

———(1991) 'The electronic cottage revisited: towards the flexible workstyle', Urban Futures 5, 27–33.

Foucault, M. (1977) Discipline and Punish, New York: Pantheon.

Frederick, H. (1993) 'Computer networks and the emerging global civil society', in L. Harasim (ed.), Global Networks, Cambridge Mass: MIT Press, 283–296.

Freeman, C. (1987) 'The case for technological determinism', in R. Finnegan, G. Salaman and K. Thompson (eds), *Information Technology: Social Issues – A Reader*, Sevenoaks: Hodder and Stoughton, 5–18.

———(1991) 'Information technology and the new economic paradigm', in H. Schutte (ed.), *Strategic Issues in Information Technology: International Implications for Decision Makers*, Maidenhead: Pergamon, 159–175.

Friendreis, J. (1989) 'The information revolution and urban life', *Journal Of Urban Affairs* 11(4) 327–337.

Frissen, V. (1992) 'Trapped in electronic cages: gender and new information technologies in the private and public domain: an overview of research', *Media Culture and Society* 14, 31–49.

Furlong, M. (1989) 'An electronic community for older adults: the Seniornet network', *Journal of Communication* 39(3), 145–161.

Gaffikin F. and Warf, B. (1993) 'Urban policy and the post-Keynesian state in the United Kingdom and the United States, *International Journal of Urban and Regional Research*, 17, 67–84.

Gale D. (1990) 'At the other end of nature', *The Guardian*, 13 April, 26.

Gandy, O. (1989) 'The surveillance society: information technology and bureaucratic social control', *Journal of Communication* 39(3), 61–76.

Gann, D. (1990) 'Intelligent buildings and smart homes', in G. Locksley (ed.), *The Single European Market and Information and Communication Technologies*, London: Belhaven.

———(1991) 'Buildings for the Japanese information economy: neighbourhood and resort offices, *Futures*, June, 469–481.

———(1992) *Intelligent Buildings – Producers and Users*, Brighton: Science Policy Research Unit, University of Sussex.

Gappert, G. (ed.) (1987) *Cities of the 21st*

Century, Sage Urban Affairs Annual Review, 23.

Garnham, N. (1994) 'Whatever happened to the information society?', in R. Mansell (ed.), *Management of Information and Communication Technologies*, London: Aslib, 42–51.

Garreau, R. (1988) *Edge City: Life on the New Frontier*, New York: Doubleday.

Geller, H.S. (1989) 'Implementing electricity conservation programs: progress towards least-cost energy services among US utilities, T. Johansson (eds), *Electricity: Efficient End-Use and New Generation Technologies, and their Planning Applications*, Lund: Lund University Press.

Gershuny, J. and Miles, I. (1983) *The New Service Economy*, London: Frances Pinter.

Giannopoulos, G. and Gillespie, F. (eds) (1993) *Transport and Communications Innovation in Europe*, London: Belhaven.

Gibbs, D. (1993) 'Telematics and urban economic development policies: time for caution?', *Telecommunications Policy*, May/June, 250–256.

Gibson, D., Kozmetsky, G. and Smilor, R. (1993) *The Technopolis Phenomenon: Smart Cities, Fast Systems, Global Networks*, Lanham Mass.: Rowman and Littlefield.

Gibson, W. (1984) *Neuromancer*, London: Harper and Collins.

Giddens, A. (1979) *Central Problems in Social Theory*, London: Macmillan.

Giddens, A. (1990) *The Consequences of Modernity*, Oxford: Polity Press.

Gille, L. and Mathonnet, P. (1994) 'Les services de Proximité', in P. Musso (ed.), *Communiquer Demain*, Mouchy: Datar, 109–122.

Gillespie, A. (1991) 'Advanced communications networks, territorial integration and local development', in R. Camagni (ed.), *Innovation Networks*, London: Belhaven, 214–229.

———(1992) 'Communications technologies and the future of the city', in M. Breheny (ed.), *Sustainable Development and Urban Form*,

London: Pion, 67–77.

Gillespie, A. and Hepworth, M. (1988) 'Telecommunications and regional development in the information economy', *ESRC Programme on Information and Communications Technologies, Working Paper 1*.

Gillespie, A. and Robins, K. (1989) 'Geographical inequalities: the spatial bias of new communications technologies', *Journal of Communication*, 39(3), 7–18.

———(1991) 'Non-universal service? Political economy and communications geography', in P. Hall and P. Newton (eds), *Cities of The 21st Century*, London: Halsted, 159–170.

Gillespie, A. and Williams, H. (1988) 'Telecommunications and the reconstruction of regional comparative advantage', *Environment and Planning A* 20, 1311–1321.

Giuliano, G. (1992) 'Transportation demand management: promise or panacea?', *Journal of the American Planning Association* 58(3), 327–335.

Goddard, J. (1975) *Office Location in Urban and Regional Development*, London: Oxford

———(1990) 'The geography of the information economy', *PICT Policy Research Papers, no. 11*, Programme on Information and Communications Technologies, Swindon: Economic and Social Research Council.

——— (1992) 'New technology and the geography of the UK information economy', in K. Robins (ed.), *Understanding Information: Business, Technology, Geography*, London: Belhaven, 165–177.

———(1994) 'ICTs, space and place', in R. Mansell (ed.), *Management of Information and Communication Technologies*, London: Aslib, 274–285.

Godfrey, D. and Parhill, D. (1979) *Gutenberg Two*, Toronto: Porcepic.

Gökalp, I. (1988) 'Global networks: space and time', in G. Muskens and J. Gruppelaar (eds), *Global Telecommunications: Strategic Considerations*, Dordrecht: Kluwer, 186–210.

———(1992) 'On the analysis of large technical systems', *Science, Technology and Human Values* 17(1), 578–587.

Gold, J. (1985) 'The city of the future and the future of the city', in R. King (ed.), *Geographical Futures*, Sheffield: Geographical Association, 92–101.

———(1990) 'A wired society? Utopian literature, electronic communications and the geography of the future city', *National Geographic Journal of India* 36(1–2), 20–29.

———(1991) 'Fishing in muddy waters: communications media and the myth of the electronic cottage', in S. Brunn and T. Leinbach (eds), (1991) *Collapsing Space and Time: Geographic Aspects of Communications and Information*, Harper: London, 327–341.

Golding, P. (1990) 'Political communication and citizenship: the media and democracy in an inegalitarian social order', in M. Ferguson (ed.), *Public Communication: The New Imperatives*, London: Sage, 84–100.

Golding, P. and Murdock, G. (1986) 'Unequal information: access and exclusion in the new communications marketplace', in M. Ferguson (ed.), *New Communications Technologies and the Public Interest*, London: Sage, 71–83.

Goldmark, P. (1972) 'Tomorrow we will communicate to our jobs', *The Futurist*, April, 55–59.

Gonzalez-Manet, E. (1988), *The Hidden War of Information*, New Jersey: Ablex.

Goodchild, B. (1990) 'Planning and the modern/postmodern debate', *Town Planning Review* 61(2), 119–137.

Gottman, J. (1982) 'Urban settlements and telecommunications', *Ekistics* 302, Sept/Oct, 411–416.

———(1983), *The Coming of the Transactional City*, Institute for Urban Studies, University of Maryland, Monograph Series, No. 2.

Gowdy, V. (1994) 'Alternatives to prison', *The Futurist*, January–February, 53.

Graham, S. (1991) 'Telecommunications and the local economy: some emerging policy issues', *Local Economy* 6(2), August, 116.

———(1992a) 'Electronic infrastructures and the city: some emerging municipal policy roles in the UK' *Urban Studies* 29(5), 755–781.

———(1992b) 'The role of cities in telecommunications development', *Telecommunications Policy* 16(3), 187–193.

———(1993) 'The changing communications environment: opportunities and threats for British cities', *Cities* 10(2), 158–166.

———(1994) 'Networking cities: telematics in urban policy – a critical review', *International Journal of Urban and Regional Research* 18(3), 416–431.

———(1995a) 'Cities, nations and communications in the global era: urban telematics policies in France and Britain', *European Planning Studies* 3(3), 357–380.

———(1995b) 'From urban competition to urban collaboration? The development of interurban telematics networks', *Environment and Planning C: Government and Policy* (forthcoming).

Graham, S. and Dominy, G. (1991) 'Planning for the information city: the UK case', *Progress in Planning* 35(3), 169–248.

Graham, S. and Marvin, S. (1994) 'Cherry picking and social dumping: British utilities in the 1990s', *Utilities Policy* 4(2), 113–119.

———(1995) 'More than ducts and wires: Post Fordism, cities and utility networks', in P. Healey, S. Cameron, S. Davoudi, S. Graham and A. Madani Pour (eds), *Managing Cities: The New Urban Context*, London: Wiley, 169–190.

Grant, W. (1994) 'Transport and air pollution in California', *Environmental Management and Health*, 1, 31–34.

Graves, J. (1986) *Liberating Technology: Steps Towards a Benevolent Society*, London: Peter Owen.

Gregory, D. (1994) *Geographical Imaginations*, Oxford: Blackwell.

Greig, J. (1987) 'Integrated building technology is here', *Architectural Journal Supplement*, 25 November, 36–39.

Gross, B. (1973) 'Introduction', in G. Gerbner, L. Gross and W. Melody (eds), *Communications Technology and Social Policy*, London: Wiley, 289–292.

Grubler, A. (1989) *The Rise and Fall of Infrastructures*, Heidelberg: Physica-Verlag.

Gurstein, P. (1991) 'Working at home and living at home: emerging scenarios', *The Journal of Architectural and Planning Research* 8(2), 164–180.

Guthrie, K. (1991) 'The politics of citizen access technology: the development of community communication and information utilities in four cities', unpublished Phd dissertation, University of Southern California.

Guthrie, K. and Dutton, W. (1992) 'The politics of citizen access technology: the development of public information utilities in four cities', *Policy Studies Journal*, 20(4), 574–597.

Gwilliam, K.M. and Geerlings, H. (1994) 'New technologies and their potential to reduce the environmental impact of transportation', *Transportation Research A*, 28a(4), 307–319.

Hägerstrand, T. (1970) 'What about people in regional science?', *Papers, Regional Science Association* 24, 7–21.

Hall, P. (1985) 'Optimism and pessimism in future planning', in J. Brotchie, P. Newton, P. Hall, and P. Nijkamp (eds), *The Future of Urban Form: The Impact of New Technology*, London: Croom Helm.

———(1987) 'The anatomy of job creation: nations, regions and cities in the 1960s and 1970s', *Regional Studies* 21, 95–106.

———(1988), *Urban and Regional Planning*, London: Routledge.

———(1991) 'Three systems, three separate paths, *Journal of the Americal Planning Association*, Winter, 16–20.

———(1992) 'New technologies, participation, integration and lifestyle', in OECD *Cities and New Technologies*, Paris: OECD, 255–262.

———(1993) 'Forces shaping urban Europe', *Urban Studies* 30(6), 883–898.

Hall, P. and Preston, P. (1988) *The Carrier Wave: New Information Technology and the Geography of Innovation, 1846–2003*, London: Unwin.

Hall, S. and Jacques, M. (1989) *New Times: The Changing Face of Politics in the 1990s*, London: Lawrence and Wishart.

Hammer, M. (1990) 'Reengineering work: don't automate, obliterate', *Harvard Business Review*, July–August, 104–112.

Hammer, M. and Champy, J. (1993) *Reengineering the Corporation*, New York: New York Books.

Hansen, U. (1990) 'Delinking of energy consumption and economic growth', *Energy Policy* 18(7), 631–640.

Harasim, L. (1993) 'Global networks: an introduction', in L. Harasim (ed.), *Global Networks: Computers and International Communication*, London and Cambridge, Mass.: MIT Press, 3–14.

Haraway, D. (1991) 'A manifesto for cyborgs: science, technology, and socialist-feminism in the late twentieth century, in D. Haraway (ed.), *Simians, Cyborgs and Women: The Reinvention of Nature*, New York: Routledge, 149–181.

Harding, A. (1991) 'The rise of urban growth coalitions, UK-style?', *Environment and Planning C: Government and Policy* 9, 295–317.

Harkness, R.C. (1977) 'Selected results from a technology assessment of tele-communication–transportation interactions, *Habitat* 2(1/2), 37–48.

Harris, B. (1987) 'Cities and regions in the electronic age', in J. Brotchie *et al.*, *The Spatial Implications Of Technological Change*, London: Croom Helm.

Harrison, B. (1994) 'The dark side of flexible production', *Technology Review*, May–June, 38–45.

Harrison, M. (1995) *Visions of Heaven and Hell*, London: Channel Four Television.

Hart, D. (1983) 'Urban economic development measures in West Germany and the United States', K. Young, and C. Mason (eds), *Urban Economic Development: New Roles and Relations*, London: Macmillan.

Harvey, D. (1985) *The Urbanization of Capital*, Oxford: Blackwell.

———(1988) 'Urban places in the "global village": reflections on the urban condition in late twentieth centry capitalism', in L. Mazza (ed.), *World Cities and the Future of the Metropolis*, Milan: Electra.

———(1989) 'From managerialism to entrepreneurialism: the transformation of urban governance in late capitalism', *Geografisker Annaler 71 (Series B)*, 3–17.

———(1989) *The Condition Of Postmodernity*, Oxford: Blackwell.

———(1993) 'From space to place and back again: reflections on the condition of post-modernity', J. Bird, B. Curtis, T. Putnam, G. Robertson and L. Tickner, *Mapping The Futures: Local Cultures, Global Change*, London: Routledge, 3–29

Hasagawa, F. (1990) 'Tokyo: a highly information-oriented city', in IC^2 Institute, *The Technopolis Phenomenon*; Austin, Texas: IC^2 Institute, 77–92.

Hausken, T. and Bruening, P. (1994) 'Hidden costs and benefits of government card technologies', *IEEE Technology and Society Magazine*, Summer, 24–32.

Healey, P., Cameron, S., Davoudi, S., Graham., S and Madani-Pour, A. (eds) (1995) *Managing Cities: The New Urban Context*, London: Belhaven.

Heilmeier, G. (1992) '"Global" begins at

home', *IEEE Communications Magazine*, October, 50–57.

Hemrick, C. (1992) 'Building today's global computer Internetworks', *IEEE Communications Magazine*, October, 44–49.

Hepworth, M. (1986) 'The geography of technological change in the information economy', *Regional Studies* 20(5), 407–424.

———(1987) 'The information city', *Cities*, August, 253–262.

———(1989a) 'Wheel and wires', *Town and Country Planning*, May, 145–146.

———(1989b) *The Geography of the Information Economy*, London: Belhaven.

———(1990) 'Planning for the information city: the challenge and response', *Urban Studies* 27(4), 537–558.

———(1991a) 'The municipal information economy', in J. Brotchie, M. Batty, P. Hall and P. Newton (eds), *Cities of the 21st Century*, London: Longman, 171–178.

———(1991b) 'Information cities in 1992 Europe', *Telecommunications Policy*, June, 175–181.

———(1992) 'Telecommunications and the future of London', *Policy Studies*, Summer, 13(2), 31–45.

Hepworth, M.E., Dominy, G and Graham, S. (1989) 'Local authorities and the Information economy in Great Britain', *Newcastle Studies Of the Information Economy*, No. 11, University of Newcastle upon Tyne.

Hepworth, M. and Ducatel, K. (1992) *Transport in the Information Age: Wheels and Wires*, London: Belhaven Press.

Hepworth, M.E., Green, A.E. and Gillespie, A.E. (1987) 'The spatial division of information labour in Great Britain', *Environment and Planning, A* 19, 793–806.

Herbert, D. and Thomas, C. (1982) *Urban Geography: A First Approach*, London: Wiley.

Herman, R., Ardekani, S.A. and Ausubel, J.H. (1989) 'Dematerialisation', in Ausubel, J.H. and Sladovich, H.E. (1989) *Technology and the Environment*, Washington DC: National Academy Press.

Hill, G.M. (1992) 'Reinventing systems in the utility industry', *Journal of Systems Management* 43(7), July, 23–35.

Hill, S. (1988) *The Tragedy of Technology*, London: Pluto.

Hillman, J. (1991) *Revolution or Evolution? The Impact of Information and Communications Technology on Buildings and Places*, London: Royal Institute of Chartered Surveyors.

———(1993) *Telelifestyles and the Flexicity: A European Study*, Dublin: European Foundation For the Improvement Of Living And Working Conditions.

Hinshaw, M. (1973) 'Wiring megalopolis: two scenarios', in G. Gerbner, L. Gross and W. Melody (eds), *Communications Technology and Social Policy: Understanding The New "Cultural Revolution"*, London: Wiley, 305–317.

Hirschl, T. (1993) 'Electronics, permanent unemployment and state policy', mimeo.

Holcomb, B. (1991) 'Socio-spatial implications of electronic cottages', in S. Brunn and T. Leinbach (eds), *Collapsing Space and Time: Geographic Aspects of Communications and Information*, London: Harper Collins, 342–353.

Holmes, S.J. and Campbell, D. (1990) 'Communicating with domestic electricity meters', in *Proceedings of Sixth International Conference on Metering Apparatus and Tariffs for Electricity Supply*, London: Institution Of Electrical Engineers, Conference Publication No. 317, 129–133.

Houlihan, B. (ed.) (1992) *The Challenge of Public Works Management: A Comparative Study of North America, Japan and Europe*, Brussels: Brussels: IIAS.

Howells, J. (1988) *Economic, Technological and Locational Trends in European Services*, Aldershot: Avebury.

Howells, J. and Wood, M. (1993) *The Globalisation of Production and Technology*,

London: Belhaven.

Howkins, J. (1987) 'Putting wires in their social place', in W. Dutton, J. Blumler, and K. Kraemer (eds), *Wired Cities: Shaping the Future of Communications*, Washington: Communications Library.

Hudson, H.E. and Parker, E.B. (1990) 'Information gaps in rural America: telecommunications policies for rural development', *Telecommunications Policy* 14, 193–205.

Hughes, T. (1983) *Networks of Power: Electrification in Western Society, 1880–1930*, Baltimore: Johns Hopkins University Press.

Hughes, T.P. (1987) 'The evolution of large technological systems, in W.E. Bijker, T.P. Hughes and T. Pinch (eds), *The Social Construction of Technological Systems*, Cambridge, Ma: MIT Press, 51–82.

Hunt, R. (1994) 'The man machine', *The Guardian Weekend*, 24 April, 40–45.

Huston, D.R. and Fuhr, P.L. (1993) 'Intelligent materials for intelligent structures', *IEEE Communications Magazine*, October, 40–44.

Huws, U. (1985) 'Terminal isolation: the atomisation of home and leisure in the wired society', in Radical Science Collective (ed.), *Making Waves: The Politics of Communications*, London: Free Association, 9–25.

IBEX (1991) *Review of Possible Roles of Teleports in Europe, Parts 1, 2 and 3*, Report to the European Commission.

IEE (1992) 'Telecommunications and the Environment', *IEE Review*, June, 212.

Imrie, R. and Thomas, H. (1993) 'The limits of property-led regeneration', *Environment and Planning C; Government and Policy* 11, 87–102.

Innis, H. (1950) *Empire and Communication*, Toronto: University of Toronto Press.

Irvine, S. (1993) 'Terminal illness', *Real World* 6, 4–7.

Irwin, M. and Merenda, M. (1989) 'Corporate networks, privatisation and state sovereignty', *Telecommunications Policy*, December,

329–335.

Itoh, S. (1988) 'Urban development by teleport', in K. Duncan and J. Ayers (eds), *Teleports and Regional Development*, North Holland: Elsevier, 235–241.

Iwama, M. and Kano, S. (1993) 'Toward the global intelligent network', *IEEE Communications Magazine*, March, 22–24.

Jacobs, J. (1962) *The Death and Life of Great American Cities*, London: Jonathan Cape.

Jaeger, C. and Dürrenberger, G. (1991) 'Services and counterurbanisation: the case of Central Europe', in P. Daniels (ed.), *Services and Metropolitan Development: International Perspectives*, London: Routledge.

Jameson, F. (1984) 'Postmodernism, or the cultural logic of late capitalism', *New Left Review* 146, 53–92.

Janelle, D. (1991) 'Global interdependence and its consequences', in S. Brunn and T. Leinbach (eds), (1991) *Collapsing Space and Time: Geographic Aspects of Communications and Information*, London: Harper, 49–81.

Janokovic, L. (1993) *Intelligent Buildings Today and in the Future*, Proceedings of a conference University of Central England, 7 October, Birmingham UK.

Jarratt, J. and Coates, J.F. (1990) 'Future use of cellular technology – some social implications', *Telecommunications Policy*, February, 78–84.

Jenkins, P. (1992) 'Eye can see you', *New Statesman and Society*, 21 February, 14–15.

Johnson, N. (1967) 'Communications', *Science Journal*, October.

——(1970) 'Urban man and the communications revolution', *Regional Urban Communications*, Detroit: Metropolitan Fund Inc.

Johnston, P. (1993) 'Teleworking as an enabler factor for economic growth and job creation in Europe', paper presented at the Telematics and Innovation Conference, Palma, Majorca, 17–19 November.

Jones, D. (ed.) (1970) *Communication and Energy*

in *Changing Urban Environments*, Sevenoaks: Butterworth.

Keating, M. (1991) *Comparative Urban Politics: Power and the City in the US. Canada, Britain and France*, Aldershot: Elgar.

Keegan, V. (1994) 'Fibre firepower', *The Guardian*, 25, February, 2–3.

Keen, P. (1986) *Competing in Time: Using Telecommunications for Competitive Advantage*, Cambridge Mass.: Harper and Row.

———(1991) *Shaping the Future: Business Design Through Information Technology*, Cambridge Mass: Harvard University Press.

Keith, M. and Pile, S. (1993) *Place and the Politics of Identity*, London: Routledge.

Keller, P. (1989) 'The Manto project: telematics – opportunities and risks for traffic and settlement, International Geographical Union, *Geography of Telecommunications and Communications*, Geneva, 7–8 November.

Kellerman, A. (1993) *Telecommunications and Geography*, London: Belhaven.

Kellner, P. (1989) 'Electronic requiem for cities', *The Independent*, 14 August.

Kemp, R. and Soete, L. (1992) 'The greening of technological progress – an evolutionary perspective', *Futures* 24(5), 437–457.

Kennett, P. (1994) 'Modes of regulation and the urban poor', *Urban Studies* 31(7), 1017–1031.

Kern, S. (1983) *The Culture of Time and Space, 1880–1918*, London: Weidenfeld and Nicolson.

King, A. (1993) 'Identity and difference: the internationalization of capital and the globalization of culture', in P. Knox (ed.), *The Restless Urban Landscape*, Englewood Cliffs: Prentice Hall, 83–97.

Knight, R. (1989) 'City development and urbanization: building a knowledge based city', in R. Knight and G. Gappert (eds), *Cities in a Global Society*, Sage: London, 223–242

Knight, R. and Gappert, G. (eds), (1989) *Cities in a Global Society*, Sage: London.

Knox, P. (1993) 'Capital, material culture and socio-spatial differentiation', in P. Knox (ed.), *The Restless Urban Landscape*, Englewood Cliffs: Prentice Hall, 1–33.

Kok, B. (1992) 'Privatisation in telecommunications – empty slogan or strategic tool?', *Telecommunications Policy*, December, 699–704.

Kraemer, K. (1982) 'Telecommunications/transportation substitution and energy conservation', Part 1, *Telecommunications Policy*, March, 39–99.

Kraemer, K. and King, J.L. (1982) 'Telecommunications/transportation substitution and energy conservation', Part 2, *Telecommunications Policy*, June, 87–99.

Kraemer, K. and King, J. (1987) 'The role of Information technology in managing cities', *Local Government Studies*, 14(2), 23–47.

Kraemer, K., King, J. and Schetter, D. (1985) *Innovative use of Information Technology in Facilitating Public Access to Agency Decision Making*, Report prepared for the Office of Technology Assessment, Congress of the United States.

Kraemer, K., King, J., Dunkle, D. and Lane, J. (1986) *Trends in Municipal Information Systems 1975–1985*, Irvine Ca: Public Policy Research Organization.

Krier, D. and Goodman, I. (1992) *Energy Efficiency: Opportunities For Employment*, Report prepared for Greenpeace.

Kroker, A. (1992) *The Possessed Individual: Technology and Postmodernity*, London: Macmillan.

Kubieck (1988) 'The technological infrastructure of home interactive telematics: ISDN and alternative systems', in F. Van Rijn and R. Williams (eds), *Concerning Home Telematics*, North Holland: Elsevier, 97–125.

Kuhn, T. (1970) *The Structure of Scientific Revolutions*, Chicago: Chicago University Press.

Kuttner, R. (1993) 'The pitfalls of the virtual corporation', *Boston Globe*, 30 July.

Lanvin, B. (ed.) (1993) *Trading in a New World Order: The Impact of Telecommunications and Data Services on International Trade in Services*, Boulder: Altwater.

Lash, S. and Urry, J. (1994) *Economies of Signs and Space*, London: Sage.

Laterrasse, J. (1992) 'The intelligent city: utopia or tomorrow's reality?', in F. Rowe and P. Veltz (eds), *Telecom, Companies, Territories*, Paris: Presses De L'ENPC.

Laterrasse, J., Chatzis, K. and Coutard, O. (1990) 'Information et gestion dynamique ou: quand les reseaux deviennent intelligents', *Flux 2*, Automne.

Latour, B. (1987) *Science in Action: How to Follow Scientist and Engineers Through Society*, Milton Keynes: Open University Press.

Laudon, K. (1977) *Communications Technologies and Democratic Participation*, New York: Praeger.

Lavocat, E. (1989) 'From networked towns to town networks: new solidarities, new territories and the impact of telecommunications', *Netcom* 2(2), November, 381–403.

Law, J. and Bijker, W. (1992) 'Postcript: technology, stability and social theory', in W. Bijker and J. Law (eds), *Shaping Technology, Building Society: Studies in Sociotechnical Change*, London: MIT Press.

Leary, T. (1994) *Chaos and Cyberculture*, Berkeley: Ronin.

Lee, M. (1991) 'Social responsibilities of the telecommunications business', *IEEE Technology and Society Magazine* 10(2), 29–30.

Lefebvre, H. (1984) *The Production of Space*, Oxford: Blackwell.

Leinberger, C. (1994) 'Flexecutives: redefining the American dream', *Urban Land*, August, 51–54.

Lerner, S. (1994) 'The future of work in North America: good jobs, bad jobs, beyond jobs', *Futures* 26(2), 185–196.

Levidow, L. and Robins, K. (1989) *Cyborg Worlds: The Military Information Society*, London: Free Association Books.

Ley, D. and Mills, C. (1993) 'Can there be a postmodernism of resistance in the urban landscape?', in P. Knox (ed.), *The Restless Urban Landscape*, Englewood Cliffs: Prentice Hall, 255–278.

Leyshon, A. (1994) 'The geography of financial exclusion', paper presented at a seminar at Newcastle University, November.

Lion, C.P. and Van De Mark, G. (1990) 'Los Angeles', in J. Schmandt *et al.* (eds), *The New Urban Infrastructure – Cities and Telecommunications*, New York: Praeger.

Lloyd, P. and Dicken, P. (1982) *Location in Space*, London: Unwin.

Locksley, G. (1992) 'The information business', in K. Robins (ed.), *Understanding Information: Business, Technology, Geography*, London: Belhaven.

Logan, J. and Molotch, H. (1987) *Urban Fortunes*, London: University of California Press.

Logan, J. and Swanstrom, T. (1990) 'Urban restructuring: a critical view', in J. Logan and T. Swanstrom (eds), *Beyond the City Limits: Urban Policy and Economic Restructuring in Comparative Perspective*, Philadelphia: Temple, 3–26.

Longhini, G. (1984) 'Coping with high tech headaches', *Planning*, American Planning Association, March 1984, 28–32.

Loukaitou-Sideris, A. (1993) 'Privatisation of public open space: the Los Angeles experience', *Town Planning Review* 64(2), 139–167.

Lovering, J. (1988) 'The local economy and local economic strategies', *Policy and Politics* 16(3), 145–157.

—— (1995) 'Creating discourses rather than jobs: the crisis in the cities and the transition fantasies of intellectuals and policy makers', in P. Healey, S. Cameron, S. Davoudi, S. Graham and A. Madani-Pour (eds), *Managing Cities: The New Urban Context*, London: Belhaven/Wiley, 109–126.

Lüthe, R. (1993) 'On the political economy of "post Fordist" telecommunications: the US experience', *Capital and Class* 51, 81–120.

Lynn, D.A. (1976) *Air Pollution – Threat And Response*, Massachusetts: Addison-Wesley.

Lyon, D. (1988) *The Political Economy of the Information Society*, Cambridge: Polity Press.

———(1993) *The Electronic Eye: The Rise of Surveillance Society*, London: Polity.

Lyotard, J. (1984) *The Postmodern Condition: A Report on Knowledge*, Manchester: Manchester University Press.

McBeath, G. and Webb, S. (1995) 'Cities, subjectivity and cyberspace', mimeo.

McGowan, F. (1993) 'Transeuropean networks: utilities as infrastructures', *Utilities Policy*, July, 179–186.

McHale, J. (1976) *The Changing Information Environment*, London: Eleck.

Machart, J. (1994) 'Roubaix Euroteleport', *Technopolis International*, March.

Mackenzie, D. and Wajcman, J. (eds) (1985) *The Social Shaping of Technology*, Milton Keynes: Open University Press.

Mckie, R. (1994) 'Never mind the quality, just feel the collar', *The Observer*, 13 November, 1.

McLuhan, H.M. (1964) *Understanding Media: The Extensions of Man*, London: Sphere Books.

McNeil, M. (1991) 'The old and new worlds of information technology in Britain', in J. Corner and S. Harvey (eds), *Enterprise and Heritage: Crosscurrents of National Culture*, London: Routledge.

Macuiszko, K. (1990) 'A quiet revolution: community on-line systems', *Online*, November, 24–32.

Madani-Pour, A. (1995) 'Reading the city', in P. Healey, S. Cameron, S. Davoudi, S. Graham and A. Madani (eds), *Managing Cities: The New Urban Context*, London: Wiley.

Madden, D. (1992) 'Light at the end of the tunnel', *Financial Times*, 30 January.

Mair, A. (1993) 'New growth poles? Just-in-time manufacturing and local economic development strategy, *Regional Studies* 27(3), 207–221.

Maisonrouge, J. (1984) 'Putting information to work for people', *Intermedia*, 12(2), 31–33.

Malecki, E. (1991) *Technology and Economic Development*, London: Longman.

Manchester City Council (1991) *Manchester: The Information City*, Manchester: City Council.

Mandelbaum, S. (1986) 'Cities and communication: the limits of community', *Telecommunications Policy*, June, 132–140.

Mansell, R. (1994) 'Introductory overview', in R. Mansell (ed.), *Management Of Information and Communication Technologies*, London: Aslib, 1–7.

Marchant, M. (1987) *Grand Aventure Du Minitel*, Paris: Larousse.

Marcou, T. (1990) 'Ville Moyenne network: a flow of information and experiences', *ELISE, On-Line Services: A Tool For Local Development*, Brussels, Belgium: Commission of European Communities.

Marsh, S. (1994) 'Competitive communication strategies', *Logistics Information Management*, 7(2), 25–31.

Martin, J. (1978) *The Wired Society*, London: Prentice Hall.

———(1981) *Telematic Society: A Challenge for Tomorrow*, Englewood Cliffs: Prentice-Hall.

Martin, M. (1991) 'Communication and social forms: the development of the telephone, 1876–1920', *Antipode* 23(3), 307–333.

Marvin, S. (1992) 'Urban policy and infrastructure networks', *Local Economy* 7(3), 225–247.

———(1993) *Telecommunications and the Environmental Debate*, Working Paper No. 20, Department Of Town And Country Planning, University Of Newcastle.

———(1994) 'Green signals: the environmental role of telecommunications in cities', *Cities*, 11(5), 325–331.

Marvin, S. and Cornford, J. (1993) 'Regional policy implications of utility regionalization, *Regional Studies* 27(2), 159–165.

Marvin, S. and Graham, S. (1993) 'Utility networks and urban planning: an issue agenda, *Planning Practice and Research* 8(4), 6–14.

Mason, R. (1983) *Xanadu*, New York: Acropolis Books.

———(1988) 'Living in tomorrow's electronic home today', in F. Van Rijn and R. Willials (eds), *Concerning Home Telematics*, North Holland: Elsevier, 165–170.

Mason, R. and Jennings, L. (1982) 'The computer home: will tomorrow's housing come alive?', *The Futurist* 16(1), February, 35.

Masser, I., Sviden, O. and Wegener, M. (1992) *The Geography Of Europe's Futures*, London: Belhaven.

Massey, D. (1984) *Spatial Division of Labour*, London: Macmillan.

———(1991) 'The political place of locality studies', *Environment and Planning A* 23, 267–281.

———(1992) 'Politics and space/time', *New Left Review* 196, 65–84.

———(1993) 'Power-geometry and a progressive sanse of place', in J. Bird, B. Curtis, T. Putnam, G. Robertson and L. Tickner, *Mapping The Futures: Local Cultures, Global Change*, London: Routledge, 59–69.

Maunder, W.J. (1989) *The Human Impact of Climate Uncertainty*, London: Routledge.

Mayer, M. (1995) 'Urban governance and the post-Fordist city', in P. Healey, S. Cameron, S. Davoudi, S. Graham and A. Madani Pour (eds), *Managing Cities: The New Urban Context*, London: Wiley.

Mazza, L. (1988) 'Introduction', in L. Mazza (ed.), *World Cities and the Future of the Metropolis*, Milan: Electra, 13–19.

Meehan, E. (1988) 'Technical capability versus corporate imperatives: towards a political economy of cable television and information

diversity', in V. Mosco, *The Political Economy of Information*, Madison: Wisconsin University Press, 167–187.

Meier, R. (1962) *A Communications Theory of Urban Growth*, Cambridge: MIT Press.

———(1985) 'High tech and urban settlement', in J. Brotchie, P. Newton, P. Hall, and P. Nijkamp (eds), *The Future of Urban Form The Impact of New Technology*, London: Croom Helm.

Melbin, M. (1978) 'Night as frontier', *American Sociological Review* 43 (1), 5–6.

Melody, W. (1986) 'Implications of the information and communications technologies: the role of policy research', *Policy Studies* 6, 46–58.

Meyer, S.L. (1977) 'Conservation of resources, telecommunications, and microprocessors', *Journal Of Environmental Systems* 7(2), 121–129.

Meyrowitz, J. (1985) *No Sense of Place: The Impact of Electronic Media on Social Behavior*, New York: Oxford.

Miles, I. (1988) *Home Informatics*, London: Pinter.

Miles, I. and Robins, K. (1992) 'Making sense of information', in K. Robins (ed.), *Understanding Information: Business, Technology, Geography*, London: Belhaven, 1–26.

Miles, I., Rush, H., Turner, K. and Bessant, J. (1988) *Information Horizons: The Long-term Social Implications of New Information Technologies*, Cheltenham: Edward Elgar.

Mill, P.A., Hartfopf, V., Loftness, V. and Drake, P. (1993) 'The challenge to smart buildings: user-controlled ecological environments for productivity', *The Technopolis Phenomenon*, 53–68.

Miller, R. (1994) 'Global R & D networks and large-scale innovations: the case of the automobile industry', *Research Policy* 23, 27–46.

Milne, C. (1991) 'Opening the debate on universal service in the UK', *Telecommunica-*

tions Policy 15(2), 85–87.

Mingione, E. (1991) *Fragmented Societies: A Sociology of Economic Life Beyond the Market Paradigm*, Oxford: Blackwell.

Mitchell, D. (1994) 'Landscape and surplus value: the making of the ordinary in Brentwood, Ca', *Environment and Planning D: Society and Space* 12, 7–30.

Mitchell, W. (1995) *City of Bits: Space, Place and the Infobahn*, Cambridge Mass: MIT Press.

Mitchelson, R. and Wheeler, J. (1994) 'The flow of information in a global economy: the role of the American urban system in 1990', *Annals of the Association of American Geographers* 84(1), 87–107.

Moeller, D.W. (1992) *Environmental Health*, Massachusetts: Harvard University Press.

Mokhtarian, P.L. (1988) 'An empirical evaluation of the travel impacts of teleconferencing', *Transportation Research-A* 22a, 283–289.

——(1990) 'A typology of relationships between telecommunications and transportation', *Transportation Research* 24a(3), 231–242.

Mokhtarian, P.L., Handy, S.L. and Saloman, I. (1994) 'Methodological issues in the estimation of the travel, energy, and air quality impacts of telecommuting', *Transportation Research A*.

Money, P. (1992) 'White collar jobs flow from Britain as data processors are lured by cheap labour', *The Guardian*, 25 August.

Moores, S. (1993) 'Satelite TV as cultural sign: consumption, embedding and articulation', *Media, Culture And Society* 15, 621–639.

Moran, R. (1993) *The Electronic Home: Social and Spatial Aspects*, Dublin: European Foundation of Living and Working Conditions.

Morgan, K. (1992) 'Digital highways: the new telecommunications era', *Geoforum* 23(3), 317–332.

Morley, D. and Robins, K. (1990) 'Non-tariff barriers: identity, diversity and difference', in G. Locksley (ed.), *The Single European*

Market and Information and Communications Technologies, London: Belhaven, 44–56.

Morley, D. and Robins, R. (1995) *Spaces of Identity*, London: Routledge.

Mosco, V. (1988) 'Introduction: information in the pay-per society', in V. Mosco and J. Wasko (eds), *The Political Economy of Information*, Madison: University of Wisconsin Press, 3–26.

Mosco, V. and Wasko, J. (1988) *The Political Economy of Information*, London: University of Wisconsin Press.

Moss, M.L. (1986) 'Telecommunications and the future of cities', *Land Development Studies* 3, 33–44.

——(1987) 'Telecommunications, world cities and urban policy', *Urban Studies* 24, 534–546.

——(1988) 'Telecommunications: shaping the future', in G. Stemlieb and J.W. Hughes (eds), *Market Geography: Nation, Region and Metropolis*, New Brunswick, NJ: Rutgers University Press, 255–275.

Motiwalla, J., Yap, M. and Hung, L. (1993) 'Building the intelligent island', *IEEE Communications Magazine*, October, 28–34.

Mouftah, H. (1992) 'Multimedia communications: an overview', *IEEE Communications Magazine*, May, 18–19.

Moulaert, F., Swyngedouw, E. and Wilson, P. (1988) 'Spatial responses to Fordist and post Fordist accumulation and regulation', *Papers of the Regional Science Association* 64, 11–23.

Moyal, A. (1992) 'The gendered use of the telephone: an Australian case study', *Media, Culture and Society* 14(1): 51–72.

Muid, C. (1992), 'New public management and information: a natural combination?', *Public Policy and Administration* 7(3), 75–79.

Mulgan, G. (1989) 'A tale of new cities', *Marxism Today*, March, 18–25.

——(1991) *Communication and Control: Networks and the New Economies of Communication*, Oxford: Polity Press.

Mumford, L. (1934) *Technics and Civilisation*, London: Routledge and Kegan Paul.

Murdock, G. (1993) 'Communications and the constitution of modernity', *Media, Culture and Society* 15, 521–539.

Murdock, G. and Golding, P. (1989) 'Information poverty and political inequality: citizenship in an age of privatized communication', *Journal of Communication* 39(3): 180–195.

Myers, N. (1994) 'Gross reality of global statistics', *The Guardian*, Monday, 2 May.

Nakicenovic, N. (1988) 'Dynamics and replacement of US transport infrastructures', in J.H. Ausubel and R. Herman, *Cities and their Vital Systems – Infrastructure Past, Present and Future*, Washington: National Academy Press, 175–221.

Nalsbitt, J. and Aburdene, P. (1991) *Megatrends 2000 – Ten Directions for the 1990s*, New York: Avon Books.

Naughton, J. (1994) 'Smile, you're on TV', *Observer, Life*, 13 November, 38–42.

Negrier, E. (1990) 'The politics of territorial network policies: the example of video-communications networks in France', *Flux*, Spring, 13–20.

Negroponte, N. (1995) *Being Digital*, London: Hodder and Stoughton.

New York Telephone (1993) *The Role of Advanced Telecommunications Technology in Government Operations Today*, New York: NY Telephone.

Newberry, D. (1990) 'Pricing and congestion: economic principles relevant to pricing roads', *Oxford Review of Economic Policy* 5(2), 22–38.

Newman, P. (1991) 'Greenhouse, oil and cities', *Futures*, May, 335–348.

Newstead, A. (1989) 'Future information cities: Japan's vision', *Futures*, June, 263–276.

Newton, P. (1991a) 'The new urban infrastructure: telecommunications and the urban economy', *Urban Futures* 5, 54–75.

———(1991b) 'Telematic underpinnings of the information economy', in J. Brotchie,

M. Batty, P. Hall and P. Newton (eds), *Cities of the 21st Century*, London: Longman, 95–126.

———(1993) 'Australia's information landscapes', *Prometheus* 11(1), 3–29.

Nicol, L. (1985) 'Communications technology: economic and spatial impacts', in M. Castells (ed.) *High Technology, Space and Society*, London: Sage, 191–209.

Nijkamp, P. (1993a) *Europe on the Move*, Avebury: Aldershot.

———(1993b) 'Towards a network of regions: the United States of Europe', *European Planning Studies* 1(2), 149–167.

Nilles, J.M. (1988) 'Traffic reduction by telecommuting: a status review and selected bibliography', *Transportation Research – A*, 22a(4), 301–317.

———(1993) 'Telework in the US today', paper presented at The Telematics And Innovation Conference, Palma, Majorca, 17–19 November.

Nilles, J.M., Carlson, F.R., Gray, P and Hanneman, G. (1976a) 'Telecommuting – an alternative to urban transportation congestion', *IEEE Transactions on Systems, Man, and Cybernetics* 6(2).

Nilles, J.M. *et al.* (1976b) *The Telecommunications–Transportation Trade-off: Options For Tomorrow*, Chichester: Wiley.

Noam, E. (1992) *Telecommunications in Europe*, Oxford: Oxford University Press.

Noll, A. (1989) 'The broadbandwagon! A personal view of optical fibre to the home', *Telecommunications Policy*, September, 197–201.

Noothoven Van Goor, J. and Lefcoe, G. (eds) (1986) *Teleports in the Information Age*, North-Holland: Elsevier.

Nora, S. and Minc, A. (1978) *The Computerisation of Society*, Cambridge: MIT Press.

Norfolk, S. (1994) 'Houston streets: a world apart', *The Independent*, 9 November, 26.

North Communications (undated) *Multimedia*

Networks, promotional brochure.

Northumbria Police (1991) *A Proposed Urban Surveillance System for Newcastle upon Tyne*, mimeo.

Northumbria Police (1995) *CCTV: Initial Findings*, Newcastle.

Nortoft, P. (1991) 'Data communication standards for power utilities: a European perspective, *IEEE Transactions on Power Systems* 7(1), 215–221.

Nowotny, H. (1982) 'The information society: its impact on the home, local community and marginal groups', in H. Bjorn Andersen, M. Earl, O. Holst and E. Mumford (eds), *Information Society, For Richer, For Poorer*, North Holland: Elsevier.

Oades, R. (1990) 'Cabling intelligent buildings', *Architectural Journal*, 22 and 29 August, 60–63.

OECD (1991) *Urban Infrastructure: Finance and Management*, Paris: OECD.

OECD (1992) *Cities and New Technologies*, Paris: OECD.

OFFER (1992) *Metering Consultation Paper*, London, January.

Office Of Technology Assessment (OTA) (1993) *Making Government Work: Electronic Delivery of Federal Services*, OTA-Tct-578, Washington DC: Government Printing Office.

Ogden, M. (1994) 'Politics in a parallel universe: is there a future for cyberdemocracy?', *Futures* 26(7), 713–729.

Ohba, R. (1992) *Intelligent Sensor Technology*, New York: Wiley.

Olalquiaga, C. (1992) *Megalopolis: Contemporary Cultural Sensibilities*, Minneapolis: University of Minnesota Press.

Openshaw, S. and Goddard, J. (1987) 'Some implications of the commodification of information and the emerging information economy for applied geographical analysis in the UK', *Environment and Planning A* 19, 1423–1439.

O'Riordan, T. (1981) *Environmentalism*, London: Pion.

Parfait, Y. (1994) 'Science parks and state-of-the-art telecommunications networks', *Technopolis International*.

Parker, E. (1976) 'Social implications of computer/telecom systems', *Telecommunications Policy* 1(1), 3–20.

Parker, S. and Cocklin, C. (1993) 'The use of geographical information systems for cumulative environmental effects assessment', *Computers, Environment and Urban Systems*, 17, 393–407.

Parkinson, M. (1992) 'City links', *Town and Country Planning*, September, 235–236.

————(1994) 'European cities towards 2000: the new age of entrepreneurialism?', mimeo.

Parsons, P. (1989) 'Defining cable television: structuration and public policy', *Journal Of Communication* 39(2), 10–26.

Pascal, A. (1987) 'The vanishing city', *Urban Studies* 24, 597–603.

Patri, P. (1987) 'The smart building: an overview', in J.M. Noothoven Van Goor and G. Lefcoe (eds), *Teleports in the Information Age*, North-Holland: Elsevier Science.

Pedersen, F. (1982) 'Power and participation in an information society: perspectives', in K. Grewlich and F. Pedersen (eds), *Power and Participation in an Information Society*, Brussels: European Commission, 249–289.

Pelton, J. (1989) 'Telepower: the emerging global brain', *The Futurist*, September–October, 9–11.

————(1992) *Future View: Communications, Technology and Society in the 21st Century*, New York: Johnson Press.

Perez, C. (1983) 'Structural change and the assimilation of new technologies in the economic and social system', *Futures*, October, 357–375.

Perry, C. (1977) 'The British experience 1876–1912: the impact of the telephone during the years of delay', in I. Pool (ed.), The Social Impact of the Telephone, Cambridge Mass.: MIP Press, 69–96.

Pickup, L. et al. (1990) 'Measuring the potential effects of road transport informatics on travel patterns in European cities', Transport Policy. PTRC Report Volume P330.

Pickvance, C. and Preteceille, E. (1991) (eds) State Restructuring and Local Power: A Comparative Perspective, London: Pinter.

Pilkington, E. (1994) 'Ghetto blaster', The Guardian Weekend, 4 June, 34–40.

Pinch, S. (1985) Cities and Services: The Geography of Collective Consumption, London: Routledge.

———(1989) 'The restructuring thesis and the study of public services', Environment And Planning A 21, 905–926.

Piore, M. and Sabel, C. (1984) The Second Industrial Divide, New York: Basic Books.

Piorinski, R. (1991) 'Télétopia: Nouvelles technologies et aménagement de territoire', Futuribles, November, 47–65.

Piperno, F. (1986) 'Innovation technologique et transformation de l'être social', in M. Tahon and A. Corten (eds), L'italie: Le Philosophe et Le Gendarme, Montral: Vlb Editeur, 126–128.

Pool, I. de Sola (ed.) (1977) The Social Impact Of The Telephone, Boston: MIT Press, 140–145.

———(1980) Communities Without Boundaries, Cambridge Mass.: MIT Press.

———(1982) 'Communications technology and land use', in L.S. Bourne (ed.), Internal Structure of the City, Oxford: Oxford University Press.

———(1983) Technologies Of Freedom, Belnap Press.

Pool, I. de Sola, Decker, C., Dizard, S., Israel, K., Rubin, R. and Winstein, B. (1977) 'Foresight and hindsight: the case of the telephone', in I. Sola Pool (ed.), The Social Impact of the Telephone, Boston: MIT Press, 127–157.

Porat, M. (1977) The Information Economy: Sources and Methods for Measuring the Primary Information Sector, Washington: US Department Of Commerce, Office Of Telecommunications.

Poster, M. (1990) The Mode of Information: Poststructuralism and Social Context, London: Polity Press.

Postman, N. (1992) Technopoly: The Surrender of Culture to Technology, New York: Vintage.

Pred, A. (1977) City Systems in Advanced Economies. London: Hutchinson.

Preston, P. (1990) 'History lessons: some themes in the history of technology systems and networks', PICT Paper, 1–2 March.

Preteceille, E. (1990) 'Political paradoxes or urban restructuring: globalization of the economy and localization of politics', in J. Logan and T. Swanstrom (eds), Beyond the City Limits, Philadelphia: Temple University Press.

Price Waterhouse (1990) 'The economic impacts of information technology and telecommunications in rural areas', final report to DGXII, Commission of the European Communities, Brussels, London: Price Waterhouse.

Pryke, M. and Lee, R. (1994) 'Place your bets: globalisation, financial instruments and the nature of competition between and within financial centres', paper presented at the conference, Cities, Enterprises and Society on the Eve of the 21st century, Lille, March.

Putnam, T. (1993) 'Beyond the modern home: shifting the parameters of residence', in J. Bird, B. Curtis, T. Putnam, G. Robertson and L. Tickner, Mapping the Futures: Local Cultures, Global Change, London: Routledge, 150–168.

Quillinan, J. (1993) 'Curse of the money's tomb', Telecom World, Spring, 13–15.

Qvortrup, L. (1988) 'The challenge of telematics: social experiments, social informatics and orgware architecture', in G. Muskens and J. Gruppelaar (eds) *Global Telecommunications Networks: Strategic Considerations*, Dordrecht: Reidel.

———(1989) 'The Nordic telecottages: community teleservice centres for rural regions', *Telecommunications Policy*, March, 59–68.

Qvortrup, L., Ancelin, C., Fawley, J. Hartley, J., Pichault, F. and Pop, P. (eds) (1987) *Social Experiments with Information Technology and the Challenge of Innovation*, Dordrecht: Reidel.

Regional Trends (1991) London: HMSO.

Reich, R. (1992) *The Work Of Nations*, New York: Simon and Schuster.

Relph, E. (1987) *The Modern Urban Landscape*. Baltimore: Johns Hopkins Press.

Rheingold, H. (1994) *The Virtual Community*, London: Secker and Warburg.

Richardson, R. (1994a) 'Telebased customer services', *Communicore*, newsletter of Newcastle PICT Centre, No. 2.

———(1994b) 'Back officing front office functions – organisational and locational implications of new telemediated services', in R. Mansell (ed.) *Management of Information and Communication Technologies*, London: Aslib, 309–335.

———(1994c) 'Finance floods out of the high street', *Planning Week* 31, March, 10–11.

Richardson, R., Gillespie, A. and Cornford, J. (1994) 'Requiem for the teleport? The teleport as a metropolitan development and planning tool in western Europe', *Newcastle Programme on Information and Communications Technologies, Working Paper* 17.

Rittner, D. (1992) *Ecolinking: Everyone's Guide to Online Environmental Information*, Berkeley: Peachpit Press.

Roarke Associates (Nd) *A Proposal for Congestion Relief on London and South East Rail Services*, London.

Robins, K. (1989) 'Global times', *Marxism Today*, December, 20–27.

———(ed.) (1992) *Understanding Information: Business, Technology and Geography*, London: Belhaven.

Robins, K. and Gillespie, A. (1992) 'Communication, organisation and territory', in K. Robins (ed.), *Understanding Information: Business, Technology and Geography*, London: Belhaven, 145–164.

Robins, K. and Hepworth, M. (1988) 'Electronic spaces: new technologies and the future of cities', *Futures*, April, 155–176.

Robins, K. and Webster, F. (1986) 'Broadcasting politics: communications and consumption', *Screen* 27(3–4), 30–44.

Robins, K. and Cornford, J. (1990) 'Bringing it all back home', *Futures*, October, 870–879.

Robson, B. (1992) 'Competing and collaborating through urban networks', *Town and Country Planning*, September, 236–238.

Roche, E. (1993) 'The geography of information technology infrastructure in multinational corporations', in H. Bakis, R. Abler and E. Roche, *Corporate Networks, International Telecommunications and Interdependence*, London: Belhaven, 181–205.

Rodenburg, E. (1992) *Eyeless in Gaia: The State of Global Environmental Monitoring*, Washington D.C.: World Resources Institute.

Roos, J. (1994) 'A post-modern mystery', *Intermedia*, August–September, 24–28.

Rose, G. (1993) 'Some notes towards thinking about the spaces of the future', in J. Bird, B. Curtis, T. Putnam, G. Robertson, and L. Tickner, *Mapping The Futures: Local Cultures, Global Change*, London: Routledge, 70–86.

Rosenberg, J., Kraut, R., Gomez, L. and Buzzard, A. (1992) 'Multimedia communications for users', *IEEE Communications Magazine*, May, 20–36.

Rosenfield, A.H., Bulleit, D.A. and Peddie, R.A. (1986) 'Smart meters and spot pricing: experiments and potential', *IEEE Technology And Society Magazine.* 5(1), 23–28.

Roszak, T. (1994) *The Cult of Information*, Berkeley: University of California Press.

Rotenberg, R. and McDonogh, G. (eds) (1993) *The Cultural Meaning Of Urban Space*, Westport Co: Bergin and Garvey

Rouse, M.J. and Cranfield, R.F. (1988) 'An overview of the trends and future direction of information technology in the water industry, *Water Pollution Research And Control* 21, 1129–1135.

Royal Society for Nature Conservation (1990) *Information and Guidelines for Bids For BT Environment City Designation*, London.

Rullani, E. and Zanfei, A. (1988) 'Area networks: telematics connections in a traditional textile district', in C. Antonelli *New Information Technologies and Industrial Change: The Italian Case*. London: Kluwer, 97–112.

Ruzic, F. (1989) 'Teleports as precursors of the 21st century's information society', *The Information Society* 6, 109–116.

Rydin, Y. (1993) *The British Planning System*, London: Macmillan.

Sackman, H. and Boehm, B. (1972) *Planning Community Information Utilities*, Montvale, NJ: Afips Press.

Sackman, H. and Nie, N. (1973) *The Information Utility and Social Choice*, Montvale, NJ: Afips Press.

Salomon, I. (1986) Telecommunications and travel relationships: a review', *Transportation Research – A* 20a(3), 223–238.

Salsbury, S. (1992) 'Emerging global computer and electronic information systems and the challenge to world stock markets', *Flux*, January–March, 27–40.

Samarajiva, R. and Shields, P. (1990) 'Integration, telecommunications, and development: power of the paradigms, *Journal Of Communications* 40(3), 84–105.

Sant, R. (1984) *Creating Abundance: America's Least Cost Energy Strategy*, Maidenhead: Mcgraw Hill.

Santucci, G. (1994) 'Information highways worldwide: challenges and strategies', *I&T Magazine*. Spring, 614–23.

Sassen, S. (1991) *The Global City: New York, London, Tokyo*, Princeton: Princeton University Press.

Savage, M. and Warde, A. (1993) *Urban Sociology: Capitalism and Modernity*, London: Macmillan.

Savitch, H. (1988) *Post-Industrial Cities: Politics and Planning in New York, Paris and London*, Princeton: Princeton University Press.

Sawhney, H. (1992) 'The public telephone network: stages in infrastructure development', *Telecommunications Policy*, September, 538–552.

Schenker, J. (1994) 'No shopping spree', *Communications Week International*, 21 February, 12–13.

Schiller, D. and Fregaso, R. (1991) 'A private view of the digital world', *Telecommunications Policy*, June, 195–207.

Schmandt, J., Williams, F., Wilson, R. and Strover, S. (eds) (1990) *The New Urban Infrastructure: Cities and Telecommunications*, London: Praeger.

Schroeder, R. (1994) 'Cyberculture, cyborg post-Modernism and the sociology of virtual reality technologies', *Futures*, June, 26(5), 519–528.

Schuler, R.E. (1992) 'Transportation and telecommunications networks: planning urban infrastructure for the 21st century', *Urban Studies* 29(2), 297–309.

Scott, A. (1988) 'Flexible production systems: the rise of new industrial spaces in North America and western Europe', *International Journal of Urban and Regional Research* 12(2), 171–185.

Shachar, A. (1994) 'Economic globalization and urban dynamics', mimeo.

Shade, I. (1993) 'Gender issues and computer networking', mimeo.

Shearer, D. (1989) 'In search of equal partnerships: prospects for progressive urban policy in the 1990s', in G. Squires (ed.), *Unequal Partnerships: the Political Economy of Urban Redevelopment in Postwar America*, London: Rutgers Press, 289–307.

Sheth, J. and Sisodia, R. (1993) 'The information mall', *Telecommunications Policy*, July, 376–389.

Shields, P., Dervin, B., Richter, C. and Soller, R. (1993) 'Who needs "POTS-plus" services? A comparison of residential user needs along the rural–urban continuum', *Telecommunications Policy*, November, 564–587.

Shulman, S. (1992) 'Pay-per-view libraries', *Technology Review*, October, 14.

Silverstone, R. (1994) 'Domesticating the revolution – information and communications technologies and everyday life', in R. Mansell (ed.), *Management of Information and Communication Technologies*, London: Aslib, 221–233.

Silverstone, R., Hirsch, E. and Morley, D. (1992) *Consuming Technologies: Media and Information in Domestic Space*, London: Routledge.

Simmons, T. (1994) 'Telecoms contribute to city's world status', *Municipal Review*. January/February, 210.

Simon, J. (1993) 'The origins of US public utilities regulation: elements for a social history of networks, *Flux* 11, 33–41.

Sioshansi, F.P. and Davis, E.H. (1989) 'Information technology and efficient pricing – providing a competitive edge for electric utilities', *Energy Policy*, 17(6), 559–607.

Sklair, L. (1991) *Sociology of the Global System*, London: Harvester Wheatsheaf.

Slack, J. (1987) 'The information age as ideology: an introduction', in J. Slack and F. Feijes (eds), *The Ideology of the Information Age*, Norwood, New Jersey: 1–12.

Slack, J.D. and Fejes, F. (eds) (1987) *The Ideology of the Information Age*, Norwood, New Jersey: Ibex.

Sleeman, J. (1953) *British Public Utilities*, London: Issac Pitman.

Smart, B. (1992) *Modern Conditions, Postmodern Controversies*, London: Routledge.

Smith, R. (1994) 'Bell Atlantic's virtual workforce', *The Futurist*, March–April, 13–14.

Sociomics (1992), *Exploratory Investigation of Employment Trends in Rural Areas Related to ECS*, Report to European Commission DG XIII.

Soja, E. (1989), *Postmodern Geographies*, London: Verso.

Sorkin, M. (ed.) (1992) *Variations on a Theme Park*, New York: Hill and Wang.

Spector, P. (1993) 'Wireless communications and personal freedom', *Telecommunications Policy*, August, 403–407.

Spooner, D. (1992) 'The Manchester host computer communications system', paper presented at the Centre for Local Economic Strategies Conference on Local Economic Development and Communications Policy, Manchester, 13 May.

Squires, J. (1994) 'Private lives, secluded spaces: privacy as political possibility', *Environment and Planning D: Society and Space* 12, 387–401.

Staple, G. (1992) *Telegeography: Global Telecommunications, Traffic Statistics and Commentary*, International Institute For Communications.

State Of California (1990) *The California Telecommuting Pilot Project, Final Report*, Sacremento: Department Of General Services, State Of California.

Stoker, G. and Young, S. (1993) *Cities in the 1990s*, London: Longman.

Storgaard, K. and Jensen, O. (1991) 'IT and ways of life', in P. Cronberg, P. Dueland, O. Jensen and L. Qvortrup (eds), *Danish Experiments: Social Constructions of Technology*, Copenhagen: New Social Science Monographs, 123–139.

Storper, M. and Scott, A. (1989) 'The

geographical foundations and social regulation of flexible production complexes', in J. Wolch and M. Dear (eds), *The Power of Geography: How Territory Shapes Social Life*, Boston: Unwin Hyman, 21–40.

Streeton, H. (1976) *Capitalism, Socialism and the Environment*, Cambridge: Cambridge University Press.

Strover, S. (1988) 'Urban policy and telecommunications', *Journal Of Urban Affairs* 10(4), 341–356.

———(1989) 'Telecommunications and economic development: an incipient rhetoric', *Telecommunications Policy*, September, 194–196.

Sui, D. (1994) 'GIS and urban studies: positivism, post-positivism and beyond', *Urban Geography* 15(3), 258–278.

Sussman, G. and Lent, J. (1991) *Transnational Communications: Wiring the Third World*, London: Sage.

Suzuki, S. (1993) 'IN roll out in Japan', *IEEE Communications Magazine*, March, 48–55.

Swyngedouw, E. (1989) 'The heart of the place: the resurrection of locality in the age of hyperspace', *Geografiska Annaler* 71(B), 31–42.

———(1993) 'Communication, mobility and the struggle for power over space', in G. Giannopoulos and A. Gillespie, *Transport and Communications in the New Europe*, London: Belhaven, 305–325.

Tarr, J.A. (1984) 'The evolution of urban infrastructure in the nineteenth and twentieth centuries, in R. Hanson, (ed.), *Perspectives on Urban Infrastructure*, Washington D.C.: National Academy Press.

Tarr, J., Finholt, T. and Goodman, D. (1987) 'The city and the telegraph: urban telecommunications in the pre-telephone era', *Journal of Urban History* 14(1), 38–80.

Tarr, J.A. and Dupuy, G. (eds) (1988) *Technology and the Rise of the Networked City in Europe and America*, Philadelphia: Temple University Press.

Taylor, G. and Welford, R. (1994) 'A commitment to environmental improvement: the case of British Telecommunications, in R. Welford, *Cases in Environmental Management and Business Strategy*, London: Longman, 60–76.

Taylor, J. and Williams, H. (1989) 'Telematics, organisation and the local government mission', mimeo.

———(1990) 'Themes and issues in an information polity', *Journal Of Information Technology*, 5, 151–160.

Taylor, R. and Rushton, S. (1993) 'The psychology of "immersion" in virtual worlds', *Intermedia*, June–July, 40–43.

Telecities (1994) *Telecities: The European Telematics Partnership*, promotional brochure.

Terasaka, A., Wakabayashi, Y., Nakabayashi, I, and Abe, K. (1988) 'The transformation of regional systems in an information-oriented society', *Geographical Review of Japan* 61 (series B)(1), 159–173.

Thompson, G. (1993) 'Fordist and post Fordist international economic relations? The globalisation of FDI and its public governance', mimeo.

Thompson, J. (1990) *Ideology and Modern Culture*, Cambridge: Polity.

Thrift, N. (1993) 'Inhuman geographies: landscapes of speed, light and power', in P. Cloke, M. Doel, D. Matless, M. Phillips and N. Thrift (eds), *Writing the Rural: Five Cultural Geographies*, London: Paul Chapman, 191–232.

Tickell, A. and Peck, J. (1992) 'Accumulation, regulation and the geographies of post-Fordism: missing links in regulationist research', *Progress in Human Geography* 16(2), 190–218.

Toffler, A. (1981), *The Third Wave*, New York: Morrow.

Toke, D. (1990) 'Increasing energy supply not inevitable', *Energy Policy* 18(7), 671–673.

Tokyo Teleport Center Inc (undated) *Toward the Realization of a Futuristic Information City*, promotional brochure.

Tompkins, R. (1994) 'Shop-till-you-drop at the touch of a button', *Financial Times*, 9 June, 11.

Tornqvist, G. (1968) 'Flows of information and the location of economic activities', *Lund Series in Geography*, Series B3.

———(1974) 'Flows of information and the location of economic activities', in M. Eliot-Hurst (ed.), *Transportation Geography: Comments and Readings*, New York: McGraw Hill.

Toth, K. (1990) 'The workless society: how machine intelligence will bring ease and abundance', *The Futurist*, May–June, 33–37.

Tucny, J. (1993) 'Space and the new technology of information–French experience', seminar paper, Centre for Research in European Urban Environments, University of Newcastle.

Tuppen, C.G. (1992) 'Energy and telecommunications – an environmental impact analysis', *Energy And Environment* 3(1), 70–81.

———(1993) 'An environmental policy for British Telecommunications', *Long Range Planning* 26(5), 24–30.

Ungerer, H. (1988) *Telecommunications in Europe*, Brussels: Commission of the European Communities.

Valovic, T. (1993) *Corporate Networks: The Strategic Use of Telecommunications*, London: Artech.

Van Meerten, R. (1994) 'Developing a European urban observatory and decision-support system', in M. de Forn (ed.), *Citytec*, conference report, Barcelona.

Van Rijn, F. and Williams R. (eds) (1988) *Concerning Home Telematics*, North-Holland: Elsevier.

Vidal, J. (1994) 'Sceptred aisle', *The Guardian*. 4 November.

Virilio, P. (1984) *L'espace critique*, Christian Bourgois, Paris [translated from the French by Astrid Hustvedt].

———(1987) 'The overexposed city', *Zone* 1(2), 14–31.

———(1988) *La Machine de Vision*, Paris: Galilée.

———(1993) 'The third interval: a critical transition', in V. Andermatt-Conley (ed.), *Rethinking Technologies*, London: University of Minnesota Press, 3–10.

Volle, M. (1994) 'Les Évolutions Technologiques', in P. Musso (ed.), *Communiquer Demain*, Mouchy: Datar, 65–82.

Ward, S. (1990) 'Local industrial promotion and development policies 1899–1940', *Local Economy*, 2 August, 100–118.

Warf, B. (1989) 'Telecommunications and the globalization of financial services', *Professional Geographer* 41(3), 257–271.

Wark, M. (1988) 'On technological time: Virilio's overexposed city', *Arena*, 83. 82–100.

Warren, R. (1989) 'Telematics and urban life', *Journal Of Urban Affairs* 11(4), 339–346.

Watson, S. and Gibson, K. (eds) (1995) *Postmodern Cities and Spaces*, Oxford: Blackwell.

Webber, M. (1964) 'The urban place and the non place urban realm', in M. Webber, J. Dyckman, D. Foley, A. Guttenberg, W. Wheaton and C. Whurster (eds), *Explorations Into Urban Structure*, Philadelphia: University of Pennsylvania Press, 79–153.

Webber, M. (1968) 'The post-city age', *Daedalus*, Fall.

Webster, F. and Robins, K. (1986) *Information Technology: A Luddite Analysis*, Norwood Nj: Ablex.

Weckerle, C. (1991) 'Télématiques, action locale et "l'espace public"' *Espaces Et Societies* 62–63, 163–211.

Westrum, R. (1991) *Technologies and Society: The Shaping of People and Things*, Belmont, Ca.: Wadsworth:

White, P. (1994) 'Fragmentation of news –

fragmentation of politics?', *Intermedia*, June/ July, 22(3), 20–22.

Whitelegg, J. (1993) 'Confusing signals on the road to nowhere', *The Times Higher*, 19 November, x–xi.

Wigan M. (1988) 'Changes in the relationship between transport, communications and urban form', *Transportation* 14, 395–417.

Williams, F. (1983) *The Communications Revolution*, London: Sage.

Williams, F. and Brackenridge, E. (1990) 'Transfer via telecommunications: networking scientist and industry', in F. Williams (ed.), *Technology Transfer: A Communications Perspective*, London: Sage, 172–191.

Wilson, E. (1991) *The Sphinx and the City*, London: Virago.

Wilson, K. (1986) 'The videotext revolution: social control and the cybernetic commodity of home networking, *Media, Culture and Society* 8, 7–39.

Wilson, M. (1994) 'Jamaica's back offices: direct dial dependency?' mimeo.

Wilson, R. (1992) 'Communications and power struggle', *Financial Times*, 30 January.

Wilson, R. and Teske, P. (1990) 'Telecommunications and economic development: the state and local role', *Economic Development Quarterly* 4(2), May, 158–174.

Winckler, M. (1991) 'Walking prisons: the developing technology of electronic controls', *The Futurist*, July–August, 34–36.

Winner, L. (1978) *Autonomous Technology: Technics Out-of-Control as a Theme in Political Thought*, Cambridge Mass.: MIT Press.

——(1986) *The Whale and the Reactor*, Chicago: University Of Chicago.

——(1993a) 'Beyond inter-passive media', *MIT Technology Review* August–September, 69.

——(1993b) 'Upon opening the blackbox and finding it empty: social constructivism and the philosophy of technology', *Technology and Human Values* 18(3), 362–378.

Wise, D. (1971) 'Exploratory analysis of the impact of electronic communications on metropolitan form', mimeo.

Wise, Deborah (1992) 'Phone home, BT, and get the curtains closed', *The Guardian*. 14 March, 37.

Wolmar, C. (1993) 'Road pricing system for London "will soon be feasible"', *The Independent*, 1 April.

Wood, J. (1994) 'Cellphones on the Clapham Omnibus – the lead-up to a cellular mass market', in R. Mansell (ed.), *Management of Information and Communication Technologies*, London: Aslib, 248–258.

Worthington, J. (1993) 'Accommodation needs of the networked corporation', paper-presented at the Telematics and Innovation Conference, Palma, Majorca, 17–19 November.

Young, J.E. (1993) *Global Networks – Computers in a Sustainable Society*, Worldwatch Paper 115.

——(1994) 'Using computers for the environment', in L.R. Brown *et al.*, *State Of The World*, New York: Norton, 99–116.

Zimmerman, J. (1986) *Once Upon The Future*, London: Pandora Press.

Zukin, S. (1991) *Landscapes of Power: from Detroit to Disneyland*, Berkeley: University of California Press.

INDEX

NLY